PROCESS CONTROL

Statistical Principles And Tools

Ronald Gulezian

Drexel University

ISBN : 0-9627468- 1 -9

Statistical thinking will one day be as necessary for efficient citizenship as the ability to read and write.

H. G. Wells

Sound understanding of statistical control is essential to management, engineering, manufacturing, purchase of materials, and service.

W. Edwards Deming

Data without concepts are blind. Concepts without data are empty.

Immanuel Kant

FOREWORD

The contemporary business environment has made quality the code word for an avalanche of organizational objectives, products and management systems. Based upon our vast experience at the Quality-Alert Institute we have long felt that there was a need to be filled within this environment with respect to the existing literature regarding SPC. Dr. Gulezian's book fulfills that need by bridging the gap between basic statistics and quality control texts and manuals. His much needed treatment of essential statistical principles and process control tools based on a unified theme provides a framework within which to learn the necessary elements of the philosophy begun by W. Edwards Deming. His thoughtful, literate presentation takes the reader from the broadest concepts to specific how-to pointers.

In the world of statistical process control, Ronald Gulezian is a jack of two trades and master of both. He is an experienced statistician. He is an extremely clear writer. Trained as a statistician at the prestigious Wharton School of the University of Pennsylvania, Dr. Gulezian brings to SPC a conceptual clarity that has been absent from the bulk of SPC books written to date. His book, *Process Control*, is likely to remain on required reading lists long after experts on quality management, excellence, one-minute management, and the like fade from best seller lists.

QualityAlert Institute is proud to publish this book and is deeply indebted to Ronald Gulezian for developing a pathway for learning how to analyze recurrent quality problems. His book covers all the topics that an informed user of SPC needs to know. If you read one book about quality improvement this year, this is the one from which you'll learn the most.

Edward E. Emanuel
Chairman
QualityAlert Group

PREFACE

The primary goal of this book is to present the fundamental concepts of statistical process control in a way that can support a variety of instructional programs and learning needs. Depending upon the orientation of a particular program, this book may be used as a self-contained document providing fundamental concepts of statistics and process control, or it may be used as a companion to a comprehensive text on quality control. It also serves as a reference guide for basic principles and tools of statistics and process control.

The book has evolved as a result of an accumulated experience based on existing training programs, teaching, industrial and government consulting, and the environment that has provided us with a new and vigorous climate for the control of quality. Moreover, more and more requests are being made to include material regarding the statistical basis for process control techniques as presented in various training programs. Thus, there is a two-fold emphasis:

1. To provide an understanding of the basic statistical principles underlying process control procedures
2. To provide a description and an understanding of the main statistical process control tools in use today

This material has been written on the premise that the more understanding one has of the underlying principles, the better able one is to utilize the tools effectively.

Numerous textbooks are available that deal with quality control and statistical quality control. These books provide the fundamentals, but with varying perspectives. Moreover, numerous strides have been made in terms of the contributions from statistics. Somehow these advances as well as the fundamentals have not adequately reached the proper audience. As a result, this book has been specially developed in order to address the fundamentals in a more readable form in order to bridge the gap between statistics and process control.

Actually, many of the concepts presented can be found either in a basic statistics textbook or in a text on statistical quality control. The unique feature of this book rests with the essential elements of these two types of texts that traditionally have been isolated; the way they have been united is unique. Again, the book has been written in a manner that effectively links statistics and process control but emphasizes the fundamental statistical principles.

Only those statistical methods that are necessary to understand the process control tools presented in the book are considered. This is not to say that other methods of statistics are not applicable to problems of quality control. Numerous other methods are available which are quite useful. These methods must, however, be applied with more knowledge of statistics and should be preceded by the fundamentals considered here. Further, the process control tools that are presented are more focused, whereas other statistical tools are much broader in scope. Since the quality control problem at its most basic level is inherently statistical, understanding of the barest fundamentals equips the reader to think through his or her own problem and effectively utilize the available literature in order to obtain a proper solution.

Numerous formulas and computational procedures are provided and illustrated in the book, in addition to the conceptual basis for these procedures. In an automated age such as the one in which we live, such emphasis on computation should not be necessary; however, the field is undergoing transition and manual procedures still are being emphasized. Consequently, at the very least, it is essential to be aware of the formulas and computational procedures in order to recognize them when encountered.

Moreover, no matter what the means of implementation, full understanding of the underlying concepts presumably is not possible without some knowledge of the accompanying algorithms. Finally, actual implementation of a full blown process control system is not feasible all at once, especially one that possesses suitable timeliness in real-time and proximity to a particular process. It is necessary to start small and build, which is best done initially by easily understood manual means. This is the best learning experience for the beginner.

In order to meet the stated needs, the book is structured in the form of four modules, A-D, each of which contains three chapters. Although maximum benefit is derived by reading all of the material presented in sequence, each module or some chapters or sections within a chapter can be read individually as a self-contained unit on any one particular area or topic. Also, combinations of modules or chapters can be selected to meet special needs.

At the very least, one may obtain a basic understanding of statistical process control by considering Modules A and C. Module B fills in prerequisite statistical principles, and Module D provides additional process control tools. The schematic below quickly indicates the modular flexibility that is built into the book.

ALTERNATE MODULE SELECTIONS

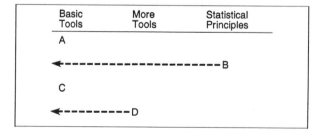

Based on this schematic, one can see that Modules A and C may be supplemented with either one or both of the two remaining modules. Actually, portions of Modules B and D can be sequenced individually.

Module A presents fundamental ideas regarding the inherent nature of data and methods and measures to describe data. The content is general and sometimes definitional in nature, and provides a backdrop for the material that follows. Measures of central tendency and dispersion that are presented in Chapters 2 and 3 are re-introduced in Chapters 5 and 7 where they actually are used.

Much of the material in Module A is similar to descriptive statistics presented in basic statistics textbooks, although some details have been eliminated and additional ideas relating to quality control have been added. Owing to the way in which many texts are written today, the concepts presented are methodological. The methods, of course, are to be used later; however, there also is a dominant theme that sets the stage for what follows. That is, statistical problems and those of process control arise because of the presence of variability, and the nature of this variability behaves in recognizable ways that can be managed or modeled. This idea, which is lost in many basic books, is so fundamental that it cannot be sufficiently emphasized in the few pages devoted to it. All of the remaining ideas should constantly be related back to Module A.

Much of the material presented in Module A is taken from the first three chapters of *Statistics For Decision Making* [57]. One distinguishing feature here, however, is in terms of the use of the sample standard deviation, which is defined in terms of a divisor of "n-1" rather than n. This notation is consistent with current usage both in statistics and statistical process control, and is consistently used throughout the remaining parts of this book.

Another important distinguishing feature of Chapter 1 that is ignored at the beginning of basic statistics texts is the introduction of the concept of variation of data over time and its relation to the main theme presented in terms of frequency distributions and histograms. Although of general importance, the concept is vital when concerned with process control.

Module B addresses basic concepts of probability, sampling, and statistical inference. The module begins with Chapter 4 discussing the fundamentals of probability and the normal curve, which is the primary tool underlying the use of the more common control charts and methods for establishing process capability. The unique part of this chapter appears at the end in terms of the introduction of various sample statistics. Random variation associated with sampled quantities is introduced for the first time and is related back to the earlier concept of a histogram and forward to material regarding time variation of these statistics. Although it is beneficial to read the entire chapter, the general concept of a probability distribution and the specifics related to the normal curve are used directly in much of the material that follows.

Chapter 5 titled Statistical Inference isolates the most relevant aspects of random sampling, estimation, and hypothesis testing that are used in process control. The material presented treats the topics in a more unified and complete way than done in basic quality control texts, yet it is briefer than presentations in basic statistics texts. For example, the concept of interval estimation has been omitted since it is not used directly in the material that follows.

Relative to the material that does follow, the chapter is self-contained and provides a complete statistical basis for understanding and using control charts. Although not used directly, the concept of power and the operating characteristic of a test are introduced in order to provide a complete presentation and to provide the basis for understanding more fully the nature of the testing process underlying the use of control charts. Complete treatments of estimation and hypothesis testing can be found in any of the basic statistics texts listed in the references presented near the end of the book. Actually, much of the material on probability, the normal curve, and hypothesis testing have been taken from *Statistics For Decision Making* [57]. Modifications have been made to suit the special needs of this document.

For anyone writing a book on statistical quality control, a real difficulty arises in terms of choices regarding concepts to be included and those not. Part of this dilemma results from the fact that statistics is a special discipline with its own language and principles that is applicable to virtually any problem area. On the other hand, quality control and quality engineering have evolved by incorporating selected portions of statistics and modifying some of the methodology on somewhat of an as-needed basis. The difficulty arises in linking the two without losing the main thread of the concepts to be presented.

The result of this, in many cases, are loose ends and an incomplete foundation for the principal tools presented. One of the most glaring problems arises with respect to estimation, since some of the underlying concepts are both subtle and difficult to understand. Consequently, optional segments have been included in Chapter 5, which discuss advanced concepts of estimation that completes the story partially told. More technically oriented readers should find this of interest.

The material in Chapter 6, titled Basic Charting, re-appears in Module C; however, it has been included here in a much simpler form. Essentially, the chapter was included as a simple bridge between the concepts of statistical hypothesis testing and control charts. On the one hand, this provides a solid conceptual basis for understanding the nature of a control chart. On the other hand, by using the Super-7 Lottery data, this chapter serves to begin to drive home Dr. Deming's fundamental message regarding the relation between chance phenomena, or lotteries, and the nature of a process.

Module C formally introduces control charts for variables data, which are the most commonly used in process control. Included here are methods of construction and ways to interpret various control chart patterns. Currently used procedures for determining process capability are considered in a separate chapter. The three chapters of this module together with those of Module A comprise the core concepts.

Chapter 7 begins with formulas and conventional methods for constructing variables control charts similar to presentations appearing in many commonly used texts. Although opposed by some, equal emphasis is given to standard deviation and range charts, whereas frequently the range is the principal measure recommended for monitoring process variation. There is no question that when implemented manually, especially in the past, that the ease of computing the range would be an

overriding consideration. With the availability of pre-programmed calculators and computer programs to perform calculations, there should be
a decreased need for the range chart. Also, a discussion unique to this book
appears at the end of the chapter relating the concept of additive sums of
squares as an aid to understanding what is measured in a control chart.
Although similar effects operate on a range chart, the alignment to the
variance and standard deviation gives reason to emphasize the standard
deviation more strongly.

Some additional points are worth noting about Chapter 7. Whenever
teaching or writing about basic statistics there are two fundamental
problems. On the one hand, everyone wants relevance in terms of their
own interests in order to be motivated. On the other hand, there is a need
for some oversimplification of examples in order to properly highlight the
methodology presented. Thus before attacking real and more complicated
problems it is necessary to present the basics in terms of simpler examples.

Unfortunately, it is impossible to meet many individuals' needs
simultaneously; concessions must be made on some common ground. In
order to accomplish this goal and satisfy various instructional requirements, data from an actual simple self-contained process were selected.
The data are associated with a filling process that contains a number of
"almost textbook-like" problems. Furthermore, it is general enough to
be understood by most everyone. The problem is used as a common
thread to link Chapters 7-9 and material presented earlier in the book.

Chapter 8, which is concerned with the interpretation of variables
control charts, presents various control patterns commonly seen in the
literature and many applications. This portion serves as a reference
guide for identifying control chart patterns.

Unique to this chapter, however, is the preceding discussion that
really sets the stage for a deeper understanding of control chart patterns. By specifically introducing the concepts of homogeneity and
heterogeneity of a process at a point in time with respect to the mean
and the variance, it is possible to develop both a clearer understanding
of various patterns and the ability to think through a problem for one's
self without having to match already published examples with one's own
problem. A clearer understanding of process control and the concept of
a change in a process also can be the result. The ideas presented are
consistent with discussions in other sources; however, greater clarity
with respect to very basic underlying statistical concepts is provided.

Chapter 9, titled Process Capability, presents four basic methods
for describing the capability of a process: (1) histogram, (2) curve fit, (3)
interval, and (4) an index. The main thrust of the chapter is methodological, though, it ends with a renewed discussion of the filling problem that
leads to unity of the ideas presented in all chapters of Module C and
material from the very beginning of the book. By using simple descriptive methods, it is possible to provide an understanding of and a need
for further analysis to improve a process beyond the description of
capability. The categorical inclusion of any one specific type of statistical
method to be used in capability studies was intentionally avoided.

Module D presents basic control charts that can be applied to qualitative process characteristics and additional variables charts. Chapter 10
introduces np and p-charts which are specifically applicable to characteristics that possess binary outcomes. These charts also are strongly tied

to product acceptability and are useful in assessing overall product quality. Commonly used Pareto charts also are presented at the end of this chapter.

Chapter 11 presents control charts that are applicable to defects, or non-conformities, and errors which represent alternative forms of attribute charts. Chapter 12 introduces moving average and moving range charts which are useful in cases where individual observations instead of samples are available. Some attention is given to other variables charts that are more responsive to process changes than the commonly used variables charts. Median charts also are given brief attention as a simple computational alternative to a mean chart. A summary chapter at the end of the book places the material covered in perspective within a broader framework.

Six appendices have been included. Appendix A provides tables of various probability distributions, which include ones usually appearing in basic statistics texts (in case additional material not covered in this book needs to be illustrated). A table of random digits is given in Appendix B, and control chart factors used to establish control limits and estimate the process standard deviation are provided in Appendix C. Appendix D provides a set of commonly used flowchart symbols which are useful for general reference. Appendix E introduces cause-and-effect diagrams, which represents a problem solving tool gaining in popularity in quality control applications. For those who have forgotten their basic algebra, Appendix F has been included as a refresher on basic operations and functions.

Acknowledgements

As is the case with any book, contributions come from many sources and from many individuals. Notable among these are a small number of people to whom I owe much gratitutde. It is not the first time that Lloyd Black has done a superb job of editing and has succeeded in cleaning-up my work. Both Mike Saccucci at Drexel University and Cyrus Mohebbi at Prudential-Bache provided valuable insights regarding the book's content.

Production of the book would not have been possible without the tireless efforts of the staff at the QualityAlert Institute. Among these, Jorge Calderon diligently translated incomplete and scrawled artwork into clear and crisp exhibits. With her unique brand of patience, conscientiousness and pride, Marie Oak got the job done. Up front, I would have had a real problem without the renewed support (and complete typing backup) from my wife Ann; I never really did understand how she unceasingly has more faith and enthusiasm than I when it comes time to write a book.

There is one person who deserves much credit but is no longer here to receive it. I owe a continued debt to the late J. Parker Bursk, the chairman of the first department of statistics in a business school, which was formed many years ago at the Wharton School of the University of Pennsylvania. Parker's appreciation and pursuit of excellence in teaching has left a lasting memory. He probably would have had a number of criticisms but I think he really would have liked this book.

Ronald Gulezian

CONTENTS

INTRODUCTION

Methods of statistical quality control have been in existence for some time and have been applied in various forms for at least fifty years. During the last decade a change in emphasis has occurred that focuses on process control and prevention as opposed to the traditional approach of acceptance sampling, defect detection, and inspection. Quite reasonably, it makes sense to continually produce higher quality output than it does to spend time to constantly produce poorer quality items that are subjected to scrap or rework. As simple as this may be in principle, it is not quite as simple to implement.

Some Historical Background

It is of interest to note some of the background of statistical quality control in order to place the material presented in some perspective. The most widely recognized methods of statistical process control, which are based on very fundamental concepts of statistics, were introduced by Walter Shewhart, [14], in the 1920's and early 1930's. Although the procedures have been expanded and supplemented over time, procedures of acceptance sampling emanated from the work of Harold Dodge and Harry Romig during the 1930's, culminating in the publication of their Sampling Inspection Tables, [7]. Both sets of efforts evolved from work done at Bell Labs in the United States. Apparently, as the story goes, it was a toss-up during the war years as to which of the two approaches to quality control should be emphasized in the prevailing military standards. Acceptance sampling was selected and became the basis for the ever-present ML-STD 105,[40], and ML-STD 414, [41], which are still required today.

Although employed effectively by a few firms, methods of statistical process control were not then universally adopted in the United States and still are having a tough time entering U.S. manufacturing today. The real thrust associated with serious applications began in post-war Japan, based on contributions made by W. Edwards Deming, a statistician, who then specialized in statistical sampling and survey methodology and

1

application. Deming introduced the ideas of statistical process control, quality circles, and market research. Today, his fourteen points of management stemming from his work are emphasized, [17]. Little realized is the fact that these management points are a logical consequence of the statistical concepts that he introduced and which also are necessary for the effectiveness of the fourteen points.

Apparently, the Japanese were faced with other problems of implementation and sought the contributions of others who are not credited with Japan's success as extensively as Dr. Deming. Joseph Juran can be credited with introducing the concepts of just-in-time manufacturing and the importance of planning and goal setting. Peter Drucker brought concepts of management, market identification, and marketing in general, while Armand Feigenbaum is attributed with introducing ideas of the usefulness of work breakdown structures, project teams, and general project management concepts. Feigenbaum is most noted for integrating the many ideas under the umbrella of total quality control, [19], which was first introduced formally in the 1950's. In essence, he is responsible for the idea that quality control does not stop at the plant floor, but involves every function within the entire organization from design to sales.

Actually, many new ideas, approaches and philosophies have been introduced since then and translated into American versions. There are two main points worth noting as they relate to the background given and the material provided in subsequent chapters. On the one hand, quality control must be multi-disciplinary in order to achieve maximum effectiveness. Secondly, since variation in output or activity lies at the heart of quality problems, understanding of the concept of variation is fundamental. Since statistics represents the field or body of knowledge, among all others, to explain or manage variation in measured observations, comprehension of basic statistical methods or concepts is essential.

Given its short but rich history the role of statistics has evolved in a rather confusing way in the minds of most people; including executives, managers, and many other types of workers. Much of this confusion is due to the subtleness of the methodology coupled with our never-ending enchantment with numbers and data, and the ease with which the methodology is abused by the untrained when attempting to analyze these data. Current methods of teaching and training also are a problem. It is near paradoxical, for example, that workers in a firm with over fifty trained statisticians in a single department neither ask their advice nor utilize their expertise in constructing a control chart. It is possible that members of the firm do not know of the existence of the statisticians nor that there is a connection between statistics and control charts. The case in point can be found within the pharmaceutical industry in which a strong incentive to use more sophisticated statistical techniques emanates from FDA requirements, whereas the motivation to use "statistical process control" or "total quality management" comes from fragmented sources that are not as single-minded or committed.

Product and Process Characteristics and Quality

The term quality can be defined in many ways, and, needless to say, there exists some confusion regarding the term. Broadly speaking, the quality of a product or service represents its fitness for use. In other words, ultimately a product or service must meet the requirements of the user or consumer of the product or service. A product, for purposes of the ideas presented subsequently, can be a final product ultimately sold by a firm or any intermediate product used in the course of final product development, including raw materials.

For purposes of clarity, we distinguish between two types of quality: quality of design and quality of conformance. Design quality corresponds to the intended level of quality built into a product. The most commonly used example is an automobile. A manufacturer may offer a low-cost compact car or a high performance, luxury model. Although both provide transportation, the latter obviously meets the needs of a consumer who requires a higher grade of car with added features of prestige, performance, workmanship, accessories, and so on.

Quality of conformance corresponds to the degree to which a product conforms to specific design specifications. This type of quality is affected by elements of the entire process used to develop or manufacture a product. Three points are important to note:

1. When emphasis is placed on defect prevention rather than inspection and defect detection, emphasis is shifted to the quality of individual items rather than the quality of a delivered shipment. If individual items are made right, the shipment will contain items that are made right.

2. Defect prevention can be accomplished by addressing both design quality and conformance quality.

3. This book is concerned with conformance quality. As such, emphasis is placed on tools that aid in controlling process characteristics, in addition to those of a product, that affect the quality of individual items produced by the process.

The Prescription

The prescription for achieving quality is very simple, and is composed of the following four components:

1. Measurement
2. Control
3. Capability
4. Improvement

Without measurement it is not possible to make objective decisions regarding quality. Consequently, in order to control a process, it is necessary to identify relevant quantifiable characteristics of the product and the process that will indicate whether a process is stable or whether it is not in control. These characteristics may be critical design specifications or they may be associated with raw materials, machines, testing instruments, intermediate parts or components, environmental conditions, or operator performance characteristics.

All measurements are subject to variation, some of which is natural and some of which is not. By using statistically based methods, it is possible to identify and separate the different types of variation. In this way, it is possible to maintain stability in a process.

Once stability is achieved, it is possible to determine whether a process is capable of meeting product specifications, or the extent to which a process may be capable of producing a better quality product. Better quality is defined in terms of a process that produces product at a targeted level with less variation from item to item. Consequently, we can think of process improvement in terms of reductions in variability. Improvement should lead to increased productivity and lower cost.

By continually considering the four components given above on a never-ending basis, essentially the concept of quality is redefined as a dynamic concept that changes over time as it improves.

Although the prescription is simple, the cure is more difficult to effect. In order to effect a cure, it is necessary to align the statistical component embodied within the above quartet with a technical, behavioral and managerial component. This involves an understanding and involvement of not just those directly producing a product or delivering a service, but of other members of an organization connected with the product or service.

Since it is vital to understand the basics before going further, this book concentrates on the fundamental principles of statistics that, in turn, underlie basic quality control tools in common use today. Emphasis is on process control, or the four components, rather than on inspection and detection. The concepts apply to any situation involving repetitive activities and tasks. Although manufactured items are emphasized, the concepts apply to repetitive services as well.

MODULE A

DESCRIBING PROCESS DATA

MODULE SUMMARY

The goals of this module are to provide the reader with an understanding of the fundamental properties of data and a familiarity with the basic methods for describing data. Key to an understanding of the nature of data is the fact that all data are subject to variation and that data behave in recognizable ways.

The first method for describing data (presented in Chapter 1) is the frequency distribution and the histogram, which is a graph of a frequency distribution. After demonstrating how to construct a frequency distribution, the concept is used to illustrate the fundamental ways in which data distribute or behave.

Since the frequency distribution is a cross-sectional concept, meaning that it is useful for describing data at a point in time, a further discussion is presented about the additional information that potentially is available by observing data over time.

Chapters 2 and 3 focus on the two main properties of data—central tendency and dispersion — and introduce ways of measuring them. Emphasized are the arithmetic mean, the standard deviation, and the sample range. Some attention is given to the median, which is an alternative measure of central tendency.

At the very end, a brief demonstration is given about the "additivity of the sums of squares," which is related to the standard deviation. At this point, the intent is to merely make the reader familiar with the idea for the first time. The concept is re-introduced in Chapter 7 where more detail is provided in order to develop a clearer understanding of the quantities used in process control to measure variation.

Actually, the concepts and procedures presented in this module are used repeatedly throughout most of the book: they are fundamental. Control charts for variables, which are presented in Chapters 6, 7, and 12, are based on the use of measures of central tendency and dispersion. In order to understand the information provided by these charts, you should have a clear understanding of an underlying distribution so as to appreciate the concept of sampling (addressed in Chapters 5, 6, 7, and 8).

HISTOGRAMS AND DATA

All meaningful efforts in quality control involve effective use of measurements associated with relevant process characteristics. The type of data and the way in which it is observed differs by the nature of the underlying problem. There is, however, one feature that is common to all data: variability. Whenever measurements are taken differences exist, no matter how small, among individual observations. Just like fingerprints, no two items can be made exactly alike. In some cases this is so even though the variations are minute, and variation among successive measurements can reflect more of the variation in the measuring instrument than of the item being measured.

An understanding of the nature of variability is essential in order to understand the various methods for analyzing process data and those used for process monitoring and control. Remarkably, variability associated with measured characteristics exhibits recognizable patterns. Basic to an understanding of these patterns are the frequency distribution and the histogram, which are presented in this chapter. In addition, these tools can be used to compare process output with tolerances and, equally important, form the basis of our knowledge of the way sample data behave, which is fundamental to the concept of a control chart.

Raw Data and Frequency Distributions

At this point, let us consider a set of measurements in the form of raw data. Raw data can be defined as a set of originally recorded observations relating to a particular problem in the order in which they are recorded. Although the time order in which observations are collected is of fundamental importance with regard to our later development, for the moment we shall assume that our data represent a cross

7

section of time and that the time order is not to be considered.

As an example, consider a seemingly simple problem in which we are interested in the color printed on a six-inch by eight-inch carton. Although color can be assessed in a number of ways, let us consider lightness or shade, which we shall assume is measured on the basis of a scale between 0.0 and 6.0. Thirty cartons are selected in sequence from the output of an inking process, and lightness is assessed in terms of the observations presented in **Exhibit 1**. The collection of the 30 observations in the exhibit can be considered as raw data. Due to the differences or variation among the individual observations, it is difficult to know much until we determine special features that are present in the data; unless something is done to this set of raw numbers, they tell us little.

EXHIBIT 1
EXAMPLE OF RAW DATA
Color Lightness

2.0	1.0	4.0	2.0	2.0	3.0
5.0	2.0	2.0	1.0	0.0	5.0
2.0	3.0	0.0	2.0	1.0	2.0
3.0	0.0	6.0	3.0	2.0	3.0
4.0	2.0	1.0	4.0	3.0	1.0

A useful way of summarizing data is in the form of a **frequency distribution**. This is a table that presents the values of a measured quantity together with the frequency, or number of times, that each of the values appears in the set of raw observations. By counting the number of times each distinct lightness reading appears in the exhibit of raw data and tabulating, we construct the table or frequency distribution shown in **Exhibit 2**. The distribution shown in the exhibit is referred to as an ungrouped frequency distribution.

EXHIBIT 2
EXAMPLE OF A FREQUENCY DISTRIBUTION
Color Lightness

Lightness	Number of Cartons
0.0	3
1.0	5
2.0	10
3.0	6
4.0	3
5.0	2
6.0	1
Total	30

In addition to being a useful summary of a set of data, the **frequency distribution**, when graphed, imparts important information about the way data behaves. For example, a graph of the color lightness distribution is presented in **Exhibit 3**. Although different sets of data do not have distributions with shapes exactly like the one displayed in the exhibit, they tend to exhibit similar tendencies. That is, low frequencies generally correspond to relatively low and relatively high measurements, whereas higher frequencies correspond to values within the center of the distribution. In other words, when viewed graphically, frequency distributions generally begin on the left with lower frequencies, rise to a single peak, and then "tail-off" to lower frequencies again.

EXHIBIT 3
GRAPHICAL REPRESENTATION OF A FREQUENCY DISTRIBUTION

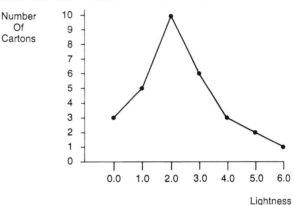

No matter how much is said, or done, it is difficult to get people to appreciate the significance of the phenomenon described in the previous paragraph. The tendency is universal and applies to most every measurable phenomenon such as people's heights, stock market prices, personal performance, as well as to characteristics associated with manufactured product and delivered services. When coupled with the concept of randomness, the concept becomes very powerful when used in process control.

Grouped Frequency Distributions and Histograms

When working with large amounts of data or data containing many distinct values, ungrouped distributions may not provide a useful description. However, by grouping observations into categories or classes corresponding to ranges of values, a **grouped frequency distribution** can be constructed that provides a satisfactory description.

An example of a grouped distribution is presented in **Exhibit 4**, together with the raw data on which it is based. In this case, we see that individual values of the measurements are not presented and that frequencies correspond to intervals of lightness. For example, the first interval in the table is 0 - 0.9 with a frequency of 5. This tells us that five cartons have a lightness reading somewhere between the class limits of 0 and 0.9. The values or the identity of the five values between the limits, however, is not available.

Grouped distributions also can be graphed. The most popular way of doing this is in the form of a *histogram*, such as the one illustrated in **Exhibit 5**. When constructing a histogram, the endpoints of the class intervals are specified on the horizontal axis. Each frequency is represented by a rectangle whose height equals the frequency and whose width corresponds to the size of the class interval. The way the endpoints of the intervals are specified on the graph, to some extent, is dependent on the way the class intervals are defined, the units of measurement, and the precision of the measuring device used to obtain measurements.

EXHIBIT 4
EXAMPLE OF A GROUPED FREQUENCY DISTRIBUTION
Color Lightness

```
2.6 1.1 3.5 2.1 2.5 3.5
4.3 2.9 1.5 0.7 1.1 4.5
1.5 2.1 0.7 2.3 1.9 1.0
2.2 0.4 5.9 2.5 2.6 3.2
4.6 2.2 0.9 3.3 2.7 0.8
```

Lightness	Number of Cartons
0 — 0.9	5
1 — 1.9	6
2 — 2.9	11
3 — 3.9	4
4 — 4.9	3
5 — 5.9	1
Total	30

A key point about grouped distributions or histograms to keep in mind is that the frequencies appear to exhibit the same pattern as in the ungrouped case: they start low, increase to a peak value, and then drop off to lower values. In other words, even when the identity of the individual observations is lost, data have a tendency, so to speak, to behave in *recognizable patterns* that are manifested in the "shape" of a corresponding frequency distribution.

EXHIBIT 5
EXAMPLE OF A HISTOGRAM
Color Lightness

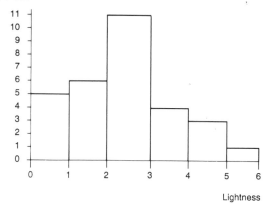

The following symbols will be used in the remainder of this chapter.

X = a numerical value of a measured quantity

f = frequency, or the number of times that a value or values of X occurs

N = the total number of individual observations

c = the size of the class interval, when all class intervals are of the same size

k = the number of class intervals

R = the range, or the difference between the largest and smallest raw observations

L = the largest raw observation

S = the smallest raw observation

p = the relative frequency

CONSTRUCTING A HISTOGRAM

In order to construct a histogram, it is first necessary to develop a grouped frequency distribution. When constructing a grouped distribution, the basic goal is to obtain a distribution that is compact in appearance and, when possible, one where the frequencies begin at low values, rise smoothly to a single peak, and then drop smoothly to lower values.

In practice, the rules for constructing a grouped distribution are general. Consequently, the steps presented below serve as a guide but do not provide a complete set of rules for constructing such a distribution. It should be kept in mind that no single distribution constructed is the only correct one. More than one alternative exists that is correct and that may satisfy the needs of a particular problem.

A general procedure for constructing a grouped distribution and a histogram is as follows:

Step 1. Calculate the range of the data using the formula:

$$R(ange) = L(argest) \; minus \; S(mallest)$$

$$= L - S$$

Step 2. Determine the number of class intervals. The general rule is that the number of classes should neither be "too few" nor "too many." In practical terms, this reduces to using between 5 and 10 classes in most cases. More detail can be obtained by adding additional classes; however, the number should not go beyond 15 or 20. A guide to selecting the number of classes based on the number of observations is given in **Exhibit 6**.

EXHIBIT 6
GUIDE TO CHOOSING NUMBER OF CLASS INTERVALS, k,
BASED ON NUMBER OF OBSERVATIONS, N

N	k
15	5
25	6
50	7
75	8
200	9
400	10
800	11
1500	12

Step 3. Determine the size of the class interval. With respect to quality control work, all class intervals should be of the same size.

Calculate the value of the class size using the formula

$$\text{Class width, } c = \frac{\text{Range}}{\text{Number of Classes}}$$

$$= \frac{R}{k}$$

The value calculated can be used directly, *or* it can be rounded up or down to a more "convenient" value.

Step 4. Establish the actual class intervals based on the result of Step 3. The value of the lower limit of the first class can be set equal to the smallest value in the data or a "convenient" number below the smallest value. If a smaller number is chosen or c is rounded up or down in Step 3, the value of k is reduced or increased accordingly.

Step 5. Tally the raw data in order to determine the number of observations that fall within each class.

Step 6. Count the number of tally bars and assign the corresponding number, or frequency, to each interval. Add the frequencies to obtain the required total. Compare this value with the original number of observations as a check on the tally procedure; these should be equal.

Step 7. Establish scales for graphing the histogram: frequencies on the vertical scale and class intervals on the horizontal scale.

Step 8. Draw rectangles such that heights correspond to frequencies and widths correspond to the class intervals.

Step 9. Label the axes and provide a title.

Note. Sufficient class intervals should be established so that all of the original observations can be placed into the intervals. No ambiguity should exist about the values of the class limits so that there is no confusion about the interval into which a particular observation is placed. One way to do this is to choose limits that do not overlap, but also consider all of the data points.

Another alternative is first to adjust the class size to be an odd-numbered multiple of the basic measuring unit that is close to the width calculated in Step 3. Then establish values of the class limits that are half a unit beyond the accuracy of the original unit of measurement. Although an overlap will exist here, an observation will fall into one interval only.

Example: Measurements of 30 items in units of one inch are given below:

1.8	2.8	5.1	2.6	4.0	3.9
4.3	5.3	3.9	4.4	6.2	2.2
4.6	4.2	2.0	3.8	3.6	3.6
3.9	3.2	5.2	6.6	1.5	3.2
5.1	5.6	8.1	4.7	7.2	5.6

The measuring device is accurate to a tenth of an inch.

Step 1. Calculate the range

Smallest value = 1.8

Largest value = 8.1

$$R = L - S$$

$$= 8.1 - 1.8$$

$$= 6.3$$

Step 2. Determine the number of class intervals.

Using Exhibit 6 as a guide, k = 6 for N = 25 and doesn't equal 7 until N = 50. Therefore, let k = 6.

Step 3. Determine the size of the class interval.

$$c = \frac{R}{k}$$

$$= \frac{6.3}{6}$$

$$= 1.05$$

This value could be used directly; however, let us reduce it to a close, but more convenient number of 1.

Step 4. Establish the actual class intervals.

As a starting value for the lower limit of the first class, we could use the smallest observed value of 1.8. Instead, let us reduce it to a more "convenient" number, 1.5. The class intervals can be written as:

1.5 - 2.4
2.5 - 3.4
3.5 - 4.4
4.5 - 5.4
5.5 - 6.4
6.5 - 7.4
7.5 - 8.4

Step 5. Tally the raw data.

1.5 - 2.4	|||
2.5 - 3.4	++++
3.5 - 4.4	++++ ++++
4.5 - 5.4	++++	
5.5 - 6.4	|||
6.5 - 7.4	||
7.5 - 8.4		

Step 6. Count tally bars and assign frequencies.

1.5 - 2.4	3
2.5 - 3.4	5

3.5 - 4.4	10
4.5 - 5.4	6
5.5 - 6.4	3
6.5 - 7.4	2
7.5 - 8.4	1

> **Note.** Consider the first interval 1.5 - 2.4 Although the difference between the limits equals 0.9, effectively the limits are the same as the following two alternatives:
>
> 1.5 - 2.4999999....
>
> 1.5 but less than 2.5
>
> such that the difference between the limits is arbitrarily close to 1. In any case, a value of 2.5 is placed in the second interval and no ambiguity exists.
>
> Also note that owing to the reduction of the class interval size and the choice of the lower limit of the first class, the number of intervals has been increased to 7 to accommodate the largest value of 8.1.

Step 7 and Step 8. Establish scales and draw rectangles. Label axes.

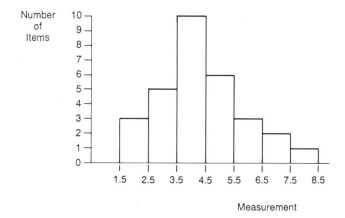

ADDITIONAL IDEAS

Distribution Shape and the Nature of Data

Although most bodies of data exhibit the same basic tendencies, differences exist. These differences are exhibited in terms of variations in the shape of the frequency distribution or the corresponding histogram. Some distributions are "taller" than others, some are "wider," while others have concentrations of observations about different values.

Consider seven distributions depicted in **Exhibit 7**. Many observed distributions have a *single peak* as depicted in diagrams A through E, whereas others may have no peak, as in G, or two peaks, as in F, or more.

Single peaked distributions sometimes are said to represent natural patterns of variation, whereas others potentially result from more than one source of variation.

Exhibit 7
FREQUENCY DISTRIBUTIONS WITH DIFFERENT SHAPES

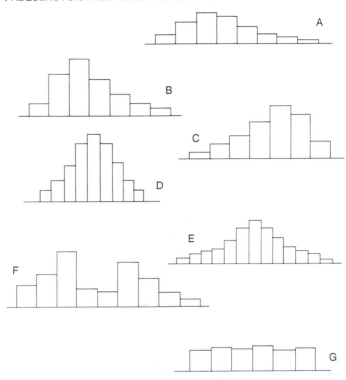

Certain properties can be associated with distributions of data, which are referred to as symmetry, kurtosis, dispersion, and central tendency. *Symmetry* occurs when each half of a distribution is the "mirror image" of the other half. The distributions labeled D and E are symmetric, and G is roughly symmetric.

Distributions that are not symmetric are said to be *skewed*; examples are depicted in A, B, and C. A and B are said to be skewed to the right, since the "tail" of each distribution points to the right. More precisely, a distribution is *skewed to the right* when relatively small frequencies are attached to large values. When a distribution has relatively low frequencies attached to low values, or has a "tail" to the left, we say that the distribution is *skewed to the left*.

Diagrammatically, we can think of *dispersion* in terms of the "width" of a distribution. For example, A appears wider than B. In such a case we say that the characteristic described in A possesses more variability than in B. The same type of comparison can be made between D and E. More variability exists in E than in D. Briefly, when more variability exists in a measured quantity when compared to another, the individual observations described by the distribution are farther apart from one another. As the variability becomes greater in both cases described above, the corresponding peaks are lower. The degree of "peakedness" or "flatness" in a distribution is referred to as *kurtosis*.

As the variability in a set of observations becomes smaller, the observations concentrate about some centrally located values. Since all sets of data exhibit some variation, it is convenient to think in terms of a single value about which the other observations tend to cluster. This value is said to be representative of the observations; in statistical terms this phenomenon is referred to as *central tendency*.

The distribution depicted in F has two peaks and is referred to as *bimodal*. Other cases can be observed in which there exists more than two peaks; such distributions are referred to as *multi-modal*. Typically, such distributions reflect unnatural patterns of variability corresponding to more than one source of variation. Multi-modal distributions typically are evidence of mixtures, which are discussed in Chapter 8.

Phenomena described by G are somewhat *uniformly* assigned: that is, various values occur with basically the same frequency. One possible situation in which this may occur is in the case where numerous sources of variation are present such that the individual distributions are masked and appear as a flat distribution.

Relative Frequency Distributions

An alternative form of a frequency distribution that is useful is the relative frequency distribution, which applies to both the ungrouped and grouped cases. The *relative frequency distribution* is obtained by dividing the total number of observations, N, into the absolute frequency, f, for each value or group of values of X. Thus, the relative frequency, p, is found by the formula

$$p = \frac{f}{N}$$

Using the distribution of color lightness, the frequency distribution and relative frequency distribution are illustrated in **Exhibit 8**.

EXHIBIT 8
A FREQUENCY DISTRIBUTION AND CORRESPONDING
RELATIVE FREQUENCY DISTRIBUTION

X Lightness	f Number of Cartons	p Relative Frequency or Proportion of Cartons
0.0	3	0.100
1.0	5	0.167
2.0	10	0.333
3.0	6	0.200
4.0	3	0.100
5.0	2	0.067
6.0	1	0.033
Total	30	1.000

In order to understand how the relative frequencies in the last column of the exhibit are obtained, consider the second value of 0.167 corresponding to a lightness rating of 1.0. This is obtained by dividing the frequency of 5 by the total of 30 (ie., 5/30 = 0.167). Hence, 0.167 or 16.7 percent of the cartons have a lightness rating of 1.0. Similar calculations and interpretations apply to the remaining values in the exhibit. The sum of all relative frequencies of any distribution must equal one.

Relative frequencies are used to compare the shapes of distributions with differing total observations. Also, by representing frequencies in relative form, a greater degree of generality is obtained. In other words, relative frequencies can be viewed as estimates applying to any number of observations in addition to the total on which they are computed. This relates closely to the problem of obtaining empirical probability estimates, which is discussed at a later point.

The Importance of Variation Over Time

Generally, when a frequency distribution in the form of a histogram is presented, it is implicitly *assumed* that the distribution represents a particular point, or cross-section, in time. If the underlying data correspond to a stretch of time, it is assumed that there are no changes in central tendency, dispersion, and skewness over time; otherwise these effects are masked and are not adequately described by a histogram.

Consider, as an example, **Exhibit 9** which provides a set of raw data together with the corresponding plotted frequency distribution and two time-plots related to the same data. Notice that the frequency distribution is skewed to the left with a long tail with a little "bump" at X = 4.

EXHIBIT 9
CROSS-SECTIONAL VS. TIME PLOT

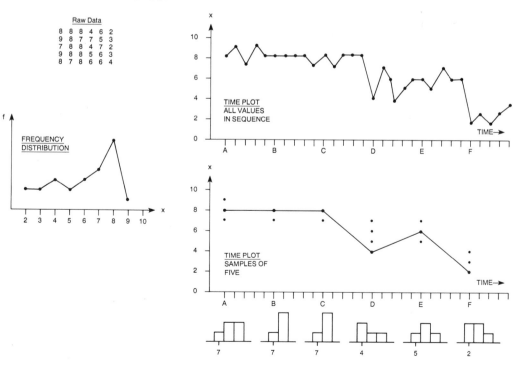

If, on the other hand, we consider the data in the time order in which they were generated, different information is provided. Assume the time order in the raw data can be read by first going down a column and then across rows. When plotted in the middle graph, it becomes apparent that the series is relatively stable throughout the first three columns and then drops in overall level and is more erratic.

By considering each column as a sample of five observations, each presumably taken at a particular point in time, we get a better picture of what is happening. The first three samples were generated under the same conditions of central tendency and dispersion, whereas these conditions were changed with respect to each of the remaining three samples. This is summarized in the third graph presented in Exhibit 9. In this case, observations from one sample are plotted at the same point in time and the "center" of consecutive samples is connected with a straight line. Notice the "central tendency" remains stable and then drops to lower values. Further, the dispersion becomes greater in the later samples, which is evidenced by the points being farther apart. Histograms at the bottom of the exhibit further demonstrate the changes described.

Essentially, the original frequency distribution is made up of samples of data from different distributions over time, whereby the data from the first three samples dominated the combined distribution. Overall, it is important to keep two ideas in mind. First, the concept of a frequency distribution is fundamental to understanding the ideas underlying process control. Second, changes in distributions occur over time and therefore observation of process data over time is essential. Ultimately, we strive for process stability, which is evidenced by similar distributions over time.

MEASURING CENTRAL TENDENCY

The frequency distribution that was introduced in the previous chapter is important to us for two basic reasons. On the one hand, it illustrates the fundamental property of data. On the other hand, it is a useful tool in assessing process output.

Although the concept of a frequency distribution should always be kept in mind as a background, it frequently is unnecessary to construct an entire distribution in order to understand the behavior of a particular process. This can be accomplished in terms of specific measures associated with the individual properties. Most often, measures of central tendency and dispersion are used. We shall discuss central tendency in this chapter.

Many measures of central tendency exist and are defined in terms of various averages, which is a familiar but elusive term. The most commonly used averages in process control are the arithmetic mean and the median. The *arithmetic mean* is found by dividing the sum of a set of observations by the number of observations. The *median* is defined as the value such that 50-percent of the observations are less than this value and 50-percent are greater. Owing to the discreteness of data, the calculated value of a median does not always conform exactly to the definition, and can be thought of as the "middle" value.

Due to the way in which various averages are treated generally and conventions within the field of quality control, some seeming inconsistencies appear in terms of notation and the way various measures are used. Here we shall present the various ways of calculating the two averages and then discuss some aspects regarding their interpretation.

The following symbols will be used in this and later chapters:

X = a numerical value of the quantity measured

f = the absolute frequency, or the number of times a value or values of X occur

N = the total number of observations, in general

n = the sample size, or the number of observations in a sample drawn from a process

μ = the mean, or the process mean

$\overline{X}$ = the arithmetic mean of a sample

$\overline{\overline{X}}$ = the arithmetic mean of a set of sample means, based on repeated samples drawn from a process

X_{50} = the median

$\tilde{X}$ = the median of a sample

$\overline{\tilde{X}}$ = the arithmetic mean of a set of sample medians, based on repeated samples drawn from a process

k = the number of samples drawn from a process

Σ = summation (read as "Sigma")

= an operator or instruction meaning that all terms following this symbol should be added

Calculating the Arithmetic Mean

The arithmetic mean of a set of numbers is found by substituting into the following formula and solving for μ.

$$\mu = \frac{\Sigma X}{N}$$

Example 1 Consider the following weights in grams of ten filled containers.

541	521	522	539	530
516	538	519	525	514

The arithmetic mean is found by substituting and solving as follows:

$$\mu = \frac{541 + 521 + 522 + 539 + 530 + 516 + 538 + 519 + 525 + 514}{10}$$

$$= \frac{5265}{10}$$

$$= 526.5 \text{ grams}$$

Note. Although the method of computation is the same as the one illustrated above, different arithmetic means can be identified that are associated with different segments of data. As a result, these are introduced here in order to distinguish among them conceptually.

When a set of data is identified as a sample, the arithmetic mean is given by the formula

$$\overline{X} = \frac{\Sigma X}{n}$$

To illustrate, consider the first row in the previous example as a sample of five. The corresponding mean equals

$$\overline{X}_1 = \frac{541 + 521 + 522 + 539 + 530}{5}$$

$$= \frac{2653}{5}$$

$$= 530.6 \text{ grams}$$

Repeating the process for the second row or sample of five, we get

$$\overline{X}_2 = \frac{516 + 538 + 519 + 525 + 514}{5}$$

$$= \frac{2612}{5}$$

$$= 522.4$$

Another mean that will be used at a later point is the mean of the individual sample means. This is given as

$$\overline{\overline{X}} = \frac{\Sigma \overline{X}}{k}$$

Using the results of the previous illustration, we have

$$\overline{\overline{X}} = \frac{530.6 + 522.4}{2}$$

Recognize that numerically this value is the same as that of μ calculated in the first example; however, conceptually the two are different. The symbol μ really is used as a theoretical construct, and, if used to represent a process mean, its numerical value is never known in practice. Therefore, it can be estimated using $\overline{X}$ or $\overline{\overline{X}}$.

Finding the Median

In general, we stated that the median can be found as the middle value in a set of data. Here, we must distinguish between an odd and an even number of observations. In the case of an odd number, the median is found as the middle value of the observations when arrayed in order of increasing magnitude. For an even number of observations, the median is found as the arithmetic mean of the middle two observations in the array.

Example 2 Consider again the weights of the ten containers given in Example 1.

541	521	522	539	530
516	538	519	525	514

In order to find the median, we first array the weights in order of increasing magnitude:

514 516 519 521 $\boxed{522\ \ 525}$ 530 538 539 541

Since we are dealing with an even number of observations, the median is the mean of the middle two:

$$X_{50} = \frac{522 + 525}{2}$$

$$= \frac{1047}{2}$$

$$= 523.4 \text{ grams}$$

Example 3 Now consider each row of the original data to be separate samples of five. The corresponding medians are found as follows:

Array first sample

521   522 $\boxed{530}$ 539   541

$$\tilde{X}_1 = 530 \text{ grams}$$

Array second sample

514   516 $\boxed{519}$ 525   538

$$\tilde{X}_2 = 519 \text{ grams}$$

Note that in both of these cases, the number of observations is odd, and the median is found as the middle value in each array.

Example 4 For reasons that will become apparent later, we may also use the mean of sample medians. This is accomplished with the formula

$$\bar{\tilde{X}} = \frac{\Sigma \tilde{X}}{k}$$

Using the results of Example 3, we have

$$\bar{\tilde{X}} = \frac{530 + 519}{2}$$

$$= \frac{1049}{2}$$

$$= 524.5 \text{ grams}$$

ADDITIONAL IDEAS

Although the concept of an average tends to be a familiar one, methods of calculating the arithmetic mean and the median were illustrated in the earlier portions of this chapter. In addition, different ways of calculating these averages were introduced that will be of use at a later point. It is useful at this point, however, to discuss the meaning of an average and to begin to obtain some insight into its usefulness.

The Meaning of an Average

Generally speaking, when a set of measurements is to be represented or summarized with a single number, an average is used. No matter which average we compute, its value must lie between the highest and lowest values of the raw observations. Typically, an average will fall somewhere in the vicinity of the center of a frequency distribution; consequently, we think of it as a measure of central tendency.

EXHIBIT 1
AVERAGES AS MEASURES OF CENTRAL TENDENCY

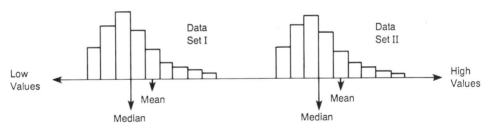

One way to understand averages is to compare frequency distributions describing different kinds of data. Consider the histograms presented in Exhibit 1 associated with two sets of data measuring the same characteristic on the same scale. Immediately, we can see that the observations described by distribution II lie above, or are higher in value than, those summarized by distribution I. By comparing a similarly computed average, whether it is the mean or median, or some other average, we are provided with the same basic information: the average associated with the distribution with higher values (II) tends to be larger than the *corresponding* average associated with the distribution with lower values (I). Although not true in all cases, typically the two values of a similarly computed average bear the same relationship to each other as do the individual values of the observations corresponding to the two distributions. In other words, an average describes the position or location of the "bulk" of the values of the observations.

When comparing sets of data using averages, the underlying frequency distributions should be of the same general shape. This is so because the mean and the median are affected by the skewness in a distribution. Both are pulled away from the mode, or value under the peak, in the direction of the skewness, or the tail of the distribution. Typically, the mean is affected more than the median. The mean and the median have the same numerical value when a distribution is symmetric.

The Weighted Arithmetic Mean

When a distinction is necessary, the mean calculated in the preceding examples can be referred to as the *unweighted* arithmetic mean. This can be contrasted with the *weighted* arithmetic mean, which is the arithmetic mean of a frequency distribution, or one such that individual values are weighted by varying frequencies of occurrence.

As an example, suppose we take 60 measurements corresponding to some characteristic and find that one of the measurements assumes a value of 1.0, five a value of 2.0, eleven a value of 3.0, thirteen a value of 4.0, and thirty a value of 5.0. If we ignore the number of times each value occurs and compute the mean, we would obtain the unweighted arithmetic mean as

$$\text{Unweighted Mean} = \frac{\Sigma X}{N}$$

$$= \frac{1.0 + 2.0 + 3.0 + 4.0 + 5.0}{5}$$

$$= \frac{15}{5}$$

$$= 3.0$$

The correct answer, however, is the mean of the following frequency distribution

X	f
1.0	1
2.0	5
3.0	11
4.0	13
5.0	30
Total	60

which is found to be a different value as

$$\text{Weighted mean} = \frac{\Sigma f X}{N}$$

$$= \frac{(1 \times 1.0) + (5 \times 2.0) + (11 \times 3.0) + (13 \times 4.0) + (30 \times 5.0)}{60}$$

$$= \frac{246}{60}$$

$$= 4.1$$

In order to explore the idea of weights further, consider **Exhibit 2** where four cases have different values of the frequencies; all of these cases are based on the same number of observations and values of X. In the first case (1), all of the frequencies are equal, and the mean has the same value as the unweighted case (ie., 3.0) computed earlier. Whenever all frequencies are identical, the weighted and unweighted means have the same numerical value. Conceptually, the two situations are alike since they correspond to distributions that are uniform; they are special forms of symmetric distributions.

EXHIBIT 2
MEAN VALUES CORRESPONDING TO DISTRIBUTIONS WITH
VARYING FREQUENCIES

Case	(1)	(2)	(3)	(4)	Combined
X	f_1	f_2	f_3	f_4	f
1.0	12	3	30	1	46
2.0	12	12	13	5	42
3.0	12	30	11	11	64
4.0	12	12	5	13	42
5.0	12	3	1	30	46
Total	60	60	60	60	240
MEAN	3.0	3.0	1.9	4.1	3.0

The distribution given as (2) also has a mean of 3.0 because it is symmetric. Although this was true of the first case, case (2) is different with respect to the way the observations are distributed about the center of the distribution. The distribution "tails-off" to lower values on both sides of the peak; the pattern of variability is different.

In the last two cases, (3) and (4), the distributions are skewed, but in opposite directions. The means are not equal to 3.0 but lie close to the points of highest frequency and also are influenced by the lower frequencies representing the extremes; consequently, in the direction of skewness.

The final column in the table presented in Exhibit 2 gives the combined frequencies of the preceding four cases. In this case, the resulting distribution is symmetric as a result of the frequencies in the individual cases, and the mean equals 3.0. If, on the other hand, we dropped case (4) and combined the frequencies for cases (1) through (3), the calculated mean would equal 2.633, which lies between 1.9 and 3.0, and is different than any of the mean values presented in the exhibit.

Essentially, the means of the combined distributions cited in the preceding paragraph are the result of different mixtures of distributions. One may liken this to the output of different streams of production, whereby the streams are not identified or isolated in the final output. The concept, therefore, of different weightings reflecting different underlying distributions becomes important when analyzing process results. More will be said about this is Chapter 8 when interpreting control chart patterns.

Before closing, it should be noted that values of the median although different numerically, will be affected by differences in frequencies or distribution shape in a manner somewhat similar to means. Medians, however, are more resistant to extremes than means. When dealing with descriptive studies in general, a choice can be made between the mean and the median based on the descriptive ability of each measure. In the case of control charting used for process control, a choice between the two generally is made merely on the basis of ease of computation.

MEASURING DISPERSION

At the end of the previous chapter we referred to different patterns of variation about the "center" of a distribution. This pattern can be thought to be composed of two components: the amount of variation, or spread, and the shape in terms of symmetry, skewness, or the direction of skewness. Although shape is important to consider, in this chapter we shall concentrate on measures of the absolute amount of variability in a set of measurements. Together, measures of central tendency and dispersion are the ones most widely used.

Just as in the case of central tendency, numerous measures of variability are available, each possessing different properties that are desirable under different circumstances or problem settings. We shall focus on the two most used in process control: the range and the standard deviation. Two main objectives stem from the material presented. On the one hand, it is necessary to know how the measures are calculated. On the other hand, understanding the measures conceptually is important, especially the standard deviation, which is used to identify different types of variation.

The *range* simply is the difference between the largest and smallest values observed in a set of data. As such, the range is easily calculated and easy to understand. The *standard deviation*, however, is not as easily calculated or understood: it is defined as the square root of the average of the squared deviations of the observations from the mean. Although we shall be able to visualize how the standard deviation measures variability, we shall not be able to demonstrate its importance fully since this is rooted in more theoretical results not considered.

The following symbols will be used in this and later chapters:

X = a numerical value of the quantity measured

N = the total number of observations

n = the sample size

μ = the mean, or the process mean

σ = the standard deviation or the process standard deviation (read as "Sigma"; lower case)

   = alternatively, a standard deviation of a sample when all of the observations, n, is used as a divisor

$\overline{X}$ = the arithmetic mean of a sample

S = the sample standard deviation

$\overline{S}$ = the arithmetic mean of a set of sample standard deviations, based on repeated samples drawn from a process

R = the range of a sample

$\overline{R}$ = the arithmetic mean of a set of sample ranges, based on repeated samples drawn from a process

k = the number of samples drawn from a process

Σ = summation (read as "Sigma")

Calculating the Range

The range of a set of observations is found as the difference between the largest and smallest values. This is given in terms of the formula

Range, R = Largest *minus* Smallest

Example 1 Consider the weights in grams of the ten filled containers of the previous chapter:

541	521	522	539	530
516	538	519	525	514

The range is found as

$$R = 541-514$$

$$= 27 \text{grams}$$

Note. Although the method of computation is the same as illustrated above, again let us distinguish between the first and second rows as separate samples, 1 and 2, respectively. The ranges of each of the two samples are found as follows:

$$R_1 = 541-521$$

$$= 20 \text{ grams}$$

$$R_2 = 538 - 514$$

$$= 24 \text{ grams}$$

Another measure that is useful in process control is the mean of the sample range. This is given as

$$\overline{R} = \frac{\Sigma R}{k}$$

Using the results of the previous illustration, we have

$$\bar{R} = \frac{20 + 24}{2}$$

$$= 22 \text{ grams}$$

Recognize that unlike the arithmetic mean of the sample means, the mean of the sample ranges does *not* equal the range of all of the observations combined. For reasons that will become clearer later, R is to be used for estimation purposes, but not directly for estimating the overall process range.

Calculating the Standard Deviation

The standard deviation of a set of observations is defined by the formula

$$\sigma = \sqrt{\frac{\Sigma(X-\mu)^2}{N}}$$

In order to find the standard deviation of N observations, the formula tells us to (a) find the mean, (b) subtract the mean from each observation, (c) square each of these differences, (d) sum them, (e) divide by the total number of observations, and (f) extract the square root.

Example 2 Again consider the weights of the containers provided in Example 1.

541	521	522	539	530
516	538	519	525	514

In order to find the standard deviation, first find the mean

$$\mu = \frac{\Sigma X}{N}$$

$$= \frac{5265}{10}$$

$$= 526.5 \text{ grams}$$

Then, the standard deviation is found by substituting into the formula and solving as

$$\sigma = \sqrt{\frac{\Sigma(X-\mu)^2}{N}}$$

$$= \sqrt{\frac{(541 - 526.5)^2 + (521 - 526.5)^2 + \dots + (514 - 526.5)^2}{10}}$$

$$= \sqrt{\frac{210.25 + 30.25 + 20.25 + 156.25 + 12.25 + 110.25 + 132.25 + 56.25 + 2.25 + 156.25}{10}}$$

$$= \sqrt{\frac{886.5}{10}}$$

$$= 9.42 \text{ grams}$$

Calculating the Sample Standard Deviation

The standard deviation presented above is the definitional form that is true when applied to any set of numbers in general. When working with samples, the result usually is employed as an estimate and, because of certain properties of the standard deviation, a modification of the above definition typically is used. Although the terminology is not truly precise, through common usage and convention the modified quantity is identified as the sample standard deviation. More will be said about the distinction in Chapter 5. Here, we shall illustrate the calculations involved.

The so-called sample standard deviation is defined by the formula

$$S = \sqrt{\frac{\Sigma(X-\overline{X})^2}{n-1}}$$

Notice that three things are different about this formula when compared to the one given above. First, the deviations are calculated about the sample mean, $\overline{X}$, rather then μ. Second, the sample size denoted by n is used in the denominator. Third a "1" is subtracted from the number of observations before dividing.

Example 3 Consider the weights given earlier as

541	521	522	539	530
516	538	519	525	514

The mean has already been found, however, we shall relabel it as $\overline{X}$ and let it equal 526.5. The sample standard deviation then is found as

$$S = \sqrt{\frac{(541 - 526.5)^2 + (521-526.5)^2 + \ldots + (514-526.5)^2}{10-1}}$$

$$= \frac{886.5}{9}$$

$$= \sqrt{98.5}$$

$$= 9.92 \text{ grams}$$

Notice that the result is a little different and larger than that of Example 2. Basically, however, the two numbers measure the same thing. On the one hand, when the number of observations is larger, the effect of subtracting one becomes negligible. On the other hand, there is a relationship between the formulas. If we let n = N and $\overline{X}$ = μ, then

$$\sqrt{\frac{\Sigma(X-\overline{X})^2}{n-1}} = \sqrt{\frac{N}{N-1}}\,\sigma$$

$$= \sqrt{\frac{N}{N-1}} \times \sqrt{\frac{\Sigma(X-\mu)^2}{N}}$$

Example 4 To show the relationship between the results of Example 3 and Example 4,

$$\sqrt{\frac{N}{N-1}}\,\sigma = \sqrt{\frac{10}{9}}\,(9.42)$$

$$= \sqrt{1.11111....}\,(9.42)$$

$$= 1.05409\,(9.42)$$

$$= 9.92$$

Note. Again consider the data given in the examples as two separate samples:

Sample 1:	541	521	522	539	530
Sample 2:	516	538	519	525	514

For each sample, we can use the formulas given above to calculate the standard deviation and the sample standard deviation. The results are as follows:

	Standard Deviation	Sample Standard Deviation
Sample 1	$\sigma_1 = 8.31$	$S_1 = 9.29$
Sample 2	$\sigma_2 = 8.64$	$S_2 = 9.66$

Based on either method of computation, the means of the standard deviations also are useful in process control. The formulas for these are given as

$$\bar{\sigma} = \frac{\Sigma\,\sigma}{k}$$

$$\bar{S} = \frac{\Sigma\,S}{k}$$

Substituting and solving, for purposes of illustration, we have

$$\bar{\sigma} = \frac{8.31 + 8.64}{2}$$

$$= \frac{16.95}{2}$$

$$= 8.48$$

$$\bar{S} = \frac{9.29 + 9.66}{2}$$

$$= \frac{18.95}{2}$$

$$= 9.48$$

Notice that neither mean equals the value of σ for the two samples combined. Later, we shall see why the averaging process associated with sample results is important when related back to the corresponding value for an entire process.

Overall, the points that should be kept in mind are that either measure can be used in the applications that follow with appropriate adjustments and that conventionally the sample standard deviation as given is becoming the standard.

Easing Calculations

Calculating the standard deviation by hand using either formula is a little tedious because of the individual squared deviations that must be computed. Many hand-held calculators, especially scientific and business versions, are specially pre-programmed to obtain the standard deviation easily. The keys used to get the final result generally are labelled as "σ_n" and "σ_{n-1}", which represent σ and S, respectively, as we have designated them here. When computer programs are employed to do process control work, the standard deviation generally is provided automatically along with other results. Usually, the "n – 1" version is furnished.

The definitional forms of the standard deviation presented and illustrated above are important in order to understand the structure and meaning of these measures. When it actually is necessary to hand-calculate either quantity, shortcut formulas are available that can be used to ease the task somewhat.

In order to introduce the shortcuts, it is instructive to introduce them in terms of the concept of a sum of squares, which is the numerator under the radical and is common to both. That is, say, in the case of the sample standard deviation,

$$\text{Sum of Squares} = \Sigma \, (X - \overline{X})^2$$

Algebraically, one can show that this can be rewritten as

$$\Sigma(X - \overline{X}) = \Sigma X^2 - \frac{(\Sigma X)^2}{n}$$

In other words, the sum of the squared deviations from the mean (or sum of squares) can be written as the sum of the squares of the individual observations $[\Sigma X^2]$ and the square of the sum of the observations $[(\Sigma X)^2]$. By using this result, it is not necessary to calculate a set of differences from the mean. The standard deviation, therefore, can be found more easily by calculating the sum of squares using the shortcut, dividing either by n or n-1 and extracting the square root. The sum of the squared observations is simplified using many calculators since an automatic cumulative feature is built-in.

Example 5 Consider Sample 1 of the previous note.

$$541 \quad 521 \quad 522 \quad 539 \quad 530$$

$$\Sigma X^2 = 541^2 + 521^2 + 522^2 + 539^2 + 530^2$$

$$= 29681 + 271441 + 272484 + 290521 + 280900$$

$$= 1408027$$

$$\Sigma X = 541 + 521 + 522 + 539 + 530$$

$$= 2653$$

$$\text{Sum of Squares} = \Sigma X^2 - \frac{(\Sigma X)^2}{n}$$

$$= 1408027 - \frac{(2653)^2}{5}$$

$$= 1408027 - \frac{7038409}{5}$$

$$= 1408027 - 1407681.8$$

$$= 345.2$$

The standard deviation, therefore, equals

$$\sigma = \sqrt{\frac{345.2}{5}}$$

$$= 8.31$$

And, the sample standard deviation equals

$$S = \sqrt{\frac{345.2}{5-1}}$$

$$= 9.29$$

Note. The preceding discussion focused on the standard deviation, σ, and the sample standard deviation, S. Generally, when these quantities are introduced the discussion is preceded by introducing quantities referred to as the *variance*, σ^2, and the sample variance, S^2, which are defined by the following formulas

$$\sigma^2 = \frac{\Sigma(X - \mu)^2}{N}$$

$$S^2 = \frac{\Sigma(X - \overline{X})^2}{n-1}$$

Variances are nothing more than standard deviations expressed in *squared units*. Consequently, as measures of variation they impart the same basic information, but in terms of squared units. Whether a variance or a standard deviation is used in a particular problem depends upon the type of application. Since standard deviations are more common in terms of the tools presented, we have emphasized it here. When referring to variation in a body of data for a particular distribution in the absence of a particular application, we may use either form interchangeably.

ADDITIONAL IDEAS

The importance of the concept of variability cannot be stressed enough. Moreover, although measures of central tendency or averages are commonly understood independently of the material presented in the previous chapter, understanding measures of variation is not quite as common. Actually, it is difficult to obtain a full appreciation of the measures by applying the formulas. Here we shall begin to deal with the interpretation of the range and standard deviation as descriptive measures and also gain some insight into the concept of averaging or pooling variations.

Interpreting the Range and Standard Deviation

In order to understand what is meant by the phrase "measuring variability," consider the two sets of numbers appearing below.

Set 1	Set 2
7 8 9 10 11	3 6 9 12 15

Both sets of numbers have the same mean and median, equal to 9; however, the differences among the values are greater in Set 2. An easy way to see this is to compare the two values on a scale in the form of a line graph as follows:

```
Set 1                  *   *   *   *   *
          3—4—5—6—7—8—9—10—11—12—13—14—15
Set 2     *       *       *       *       *
```

The asterisks represent the position of the values for the two sets of values on a common scale. We immediately can see that the values corresponding to Set 2 are spread more than those in Set 1. In other words, there is more variability or dispersion in Set 2.

The simplest way to measure the variability described is in terms of the ranges, which equal 4 and 12, respectively. These values confirm what we established visually, that Set 2 is more variable than Set 1. By calculating the range, we have described the variability in terms of a single summary, or descriptive, measure.

Now consider two sets of data, A and B, compared directly in terms of line graphs.

```
                              *
                          *   *   *
Set A (R = 12)  *         *   *   *   *
              1—2—3—4—5—6—7—8—9—10—11—12—13
Set B (R = 12)  *   *   *   *   *   *   *   *   *   *   *   *   *
```

Because both sets of values described in the diagram have the same extremes, their ranges are identical and equal 12. It is obvious, however, that the pattern of variability among the points in Set A is different than those in Set B. The points in A cluster between the values of 5 and 9, while those in B are evenly distributed between 1 and 13 and are "more spread apart." Comparing the ranges without looking at the data really does not give us much information about the comparative variation.

The standard deviation, which is not as easy to visualize and calculate, does account for the differences in the way the observations are distributed and is weighted in favor of the bulk of the observations rather than the extremes. In the case of the first two sets of values presented, the values of the standard deviation and sample standard deviation are as follows:

	Set 1	Set 2
σ	1.41	4.24
S	2.90	4.27

The corresponding results for the remaining two sets of values are

	Set A	Set B
σ	2.54	3.47
S	2.90	4.27

Notice in both cases, the standard deviations for the second set is greater than that of the first. When making the comparison for descriptive purposes, it doesn't matter which is compared (ie., σ or S) as long as the same measure is used for each data set. Obviously, S will be greater than the corresponding σ in each case. On a comparative basis, either imparts the same basic information.

In general, a measure of variability should assume larger values for a set of highly dispersed values and, at the other extreme, when all values are identical, a measure of variation should equal zero. To further understand the nature of the measures introduced, consider the three data sets in **Exhibit 1** presented in the form of graphed frequency distributions with the same mean. By examining the three diagrams, we obtain an impression about the standard deviation and range as descriptive measures of variation.

The first case deals with data where all 15 observations are identical. When graphed, the data plot as a single point, which indicates that there is no variation. Consequently, both the standard deviation and the range equal zero. The second case also deals with 15 observations. Although the most frequently occurring value remains at six, some observations are different. This results in a frequency distribution that is somewhat peaked and narrow, indicating a concentration of observations about six. The corresponding standard deviation and range equal 0.93 and 4, respectively.

As the observations become more dispersed or differ more from six, we see that the distribution that is described in the third diagram is less peaked, flatter and wider, or more spread-out, than the previous case. The standard deviation, equal to 2.8, is greater than the standard deviation in the second case. Although numerically different, the same is true of the range. Also reported is the sample standard deviation, which assumes values that bear the same relationship to one another as the other measures. Numerically, the values in the second and third case are slightly larger than the corresponding standard deviation cited above.

EXHIBIT 1
COMPARISON OF MEASURES OF VARIABILITY FOR DIFFERENT PATTERNS OF VARIATION

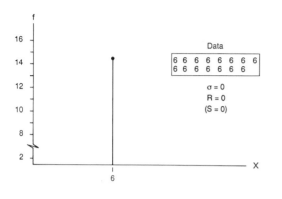

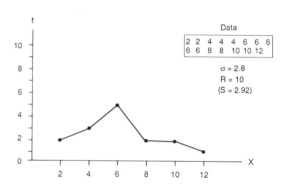

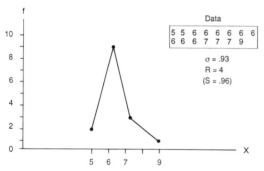

> **Note.** Implied in the above discussion is the fact that the standard deviation considers more information in the data than the range when used as a descriptive measure. By using the term descriptive it is meant that the variability in a given set of data or in a process is described or represented by a single measure. When sample data are used to estimate process variability, process variability is considered in terms of the standard deviation however both the range and the sample standard deviation can be used in the estimation process. In other words, when not used in a descriptive sense but for purposes of estimation, it is possible to use both measures despite the discrepancy between the two with respect to descriptive ability. This note has not been introduced to provide full understanding at this point, but merely to distinguish between different types of applications and to begin to resolve certain seeming inconsistencies within the field. More will be said about the estimation problem in Chapter 5.

Pooled Sums of Squares and the Standard Deviation (Optional)

It was stated in an earlier note that the mean of the standard deviation calculated for separate samples is not equal to the standard deviation of all of the observations obtained by combining the samples into one group. This difference in values results from the algebraic properties of the standard deviation, the fact that potentially different sources of variation are being measured, and that the calculations for each measure are computed about different means. The concept can become important when attempting to obtain an appropriate measure of variation in order to assess the capability of a process and, also, in order to understand the nature of a process in terms of different sources of variation.

In order to understand the above statements, we shall first consider the concept in terms of sums of squares, already presented with respect to the shortcut formulas, and then illustrate it in terms of a numerical example. The *total* sum of squares, or sum of squares for an entire group of observations, can be written as

$$\Sigma (X - \bar{\bar{X}})^2 = \Sigma \Sigma(X - \bar{X})^2 + n\Sigma(\bar{X} - \bar{\bar{X}})^2$$

In other words, variation of the entire group of observations measured in terms of the sum of squares, $\Sigma(X - \bar{\bar{X}})^2$, can be expressed as the sum of two components of variation. The first of these, $\Sigma \Sigma(X - \bar{X})^2$, represents the sum of squares *within* each sample, or subgroup, added together. The second, $n\Sigma(\bar{X} - \bar{\bar{X}})^2$, is a sum of squares not introduced earlier, which is based on the differences between the subgroup means, $\bar{X}$, and the overall mean, $\bar{\bar{X}}$. The first measures variation within samples and is referred to as the *pooled* within sum of squares; the second measures variation *among* samples. Together, they comprise total variation, or the total sum of squares. The quantity S introduced earlier is used in process control and is based solely on the measure within samples.

Example 6 This example will demonstrate the additivity of the sums of squares, and re-calculate and compare the various standard deviations based on the results.

Consider the observations given earlier:

(Sample 1)	541	521	522	539	530
(Sample 2)	516	538	519	525	514

The total sum of squares is calculated directly as

$$\Sigma (X - \bar{\bar{X}})^2 = (541 - 526.5)^2 + \dots\dots + (514 - 526.5)^2$$

$$= 886.5$$

The within sample, or subgroup, sum of squares for each sample individually is found as

$$\Sigma(X_1 - \bar{X}_1)^2 = (541 - 530.6)^2 + \dots + (530 - 530.6)^2$$

$$= 345.2$$

$$\Sigma(X_2 - \overline{X}_2)^2 = (516 - 522.4)^2 + \ldots + (514 - 522.4)^2$$

$$= 373.2$$

Based on these results, the pooled within subgroup sum of squares is the sum of the two, which is calculated as

$$\Sigma\Sigma(X - \overline{X})^2 = 345.2 + 373.2$$

$$= 718.4$$

The among subgroup sum of squares is found as

$$n\Sigma(\overline{X} - \overline{\overline{X}})^2 = 5[(530.6 - 526.5)^2 + (522.4 - 526.5)^2]$$

$$= 5(33.62)$$

$$= 168.1$$

By adding the pooled within and among subgroup sum of squares, we can see that the sum represents the total sum of squares found initially. That is,

$$\Sigma(X - \overline{\overline{X}})^2 = \Sigma\Sigma(X - \overline{X})^2 + n\Sigma(\overline{X} - \overline{\overline{X}})^2$$

$$= 718.4 + 168.1$$

$$= 886.5$$

The standard deviation based on this total, therefore, is given as

$$S = \sqrt{\frac{\Sigma(X - \overline{\overline{X}})^2}{nk - 1}}$$

$$= \sqrt{\frac{886.5}{5(2) - 1}}$$

$$= \sqrt{\frac{886.5}{9}}$$

$$= 9.92$$

Using the individual subgroup sums of squares, the standard deviations of the individual samples are found as

$$S_1 = \sqrt{\frac{\Sigma(X_1 - \overline{X}_1)^2}{n - 1}}$$

$$= \sqrt{\frac{345.2}{4}}$$

$$= 9.29$$

and

$$S_2 = \sqrt{\frac{\Sigma(X_2 - \overline{X}_2)^2}{n - 1}}$$

$$= \sqrt{\frac{373.2}{4}}$$

$$= 9.66$$

Based on the last two results, the mean of the individual sample standard deviations equals

$$\overline{S} = \frac{9.29 + 9.66}{2}$$

$$= 9.48$$

Two points are illustrated in terms of the above calculations. First, the sum of the within and among sum of squares equals the total sum of squares. Second, the standard deviation, $\overline{S}$, obtained as the average of the two sample standard deviations is not equal to the standard deviation calculated on the basis of the total sum of squares.

Typically, the ideas just presented are considered in more advanced presentations of statistical methods. Actually, they are quite fundamental since they illustrate the essential components of the process control problem. Later, we shall see that if a process is in control, among-subgroup and pooled-within-subgroup variation should be similar except for chance variation. When a process is not stable, or out of control, the contribution made by the among-subgroup component of variation will reflect non-chance factors, and the two will not be of the same order of magnitude. The concept has been introduced here as a computational exercise merely to provide an early exposure to some properties of the standard deviation.

PROBABILITY AND SAMPLING

MODULE SUMMARY

The preceding module introduced the concept of variation in data and presented ways to describe this variation. Use of measurements and ways of dealing with the variation in measurements are critical when dealing with process control problems statistically. By using the term statistically we mean that probabilistic based methods are used to analyze data in order to reach conclusions in the presence of variation. In other words, ideas about chance and randomness can be related to the way data behave and are useful in drawing conclusions in the presence of the variation in these measurements.

MODULE B

**PROBABILITY
AND SAMPLING**

This module begins in Chapter 4 by introducing certain fundamentals about chance and probability. Mainly, it is important to understand the meaning of the probability of an event. Based on this definition, we are able to develop the concept of a probability distribution, that is the fundamental statistical tool used to model the way data behave, or to model a frequency distribution.

Many special probability distributions are available that can be matched to a set of underlying assumptions in order to solve specific problems. Of these distributions, the uniform, binomial, and Poisson are illustrated since they are referenced in later sections. Emphasis, however, is placed on the normal distribution which is used as a foundation for many of the procedures in process control and the material that follows. Chapter 4 concludes with a brief discussion of the outcome of a simulated set of sample results, whose underlying principles are the basis for the methods of inference presented.

Chapter 5 introduces the fundamental concepts of statistical inference that form the basis for understanding why control charts are constructed the way they are and the meaning of control chart results. Included in this section are fundamental concepts of sampling and sampling distributions, which are the basic tools underlying all methods of statistical inference. Related methods of estimation are presented, and the concept of a test of statistical hypothesis is introduced. A statistical test of an hypothesis forms the basis for the construction of control charts.

Although many different types of tests are available that are applicable to a variety of problems, the principles of testing are presented in terms of means in order to present the underlying concepts. These concepts, with the aid of different sampling distributions, are applicable to other cases.

The module concludes with Chapter 6, which introduces control charts simply in order to bridge the gap between concepts of chance variation, sampling, estimation, hypothesis testing, and control charts. The material is repeated but with more detail in Module C, which presents a complete development of control charts for variables.

BASIC PROBABILITY AND THE NORMAL CURVE

The words chance and random are quite common. Most of us know they have something to do with probability, gambling, and lotteries. Moreover, there is a growing awareness that the concepts are important in terms of process control and improvement. There still is, however, a barrier in many people's minds when they are told that their process is stable or in control when it "behaves like a lottery!" If "random" implies lack of predictability, then how can a process that is unpredictable be stable?

In this section we shall introduce the concepts of randomness and probability more formally than they are treated in everyday encounters. We shall then introduce some techniques and methods for calculating probabilities, which at a later point can be used as a basis for action regarding the control of a manufacturing process.

The symbols used in this section are as follows:

S	=	sample space
E	=	event
N(S)	=	the number of possible outcomes or sample points in the sample space
N(E)	=	the number of outcomes or sample points in the event E
P(E)	=	the probability of the event E
P(A & B)	=	the joint probability of events A and B, or the probability that A and B occur simultaneously
P(A|B)	=	the conditional probability of event A given event B

43

x	= a value of a variable or measured quantity that varies at random
$f(x)$	= a function or rule used to assign probabilities to the values of x
$P(x = a)$	= the probability that the random variable assumes a numerical value equal to a; sometimes abbreviated as $P(a)$
$F(x)$	= the cumulative probability
μ or $E(x)$	= the mean or expected value of a probability distribution
σ^2	= the variance of a probability distribution
σ	= the standard deviation of a probability distribution
n	= the number of repetitions, or the sample size
P	= the probability that an outcome will occur on a single repetition
Q	= the probability that an outcome will not occur on a single repetition
	= $1 - P$
e	= the mathematical constant that is the base of natural logarithms
	= $2.718281828\ldots\ldots$
π	= a mathematical constant
	= $3.141592654\ldots\ldots$
$P(a \leq x \leq b)$	= the probability that a random variable assumes a value in an interval between two specific numbers, "a" and "b"
$P(\mu \leq x \leq a)$	= the probability that a random variable assumes a value in an interval between the mean and a specific number "a"
z	= the "standardized normal variate," needed when using Table A-1 in the Appendix to find normal probabilities

Randomness and the Probability of an Event

The easiest way to introduce the concept of randomness is to define a *random experiment*. This is a phenomenon having the properties that (1) all possible outcomes can be specified in advance, (2) it can be repeated, and (3) the same outcome does not necessarily occur on various repetitions, so that the actual outcome is not known beforehand. For example, measurement of the blood pressure of a group of individuals constitutes a series of random experiments. Checking a car's gas mileage is another example of such an experiment. The simplest and most popular illustration is flipping a coin.

It should be clear that the word experiment is used in a very broad sense. Typically, an experiment is thought of in terms of a test or set of trials under controlled laboratory conditions. The term is used here to refer to any type of situation that can be experienced on a repeated basis. Consequently, recording the value of a critical dimension on a series of manufactured items, in the above context, constitutes a random experiment.

The essential feature of a random experiment is that the outcome of a single repetition is not predictable in advance. In other words, there is uncertainty surrounding the outcome. Although an individual occurrence associated with a random experiment cannot be predicted exactly, it can be shown that something can be said about the frequency of occurrence in a large number of repetitions. It is through probability that this can be accomplished, and becomes the basis for resolving problems dealing with uncertainty.

In order to develop our understanding, the following concepts are defined:

> A **sample space** is a set or collection of all possible outcomes associated with a random experiment.

> An **event** is defined as any collection of elements or outcomes within the sample space.

> An event is said to **occur** if the outcome of a random experiment, once performed, is contained within the given event.

Consider a very simple case in order to illustrate the three definitions. Suppose we are interested in an experiment in which a single coin is flipped twice. The sample space can be designated in the following way:

$$\{HH, HT, TH, TT\}$$

We can obtain two heads in a row, or a head on the first flip and a tail on the second, and so on. In this example each possible outcome or *sample point* is considered separately in characterizing the sample space.

Sample spaces may be characterized in ways that conceal the identity of the sample points. For example, another legitimate way to specify the space corresponding to the two flips of a coin would be in terms of the number of heads as

$$\{0, 1, 2\}$$

The essential points to keep in mind are that a properly designated sample space considers or exhausts all possible outcomes, and that there is no overlap among the elements within the space; namely, that they are mutually exclusive. Moreover, there is no single correct way of specifying a sample space. The specification depends on the ultimate use in a particular problem.

Based on the sample space {HH, HT, TH, TT}, there are a number of events that can be considered. For example, two possibilities are {HT, TT} and {HH, HT, TH}. Events can also be characterized in other ways, depending on the way the sample space is specified.

Once an experiment is performed, the outcome will always be represented by an individual possibility or single sample point. An event is said to occur if the outcome is contained within the event, regardless of the way the event is characterized or the number of sample points contained in the event. For example, if we obtain two tails after flipping a coin twice, we can say that the event {HT, TH, TT} has occurred since it contains the outcome, TT. Furthermore, any other event that contains the outcome, TT, also is said to occur.

Based on the above definitions, we can define the probability of an event as follows:

> The **probability of an event** is a number, between zero and one, that represents the proportion of times the event occurs in an infinite sequence of trials of the same random experiment.

In order to understand this definition, imagine repeating a random experiment endlessly. You define a particular event, and each time the experiment is performed you record whether the event has occurred or not. As the number of trials of the experiment becomes large, begin computing the ratio of the number of times the event occurs to the total number of trials, or compute the relative frequency of occurrence. Loosely speaking, the larger the number of trials the closer the ratio or relative frequency will be to the probability of the event. Strictly, the computed ratio approaches the probability in "the limit" as the number of trials approaches "infinity."

If, for example, we are told that the probability of selecting "a spade" on a single draw from a well-shuffled card deck is 0.25, how should we interpret this number? Based on the definition, we can say that 0.25 or one fourth of an "infinite" number of draws would yield a spade. In more practical terms, we can say that if a large number of trials were performed, "close" to 25 percent would yield a spade. Recognize that the probability statement tells us nothing about which of the trials compose the 25 percent!

Consider another situation in which an assembly line item is produced. We are told the probability that a defective item will be produced is 0.05. Using the given definition, we can say that 0.05 or 5 percent of all the items produced will be defective, under the assumption that the manufacturing process continues endlessly into the future. More practically, we can say that large shipments taken from this process will contain nearly 5 percent defective items.

The above definition states that the probability of an event is a number between zero and one. A *zero* probability implies that an event is impossible, whereas a probability of *one* signifies that an event will occur with certainty, or 100 percent of the time. If zero, we know the event will not occur; if one, we know it must occur. Any other value provides information about the relative frequency of occurrence over a long sequence of trials, and it is really toward these values that we concentrate our interest.

The simplest method for computing the probability of an event follows directly from its definition. In order to apply the definition however, it is necessary to introduce two *assumptions*. For the present, we shall consider cases where the sample space is finite, or spaces for which we can actually count the exact number of outcomes or sample points in the space. On the other hand, we shall assume that the random experiment is such that each of the sample points is equally likely, which means that each possible outcome has the same probability of occurrence attached to it.

Based on the two assumptions provided, we can *compute* the probability of an event according to the formula

$$P(E) = \frac{N(E)}{N(S)}$$

This formula tells us to compute the ratio of the number of ways an event can occur to the total number of possible outcomes.

Example 1   Suppose a single coin is flipped twice, and we are interested in finding the probability of obtaining "at least one tail."

The sample space for the experiment is {HH, HT, TH, TT}. "As least one tail" can be obtained in one of three ways: HT, TH, or TT. The probability then becomes

$$P(\text{At least one tail}) = \frac{3}{4}$$

$$= 0.75$$

Example 2   A shipment of 5000 auto parts contains 250 that are defective. If one part is drawn from the shipment at random, what is the probability that it will be defective? The answer is found as

$$P(\text{Defective}) = \frac{\text{Number of Defectives}}{\text{Number of Parts}}$$

$$= \frac{250}{5000}$$

$$= 0.05$$

The preceding treatment of probability was quite simple. Ultimately, our objective is to move away from these simple idealized cases, but we must first obtain the basis for dealing with more complicated and realistic problems. What we need are ways of establishing theoretical probabilities that describe different kinds of problems.

The last statement suggests that, to be useful, the theoretical probability should describe the relative frequency that would be observed if we were actually confronted with random phenomena a "very large number of times."

As an alternative to determining the probability of an event theoretically, we may estimate it empirically by actually performing a large number of trials and computing the observed relative frequency. Since it is obvious that we are not able to perform a random experiment forever, the trials we do make constitute a sample of all possible trials. This would, therefore, result in an estimate of the required probability.

Regardless of the method of computation, we shall use the definition of probability presented in this section throughout the remaining portions of this book. As the kind of problem varies, different methods of computation are required. One procedure has been given. Others will be considered. Some will apply to sample spaces that assume different forms.

Other Types of Probabilities

We have introduced the idea of a probability in terms of a single event. By introducing a second event, it is possible to extend the idea of a probability of an event to events relating to both events. These we shall identify as joint and conditional probabilities. (It is not absolutely essential to read this material. It is being introduced in order to present definitions that are used in the next section of this chapter. The rules are drawn on briefly in a portion of Chapter 8 regarding the interpretation of control charts.)

The following definitions can be used to distinguish among the different types of probabilities:

1. The **joint probability** of two events is the probability that the two events occur simultaneously, or at the same time.

2. A **conditional probability** is the probability that a particular event will occur, *given* that another event occurs or is assumed to occur.

3. When more than one event is isolated, the probability of one particular event, irrespective of the others, is named a **marginal probability**.

In order to illustrate the meaning of the different types of probabilities, consider the following record of cracked caps received from three suppliers.

BOTTLE CAP RECORD BY SUPPLIERS X, Y, and Z

	X	Y	Z	Total
Cracked Caps(C)	15,000	10,000	13,000	38,000
Uncracked Caps (U)	85,000	140,000	112,000	337,000
Total	100,000	150,000	125,000	375,000

Based on the information given, we can now identify different events as C, U, X, Y, and Z. Using the above information, various joint, conditional, and marginal probabilities are presented in **Exhibit 1**, based on the frequencies provided.

Assuming the bottle cap record reflects the performance of the various suppliers, the above values were used as inputs in order to calculate the probability estimates in Exhibit 1. As an example, consider the value 0.027 corresponding to the cell identified by C and Y. The joint probability is found as

$$P(CY) = \frac{10000}{375000}$$

$$= .027$$

This number tells us that .027 or 2.7 percent of all caps are simultaneously cracked and originated from Suppler Y.

EXHIBIT 1
EXAMPLES OF JOINT, CONDITIONAL, AND MARGINAL PROBABILITIES
Bottle Cap Record By Supplier

	Joint Probabilities			
	X	Y	Z	Marginal Probabilities
C	.040	.027	.035	.101
U	.227	.373	.299	.899
Marginal Probabilities	.267	.400	.333	1.000

	Conditional Probabilities		
	X	Y	Z
C	.150	.067	.104
U	.850	.933	.896
	1.000	1.000	1.000

Instead, if we were interested in the probability of a cracked cap irrespective of the supplier, this would be found as

$$P(C) = \frac{38000}{375000}$$

$$= .101$$

which tells us that 10.1 percent of all caps are cracked.

The conditional probability of a cracked cap given Supplier Y equal to 0.067 is found as

$$P(C \mid Y) = \frac{10000}{150000}$$

$$= .067$$

In this case, the space is restricted to the 150,000 caps from Supplier Y instead of the total number of caps equal to 375,000. The remaining probabilities in the two tables presented in the exhibit can be found in a similar fashion.

Notice that the marginal probability of a cracked cap, $P(C)$, is not equal to the conditional probability, $P(C \mid Y)$. That is,

$$P(C) = .101$$

$$\neq P(C \mid Y) = .067$$

In such a case we say that the events C and Y are statistically dependent. If the marginal probability equalled the conditional, they would be independent.

In general, two events, A and B, are said to be **independent** if information about B tells nothing about the probability of occurrence of A. In terms of the probability of events, this occurs when

$$P(A \mid B) = P(A)$$

In other words, two events are independent when the conditional probability of one event, A, given another event B, is equal to the marginal probability of the first event. In the cracked cap example, independence between cap status and supplier would occur if the probabilities in each row of the second table in Exhibit 1 were equal to the corresponding marginal probability. If this were the case, the probability of selecting or generating a cracked cap would be the same regardless of the supplier.

Since we can see that the conditional probabilities in the exhibit vary and do not equal the corresponding marginal probabilities, cap status and supplier are said to be dependent. In general, two events, A and B, are said to be **dependent** if information about B tells something about A. In terms of probability, this means

$$P(A \mid B) \neq P(A)$$

In other words, the events, A and B are dependent when the conditional probability of A given B is not equal to the marginal probability of A. Dependence between characteristics implies that a relationship exists and, therefore, knowledge about one characteristic is useful in assessing the probability of the other.

Some General Rules

Many of the results from the field of probability can be reduced to simple rules and formulas. Emphasis here is on some commonly used rules that apply in cases where two events are considered. Although the material draws on definitions from material presented earlier in this chapter, it is presented in general terms. By this we mean that each of the rules applies not only to finite sample spaces, with which we already are familiar, but to other kinds of spaces as well. The strength of the rules lies in their usefulness in computing probabilities of events of interest directly in terms of probabilities of other events, without resorting to the direct counting method used so far. In many cases the direct counting method does not apply.

Two basic rules are considered: the addition rule and the multiplication rule. The *addition rule* for two events is employed where it is of interest to find the probability that either one of two events or both events will occur. In other words, it applies when one is interested in finding the probability that one event *or* another event will occur. A probability using the addition rule is found by adding probabilities associated with individual events defined within the sample space.

The *multiplication rule* is nothing more than a general procedure to find the joint probability of two events; it applies when one is interested in finding the probability that the events occur at the same time and involves the product of probabilities. An alternative way of defining the concept of independence of events is directly related to a special case of the multiplication rule.

The procedures for finding probabilities presented at this point are concerned with cases where any two events are involved. For convenience, we designate these as A and B. Each of the rules considered are general and require no assumptions about the way the individual probabilities are computed.

In order to find the probability that either event A or B (or both A and B) will occur use the **addition rule**, which takes the form

$$P(A \text{ or } B) = P(A) + P(B) - P(A \& B)$$

In other words, the probability that A or B will occur is found by adding the marginal probabilities and subtracting the joint probability from the sum of the marginals of the two events.

When A and B are *mutually exclusive*, the addition rule reduces to the "straightforward addition" of the marginal probabilities. That is,

$$P(A \text{ or } B) = P(A) + P(B)$$

Two events are mutually exclusive when they have no common elements or sample points; in other words, they cannot occur at the same time, or $P(A \& B) = 0$.

In order to find the probability that events A and B occur simultaneously, the **multiplication rule** for joint probabilities is used. The formula is

$$P(A \& B) = P(A)P(B \mid A)$$

In other words, the joint probability can be found by multiplying the marginal probability of one event by the conditional probability of the other event. By interchanging A and B in the above expression, an equivalent result can be obtained. That is, $P(A \& B) = P(B)P(A \mid B)$.

When A and B are *independent*, the multiplication rule reduces to the "straightforward multiplication" of the marginal probabilities

$$P(A \& B) = P(A)P(B)$$

Recall, two events are independent when the conditional probability equals the marginal.

Example 3   Many problems arise when probabilities are given rather than actual frequencies. In such cases, the addition and multiplication rules are useful to find other probabilities based on the ones given. To illustrate the point, suppose we are told that two key components of an electronic device fail with probabilities $P(F_1) = 0.1$ and $P(F_2) = 0.2$. The symbol F indicates failure and the subscript denotes the component number. Further, assume we know that the second component fails 70 percent of the time when the first component fails, or $P(F_2 | F_1) = 0.7$. The original data regarding failures are not available and only the above probabilities are known from reported specifications.

Suppose the device to be built with the two components experiences a total breakdown only if *both* parts fail. The probability of total breakdown, therefore, is found as a joint probability of failure of both components using the multiplication rule.

$$P(\text{Device breakdown}) = P(F_1 \ \& \ F_2)$$

$$= P(F_1)P(F_2 | F_1)$$

$$= (0.1)(0.7)$$

$$= 0.07$$

Hence, the chance that the device will fail is 0.07, or 7 times in 100.

On the other hand, suppose the device will break down if *either* of the components fails, where the probability of total device failure is given by the addition rule as

$$P(\text{Device breakdown}) = P(F_1 \text{ or } F_2)$$

$$= P(F_1) + P(F_2) - P(F_1 \ \& \ F_2)$$

$$= .1 + .2 - .07$$

$$= 0.23$$

Consequently, there is a higher chance, equal to 0.23, that the device will fail if a breakdown can occur as a result of either part failing rather than both; in other words, if there is no backup.

Sampling and the Multiplication Rule

It was stated without illustration that the multiplication rule reduces to the product of marginal probabilities when events are independent. An important way to illustrate this is in connection with the distinction between sampling with and without replacement. By sampling *with replacement* we mean that when, say, a second item is drawn from a population, the first is replaced before the second is drawn. When sampling *without replacement*, the first item is not replaced. The following two examples illustrate the calculations using the multiplication rule in both cases.

Example 4 Find the probability of drawing two defective items from a random sample without replacement from a shipment of 500 parts containing 40 that are defective.

P(Defective on first selection) = $P(D_1)$

$$= \frac{40}{500}$$

$$= 0.080$$

P(Defective on second selection
given defective on first) = $P(D_2 \mid D_1)$

$$= \frac{39}{449}$$

$$= 0.087$$

The denominator equal to 449 in the second calculation is based on the fact that the first item is not replaced. The joint probability of drawing two defectives is obtained by the multiplication rule as

P(Defective on first *and* defective
on second)= $P(D_1D_2)$

$$= P(D_1)P(D_2 \mid D_1)$$

$$= .080(.087)$$

$$= 0.007$$

Example 5 If the above problem were considered with replacement, the composition of the shipment on the second selection would be the same as on the first, and the events would be independent. The probability of two defectives is

$$P(D_1D_2) = P(D_1)P(D_2)$$

$$= \frac{40}{500} \left[\frac{40}{500} \right]$$

$$= .08(.08)$$

$$= 0.006$$

Since the two events are independent, the joint probability reduces to the product of the marginal probabilities. Also, one can see that there is a slight reduction when sampling with replacement.

In most practical applications, sampling is done without replacement, yet the probabilities based on the independence assumption are calculated. The distinction only becomes important when the number sampled is "large" relative to the size of the shipment.

PROBABILITY DISTRIBUTIONS

Our treatment in Chapter 1 began with the frequency distribution, a basic tool for describing the way data behave. Early in the history of probability and statistics it was discovered that frequency distributions could be represented mathematically and explained as a consequence of the laws of probability. It was this discovery that led to the development of modern statistical methods. The mathematical model used to represent frequency distributions is called a probability distribution. The probability distribution is the main device underlying statistical inference. It is the basis for much of the subsequent material related to statistics and its application to process control methods.

General Concept of a Probability Distribution

A **probability distribution** can be defined as a rule that assigns probabilities to every value of a variable, x, where the values of x correspond to individual outcomes related to some random phenomenon; probability distributions can appear either in the form of a table or as a formula. Instead of dealing with individual events corresponding to random phenomena as before, we shall focus on the entire sample space and work with numerical values attached to all possible outcomes.

In order to develop the idea of a probability distribution, consider the table appearing in **Exhibit 2**. The table is comprised of x-values together with a set of values designated as f(x). The f(x)-values are positive numbers that lie between zero and one where their sum equals one. The f(x) values possess the same properties as probabilities of events described earlier. Therefore, the x's can be viewed as points in a sample space that assume numerical values, and the table can be considered a probability distribution since it serves as a rule or way of assigning probability values to every value of x.

In appearance, the table looks very much like a relative frequency distribution, except f(x) is used instead of "p". This is done to distinguish between theoretical probabilities [i.e., f(x)] and observed relative frequencies (i.e., p). The resemblance between the two types of distributions is important, since ultimately the probability distribution is used as a model of empirical results or of frequency distributions.

EXHIBIT 2
AN EXAMPLE OF A PROBABILITY DISTRIBUTION

x	f(x)
1	0.05
2	0.45
3	0.25
4	0.15
5	0.10
	1.00

When dealing with chance phenomena using probability distributions, special terminology is used when referring to the numerical valued quantity, whose probabilities are provided by the distribution. We call it a *random variable*. In other words, the value of x is unknown before observing the result of a random experiment, and its numerical value is determined by chance. The probability associated with an individual value of x is obtained from the probability distribution. The x-values could represent any measured quantity such as accident rates or the number of critical defects in an assembled unit.

Some examples of probability distributions are presented next. The way these distributions are used to find probabilities also is illustrated. Related concepts are then considered.

Example 6 Given the following table:

x	f(x)
1	0.1
2	0.2
3	0.3
4	0.4
Total	1.0

Since each of the f(x)-values are non-negative, less than one, and sum to one, the table constitutes a probability distribution. Each f(x)-value represents a probability. For instance, if we want to find the probability that 3 will occur, it is read directly from the table as 0.3. That is,

$$P(x = 3) = f(3)$$

$$= 0.03$$

Probabilities for the remaining values of x are found in the same manner.

Example 7 Given the following function in formula form:

$$f(x) = \frac{x}{10}; \ x = 1,2,3,4$$

The values of x following the formula represent the values for which the function is defined.

Since all the x-values are positive, substitution into x/10 will yield positive f(x)-values. The sum of the f(x)'s is determined as

$$\Sigma f(x) = \Sigma \frac{x}{10}$$

$$= \frac{1}{10} + \frac{2}{10} + \frac{3}{10} + \frac{4}{10}$$

$$= \frac{10}{10} = 1$$

Notice that this distribution is identical to the one presented in Example 6 except that it is expressed algebraically.

Example 8   Given the following probability distribution:

$$f(x) = \frac{x^2}{90} \; ; \; x = 2, 3, 4, 5, 6$$

In order to find the probability that the value 5 will occur, we substitute 5 into the formula as follows:

$$P(x = 5) = f(5)$$

$$= \frac{(5)^2}{90}$$

$$= \frac{25}{90} = 0.28$$

Note. Consider the distribution given in Example 8 and suppose we are interested in finding the probability that x assumes a value that does not exceed 4. This is found by applying the special case of the addition theorem. That is, we add the probabilities of the x-values less than and equal to 4 as follows:

$$P(x \le 4) = \sum_{x=2}^{4} \frac{x^2}{90}$$

$$= \sum_{x=2}^{4} \frac{x^2}{90}$$

$$= \frac{2^2}{90} + \frac{3^2}{90} + \frac{4^2}{90}$$

$$= .044 + .100 + .178$$

$$= 0.322$$

In general, the probability that x will assume a value within some range is found by *adding* the probabilities of the x-values within that range.

Example 9   Consider the probability distribution

$$f(x) = \frac{1}{5} \; ; \; x = 1, 2, 3, 4, 5$$

In this case, each value of x is assigned the same value of f(x), equal to 1/5. Hence all values of x have a probability of 1/5, or 0.20. For example, the probability that x equals 3 is

$$P(x = 3) = f(3)$$

$$= \frac{1}{5} \text{ or } 0.20$$

The distribution in this example is a special case of what is referred to as the discrete *uniform distribution* and is given by the more general formula

$$f(x) = \frac{1}{k} \; ; \quad x = 1, 2, 3, \ldots, k$$

In this case, k can be any integer. Later, we shall reintroduce the distribution as it applies to die rolls and lotteries.

Example 10 Suppose a random sample of four items (i.e., n = 4) is drawn from a continually operating manufacturing process that has been demonstrated to produce 10 percent defective items (i.e., P = 0.10). Find the probability that exactly two of the four sample items will be defective.

The underlying distribution is of the form

$$f(x) = \begin{bmatrix} 4 \\ x \end{bmatrix} (0.1)^x (0.9)^{4-x} \; ; \quad x = 0, 1, 2, 3, 4,$$

where the expression $\begin{bmatrix} 4 \\ x \end{bmatrix}$ denotes the number of *combinations* of four things grouped "x" at a time and is given by the formula

$$\begin{bmatrix} 4 \\ x \end{bmatrix} = \frac{4!}{x!(4-x)!}$$

The exclamation symbol after an integer stands for "factorial" and represents the product of that integer and all non-negative integers less than it.

To find the specified probability, we substitute and solve as follows:

$$P(x = 2) = f(2) = \begin{bmatrix} 4 \\ 2 \end{bmatrix} (.1)^2 (.9)^2$$

$$= \frac{4!}{2!(4-2)!} (.1)^2 (.9)^{4-2}$$

$$= \frac{4 \times 3 \times 2 \times 1}{(2 \times 1)(2 \times 1)} (.01)(.81)$$

$$= 6(.0081)$$

$$= .049$$

This number can be interpreted to mean that roughly 4.9 percent of all samples of size four taken from a 10- percent defective process will contain 50-percent defective.

The above distribution is a special case of what is referred to as the *binomial probability distribution*, which is given by the general expression

$$f(x) = \begin{bmatrix} n \\ x \end{bmatrix} P^x Q^{n-x}; \, x = 0, 1, 2, ..., n$$

This is a special distribution that applies to probabilities associated with the number of times, x, that one of two possible outcomes occurs in n repetitions, or a sample generated from situations where the probability that an outcome on a particular repetition is constant and equals P. Q (= 1 – P) is the probability that the outcome will not occur.

Note. The above example was solved by direct substitution into the formula for the binomial distribution. Binomial probabilities for selected values of P and values of n between 1 and 20 can be found alternatively in **Table A-2** in the Appendix.

Example 11 A certain part is produced by a process with an average of two minor defects per part. What is the probability that any part produced will contain three minor defects?

The underlying distribution that applies in this case is of the form

$$f(x) = \frac{2^x e^{-2}}{x!}; \; x = 0, 1, 2, ...$$

where *e* is the mathematical constant that is the base of the natural logarithm and x! again is the factorial of x.

To find the required probability, we substitute and solve as follows:

$$P(x = 3) = f(3)$$

$$= \frac{2^3 (2.71828....)^{-2}}{3!}$$

$$= \frac{8(.135335.....)}{3 \times 2 \times 1}$$

$$= 0.180$$

The distribution used to solve this problem is a special case of what is referred to as the *Poisson probability distribution*, which is given by the general expression

$$f(x) = \frac{\mu^x e^{-\mu}}{x!}; \; x = 0, 1, 2, 3,....$$

This is a special distribution that applies to probabilities associated with the number of times, x, that an event can occur when "non-occurrences" cannot be enumerated, such that events occur independently with a constant average, μ. In our example above, defects can be documented and enumerated whereas "non-defects" cannot.

Note. The above example was solved by direct substitution into the formula for the Poisson distribution. Poisson probabilities for selected values of can be found alternatively in **Table A-3** in the Appendix.

Characteristics of Probability Distributions

Just as we defined measures describing properties of frequency distributions, we can do the same for probability distributions. The most important of these measures are the mean, the variance, and the standard deviation. Procedures for finding each are the same as those for a relative frequency distribution. Their interpretations in terms of location and distribution spread also are similar. The mean and variance (and standard deviation) also can be interpreted in terms of probabilities. This is especially important when explaining the way chance phenomena occur.

The mean of a probability distribution is referred to as the mathematical expectation, or the *expected value*. The expected value is designated by one of two symbols, μ (read "Mu") or E(x), and is defined by the following formula:

$$\mu \text{ or } E(x) = \Sigma x f(x)$$

The expected value is found by multiplying every value of x by its probability of occurrence and summing. Procedurally, computation is the same as calculating a weighted average, except the weights are probabilities rather than frequencies.

Suppose we have a probability distribution associated with a particular random phenomenon. Visualize the random phenomenon repeated endlessly. Each repetition will yield values that recur according to the relative frequency or probability of occurrence of each value. The expected value of a random variable is the value obtained as the arithmetic mean of the observations resulting from the endless repetition of the random phenomenon. The larger the number of repetitions, the closer the computed mean will be to the expected value, or theoretical mean. In other words, we "expect" the average result of a large number of repetitions of a random phenomenon will be close to the expected value or mean.

The *variance* of a probability distribution is designated by the symbol σ^2 (read Sigma-squared), and is defined by the formula

$$\sigma^2 = \Sigma(x - \mu)^2 f(x)$$

The variance is defined as the sum of the squared deviations of the x-values from the mean weighted by their probability of occurrence. Note that this procedure is the same as calculating a weighed average of squared deviations where the weights are probabilities of occurrence. The *standard deviation*, designated as σ (or "Sigma"), is obtained as the square root of the variance. For our purposes, however, it is not necessary to compute a variance or standard deviation of a probability distribution. It is important to understand what they mean.

As a measure of the spread or width of a probability distribution the variance indicates the amount of concentration or variability of the x-values within the vicinity of the mean or expected value. This is similar to the interpretation of the standard deviation and the variance of a frequency distribution presented in Chapter 3. The concentration of x-values about the mean also is reflected in the probability associated with these x-values. Hence the smaller the variance, or the standard deviation, of a probability distribution, the larger will be the probability that an x-value will fall within a specific interval about the mean. Relatively large variances are associated with greater uncertainty, whereas small variances are associated with less uncertainty. A variance of zero corresponds to absolute certainty.

Continuous Probability Distributions

The probability distributions illustrated in this section all exhibited a certain property; they are associated with x-values that are not connected when considered on a numerical scale. This becomes evident when the distributions are graphed, since spaces exist between the plotted points. In other words, the distributions are defined for a distinct set of values rather than everywhere along an interval. Distributions exhibiting this property are called *discrete*.

While the concept of a probability distribution is understood by examining the discrete case, many methods rely on another type of distribution, referred to as a continuous probability distribution. A *continuous* distribution is associated with x-values that may assume any numerical value in an interval. Probabilities are obtained as *areas* under the distribution that are associated with a set of values, rather than by direct substitution. Examples of some continuous distributions are depicted in **Exhibit 3**. Each of the distributions shown in the exhibit is

EXHIBIT 3
GRAPHIC ILLUSTRATION OF SOME CONTINUOUS
PROBABILITY DISTRIBUTIONS

Shaded Areas Represent A Probability, P(a ≤ x ≤ b)

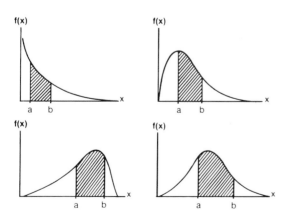

represented by a smooth curve associated with a set of x-values plotted on the horizontal axis. This is true of all continuous probability distributions regardless of their shape.

The *area* under all continuous probability distributions must equal one. Consequently, the area under each curve associated with a sub-interval of x-values is less than one, and therefore can represent a probability. The shaded regions are included in each diagram appearing in Exhibit 3 to indicate that a probability associated with any interval is represented by an area under the curve. For example, the probability that a value of x falls between any two numbers "a" and "b" is represented as P(a ≤ x ≤ b).

Whenever continuous distributions are used, probabilities always must be found as areas; the probability associated with a single x-value is equal to zero and has no meaning. It should be emphasized that, unlike the discrete case, substitution directly into the formula for f(x) is *not* appropriate in order to find probabilities for continuous variables. In general, probabilities can be more difficult to calculate when using continuous probability distributions, requiring one to rely on mathematical methods derived from calculus. Fortunately, many distributions are available in the form of tables and more and more are pre-programmed into hand-held calculators and computer software. Hence, there is no need for us to deal with the mathematical considerations at all.

Continuous probability distributions are used widely. Surprisingly, they can be used to approximate discrete distributions. In addition, continuous probability distributions are applied directly to the behavior of various measurements. In such cases we can think of continuous distributions as models of smoothed histograms.

THE NORMAL CURVE

Of the many continuous probability distributions available, the normal distribution is more widely used than most other individual distributions. In many cases it is used correctly and as an approximation; sometimes it is applied to problems where it is inappropriate. Unlike the binomial and Poisson illustrated earlier, which are based on specific assumptions, the normal is applicable to a variety of situations. It is extremely useful as an approximation with respect to many methods employed in process control, both in terms of the way measurements are distributed and with respect to characteristics of samples. Here, we shall introduce the normal and illustrate a procedure for obtaining probabilities from the normal. This will serve as a basis for procedures presented in subsequent sections.

The *normal distribution* is characterized by the following formula:

$$f(x) = \frac{1}{\sqrt{2\pi}\,\sigma}\, e^{-\frac{1}{2}[x-\mu]^2}; \; -\infty < x < \infty$$

The formula represents the probability distribution of x, which stands for some measured value that varies at random according to the normal curve. Therefore, we can refer to x as a normally distributed random variable. The formula for f(x) is the basis for generating probabilities of variables that are normally distributed.

The formula for the normal curve appears complicated. Although we do not have to work with it directly in order to find probabilities, it is instructive to describe each of the components of the formula. The symbols "π" (read as Pi) and "e" are natural constants arising in mathematics. Numerically, they are given as

$$\pi = 3.14159...$$
$$and$$
$$e = 2.71828...$$

The dots trailing after each value signify unending decimals, where additional terms in each series provide a closer approximation to the true value.

Once it is established that a distribution is normal, a particular normal distribution is identified by specific values of μ and σ (read as "Mu" and "Sigma"). Different values of μ and σ represent different normal distributions and are referred to as the *parameters* of the normal. These parameters are the *mean* and the *standard deviation.**

When plotted in the form of a graph, the normal appears as a smooth, unbroken curve, as shown in **Exhibit 4**. This smooth appearance of the normal when graphed represents a key difference between the normal and distributions such as the binomial. The binomial is associated with x-values representing integer values or counts, whereas the normal is associated with all numerical values along the horizontal scale. Therefore, the normal is referred to as a continuous probability distribution rather than as a discrete one.

EXHIBIT 4
GRAPH OF A NORMAL CURVE

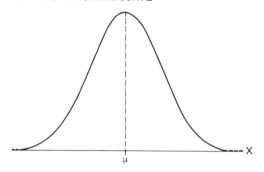

The normal is "bell shaped" and is defined for values of x assuming any real number. The distribution has a single peak in the center and drops smoothly in either direction of the peak, assuming lower and lower

---

* A seeming inconsistency exists regarding the way symbols μ and σ are used, based on certain conventions in the field of statistics. In general, they are used to denote the mean and standard deviation of any probability distribution, but also have been retained as the parameters of the normal curve. Mathematically, it can be shown that the two parameters are the mean and the standard deviation. Later, the same symbols will be used to denote the mean and standard deviation of universes or processes.

values as it approaches the limits of plus-and-minus infinity. Another way of describing this property is to say that it is asymptotic to the x-axis, or that it continually approaches the x-axis but "never" touches it. As a result, it is not possible to draw the entire curve. The normal distribution is *symmetric* about the mean, μ.

The entire area under any normal curve is equal to *one*. This is essential for finding probabilities. Instead of substituting into the formula for the normal to find probabilities, probabilities are found as areas under the curve that are assigned to intervals rather than individual values of x. Since the total area under the curve is unity, the area under the curve corresponding to an interval of x-values is less than one and represents a probability.

Calculating Normal Probabilities

Due to the nature of the normal distribution, probability computations have been simplified and have been reduced to an elementary calculation that is used with a set of tabulated values given in the Appendix. The procedure for finding normal probabilities basically involves the following three steps:

Step 1. For given values of μ and σ and a specific numerical value of x, compute the *standardized variate*, z, by substituting into the following formula:

$$z = \frac{x - \mu}{\sigma}$$

In words, z is found by subtracting the mean from a value of x and dividing by the standard deviation.

Step 2. Refer to the table of "Areas of the Normal Curve" in the Appendix (**Table A-1**); the table provides areas corresponding to *positive* values of z.

Step 3. From the table, determine the area corresponding to the value of "z" computed in Step 1.

> **Note.** Due to the way Table A-1 is constructed, the three steps may have to be repeated a second time in order to determine a required probability.

The following points should be kept in mind:

a. The entire area under the normal curve is equal to one.
b. Because the normal distribution is symmetric, the area under the curve on either side of the mean is equal to 0.5.
c. The table of Areas of the Normal Curve is tabulated only for the right half of the normal curve.
d. All probability computations involving the normal are reduced to finding the probability

$$P(\mu \le x \le a)$$

**BASIC PROBABILITY
AND THE
NORMAL CURVE**

In other words, areas in the table correspond to probabilities for intervals between the mean, μ, and a number, "a", which is greater than the mean. The area tabulated corresponds to the shaded region in the diagram appearing in **Exhibit 5**. The value of "z" computed in Step 1 provides the area in the shaded region from the table directly. When "a" is to the left of the mean, the value of z will be negative; if you ignore the negative sign, the value of z also will provide the required probability directly. In all other cases, the three steps listed above must be performed twice. Two values of z are obtained, and the two probabilities must be added or subtracted according to the specific probability required. This becomes clear by examining the examples that follow.

EXHIBIT 5
AREA TABULATED IN THE TABLE OF "AREAS
OF THE NORMAL CURVE"

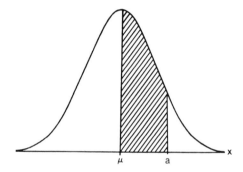

Example 12 Given that x is normally distributed with $\mu = 16$ and $\sigma = 4$. Find $P(16 \leq x \leq 21)$ or the shaded area in the diagram below.

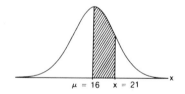

Step 1. Compute, $z = \dfrac{x - \mu}{\sigma} = \dfrac{21 - 6}{14}$

$$= \frac{5}{4}$$

$$= 1.25$$

Step 2. From Table A-1, z = 1.25 gives an area of 0.3944.

Step 3. Therefore, from the table

$$P(16 \leq x \leq 21) = 0.3944$$

Example 13 Given that f(x) is normal with $\mu = 3$ and $\sigma = 0.5$. Find $P(2 \leq x \leq 3.5)$.

Owing to the way the table of Areas of the Normal Curve is constructed, this problem must be handled in two parts, each corresponding to a different side of the mean labeled as (a) and (b) in the diagram below.

(a) $z_1 = \dfrac{x_1 - \mu}{\sigma} = \dfrac{2 - 3}{0.5}$

$= -2.00$ (left side of mean)

(b) $z_2 = \dfrac{x_2 - \mu}{\sigma} = \dfrac{3.5 - 3}{0.5}$

$= 1.00$ (right side of mean)

From Table A-1, $z_1 = -2.00$ gives an area of 0.4772 (minus sign ignored when using table), and $z_2 = 1.00$ gives an area of 0.3413. These two areas must be *added* to obtain the desired probability. That is,

$P(2 \le x \le 3.5) = 0.4772 + 0.3413$

$= 0.8185$

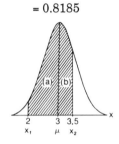

Example 14 Given that $f(x)$ is normal with $\mu = 33$ and $\sigma = 8$. Find the probability that x is less than 21.

$z = \dfrac{x - \mu}{\sigma} = \dfrac{21 - 33}{8}$

$= -1.50$

The required probability lies in the "lower tail" of the distribution, which is shown in the diagram below.

From Table A-1, $z = -1.50$ (ignoring sign) yields an area of 0.4332. *Subtracting* this from 0.5, we get

$P(x < 21) = 0.5 - .4332$

$= 0.0668$

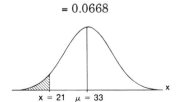

Example 15 Suppose the distribution of weights (x-values) of items produced by a manufacturing process is "closely approximated" by a normal curve with mean weight equal to 13 pounds and standard deviation equal to 3 pounds. What proportion of the items produced have weights between 7 and 19 pounds?

Based on the definition of probability, the proportion required in the problem is the same as the probability that an item has a weight between 7 and 19 pounds. Hence, we are to find the shaded area in the diagram appearing below.

$$z_1 = \frac{7 - 13}{3} = -2.00$$

$$z_2 = \frac{19 - 13}{3} = 2.00$$

Both $z_1 = -2.00$ and $z_2 = 2.00$ yield an area of 0.4772 from Table A-1. Therefore,

$$P(7 \le x \le 19) = 0.4772 + 0.4772$$

$$= 2(0.4772)$$

$$= 0.9544$$

Hence, the proportion of items with weights between 7 and 19 pounds equals 0.9544.

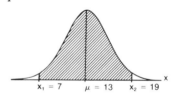

$x_1 = 7 \qquad \mu = 13 \qquad x_2 = 19$

Note. Since the absolute difference between 7 and 13, and 19 and 13 is the same, equal to 6, the problem could be restated as finding the probability that an item has a weight within 6 pounds of the mean weight. Hence, only one z-value is necessary; the answer is obtained by multiplying the corresponding tabulated probability by 2.

Example 16 A Related Problem Suppose we consider a normal curve with $\mu = 400$ and $\sigma = 75$, and we want to find the value of x below which the lowest 33 percent of x-values falls.

This is an exercise in the use of Table A-1 that does not require a probability computation but is nonetheless useful in later work. In this case the area is given and you must find the value of x. Consider the diagram below.

You are told that the area to the left of the unknown x-value is 0.33. Since the table of areas of the normal curve deals only with the cross-hatched region, or 0.17 (i.e., $0.5 - 0.33 = 0.17$), it is this value that is found in the body of Table A-12. Referring to the table, an area of 0.17 yields a value of z equal to 0.44. Since x is below the mean, z must be changed to –0.44. If we substitute all the known values into the expression of z, we can solve for x. That is.

$$z = \frac{x - \mu}{\sigma}$$

$$-0.44 = \frac{x - 400}{75}$$

$$x = 400 - 0.44(75)$$

$$= 400 - 33$$

$$x = 367$$

Hence, 33 percent of the x-values lie below 367.

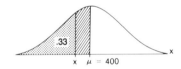

Note. Recognize that substituting and solving for x is equivalent to subtracting the standard deviation multiplied by z from the mean. The formula for x *below* the mean takes the form

$$X = \mu - Z\sigma$$

When x is *above* the mean, the formula to find x for a given probability is given as

$$X = \mu + Z\sigma$$

Here, the standard deviation multiplied by z is added to the mean.

Properties of the Normal

The use of the word "normal" to describe a probability distribution seems to imply that most random phenomena are distributed according to this distribution. This belief is supported by the frequency with which the normal curve appears in introductory textbooks as well as in a large number of applied problems. Although its usefulness is widespread, it should be emphasized that there are many phenomena that are described by other probability distributions. With this caution in mind, we shall explore some additional points about the normal.

We stated that the parameters of the normal distribution are the mean and standard deviation. It is instructive to examine these charac-

**BASIC PROBABILITY
AND THE
NORMAL CURVE**

teristics in more detail. If we assume for the moment that the standard deviation is held constant, a change in the value of the mean, μ, induces a shift in the peak of the distribution. The overall appearance of the curve remains the same, however. For example, consider the diagrams appearing in **Exhibit 6**. The three curves are normal with the same value of σ, but with different mean values of μ_1, μ_2, and μ_3. The different values of the mean correspond to different *locations* of the peak of the three distributions, which means the "bulk" of the area of the three curves also has a different position on the axis. This interpretation of the mean is similar to the one for frequency distributions presented in Chapter 2.

EXHIBIT 6
NORMAL CURVES WITH DIFFERENT MEANS

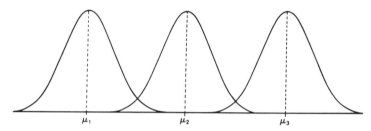

Now consider the case where the mean is held constant and we vary σ. This is depicted in **Exhibit 7**. Of the two distributions shown in the exhibit, the "narrower" curve has the smaller standard deviation, or σ_1 is less than σ_2. Also, we can say that the probability of an x-value falling within any set of fixed limits centered on the mean μ is less for the distribution with the smaller standard deviation. In other words, it is more likely that a value falls within the vicinity of the mean when the standard deviation is smaller.

EXHIBIT 7
NORMAL CURVES WITH DIFFERENT STANDARD
DEVIATIONS, $\sigma_1 < \sigma_2$

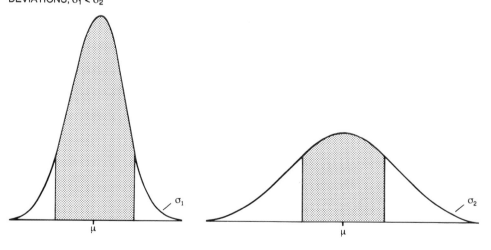

A few words also are needed to clarify the notion of a "normal phenomenon." We have seen that the normal curve is defined for all the real numbers, or between plus-and minus infinity. Regardless of any set of measurements taken, no matter how large or how small, the range of values for which the normal is defined obviously is unrealistic. Then how can we say that a variable is normally distributed and apply it to a realistic or practical problem?

If you think about the diagrams of the normal distribution used in this section, you will recall the "tails" of the distribution are very "close" to the x-axis. In terms of probability, this means that a "very small" proportion of the x-values occurs in the tails. In order to emphasize this point, consider the area or probability value in the table of Areas of the Normal Curve corresponding to a value of z equal to 3. The area given is 0.4987, which tells us that 49.87 percent of the x-values lie between the mean and some value that is 3 standard deviations from the mean. Furthermore, if we double the probability found in the table, we find that 99.74 percent of the x-values fall within 3 standard deviations of the mean. Diagrammatically, this is shown in **Exhibit 8**. Virtually all of the observations or x-values in the diagram lie within the "3-sigma" limits. Less than 0.3 percent lies outside these limits, which is the case for *any* normal distribution.

EXHIBIT 8
AREAS UNDER A NORMAL CURVE CORRESPONDING TO 3 STANDARD DEVIATIONS OF THE MEAN

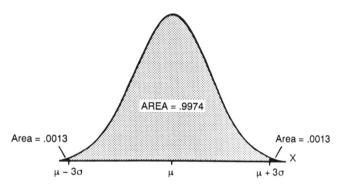

In practical terms, this result suggests that such a small percentage of observations lies in the tails of the normal that, realistically speaking, we can ignore the tails. In other words, within limits generally encountered in practice, there are phenomena that can be approximated satisfactorily by a normal distribution well enough that it is useful.

Another point about normal distributions worth noting is that regardless of the magnitudes of μ and σ, fixed proportions of the area under the curve, or probabilities, can be associated with values that are fixed multiples of the standard deviation from the mean. The results become important in relating the use of control charts presented later and parts per million (ppm) goals regarding defectives and defective reduction. **Exhibit 9** presents probabilities within various multiples of the standard deviation as well as tail-area probabilities for selected multiples.

The table presented in Exhibit 9 also provides the corresponding parts per million (ppm), or its equivalent, corresponding to each of the tail area probabilities. Essentially, when the ppm concept is applied, it is assumed that the distribution of a product characteristic is normally distributed and the parts per million correspond to the number of parts per million outside the "sigma-limits". If, say, a particular set of sigma limits represents a set of specifications, the ppm represents the number of defectives per million parts produced. Looking at it another way, given a set of specifications, reduction of process variation, or reduced σ, leads to a smaller ppm.

EXHIBIT 9
PROBABILITIES ASSOCIATED WITH NORMAL
MULTIPLES OF THE STANDARD DEVIATION

Number of Standard Deviations	Centered Probability	Combined Tail Area Probability	Parts Per Million (ppm)
μ ± 1 σ	.6827	.3173	317300
μ ± 2 σ	.9545	.0455	45500
μ ± 3 σ	.9973	.0027	2700
μ ± 4 σ	.99994	.00006	60
μ ± 5 σ	.9999994	.0000006	.6(or 6/10 million)
μ ± 6 σ	.999999998	.000000002	.002(or 2/billion)

ADDITIONAL IDEAS

Probability and probability distributions represent basic tools that are used as models of empirical observations. In this section, we introduced the idea of a theoretical distribution and presented some special cases. Much work has been done in the areas of applied probability and statistics that has resulted in distributions that apply to various situations, based on given assumptions. When dealing with a particular problem, we first look at our assumptions or given conditions, and, in many cases, we are able to select an established model or distribution off the shelf, so to speak, that applies.

Other problems exist in which actual data must be collected in order to uncover an appropriate distribution to use. This requires the construction of a frequency distribution and finding a theoretical distribution that adequately represents the observed distribution. Before addressing applications of individual distributions in later sections, it is useful to discuss some related issues at this time.

The Concept of Randomness

The term random is used in several different ways in probability and statistics; already we have been exposed to the terms random experiment, random variable and random sample. Based on our discussion, we should have an intuitive feeling for the meaning of the term.

The term random applies to possible outcomes that cannot be predicted before the occurrence of an event but whose long-run relative frequency of occurrence is known in advance and corresponds to any probability distribution. Hence, "random occurrences" are completely unpredictable in the sense that it is impossible to know beforehand what the exact outcome will be. What is known, however, is the frequency with which outcomes will occur in a "large number of trials."

When dealing with problems of process control, the problem of randomness becomes important in two distinct ways. On the one hand, reducing process measurements to ones that are completely random renders a process stable, which means that it is free of immediately correctable problems and can be characterized by an identifiable signature, so to speak, in terms of an underlying probability distribution. On the other hand, when selected properly random samples taken from a process can be used to great benefit in establishing when a process is free of such problems and when to hunt for problems and correct them. Interestingly, the characteristics of these samples also can be described in terms of the laws of chance, which forms the basis for the methods presented in subsequent chapters.

Parameters, Statistics, and Other Random Sequences

In general, when problems are dealt with using statistical methods, an important distinction is made, either implicitly or explicitly, between a universe and a sample. A **universe**, or population, is defined as the totality of elements about which a conclusion is to be drawn or a decision to be made, as it relates to an overall problem. In the case of process control, the so-called statistical universe is defined as the set of items or individual components of output of a process. When dealing with problems of acceptance sampling, typically the universe represents the set of items contained within a particular lot or shipment. The process mean associated with a critical dimension or the shipment percent defective would be considered as "parameters," since they represent identifying characteristics of the universe. The actual numerical value(s) of process parameters really are never known and are approximated or estimated with sample information.

A **sample** is defined as a subset or a portion of a universe. Although uncertainty is associated with sample information, much can be done using sample data. Characteristics of samples, which correspond to universe parameters, are called **statistics**. Such quantities as the mean, standard deviation, median, and range calculated from sample data are referred to as sample statistics.

Based on the concept of randomness, values of a sample statistic calculated from repeated samples represent a random sequence. This is based on the fact that functions of basic observations that are random also are random. The pattern of randomness, so to speak, will vary, however, depending on the particular statistic.

In order to obtain some appreciation for these statements before formally introducing methods of inference in Chapter 5, let us introduce the idea of random variation in sampled quantities in terms of a simple simulation.

Strictly speaking, simulation as opposed to trial and error and sensitivity analysis is a method by which an artificial numerical history is generated at random in order to be used as a basis for understanding or solving problems. It is the random component that distinguishes simulation from the other methods.

The simulated results that we shall use simply are based on the results of 500 actual rolls of a die that originally were presented in Gulezian [57]. There, the die rolls are used to demonstrate how the concepts of probability we have discussed actually work and, further, to link concepts of descriptive statistics and probability to sampling methods. Here, we shall pick up from that discussion and introduce some additional ideas related to the theme of this book. The results of the die rolls have been reproduced in **Exhibit 10**.

EXHIBIT 10
RESULTS OF 500 ROLLS OF A SINGLE DIE

4 4 5 6 5	4 2 6 2 3	1 6 5 4 4	1 4 1 4 3
5 2 2 2 3	3 2 1 3 3	6 3 4 5 2	1 1 2 6 5
3 2 6 3 4	5 5 1 3 3	6 1 2 4 5	4 3 6 6 2
6 3 6 2 5	4 5 2 5 3	1 1 3 3 1	5 4 5 3 6
1 1 2 1 2	4 4 2 6 6	1 2 4 6 3	4 2 5 4 1
4 3 4 3 1	2 4 4 2 2	3 2 3 4 5	2 6 2 5 1
4 5 4 5 3	6 3 3 6 2	5 4 1 5 1	1 6 1 6 4
5 6 4 2 2	5 2 4 4 4	2 6 2 3 3	1 6 5 3 4
4 1 6 4 1	4 2 2 5 3	2 6 5 2 6	4 3 2 3 5
4 5 3 6 1	3 4 4 2 3	2 6 6 5 5	2 5 5 2 6
5 2 1 2 5	1 3 2 4 3	5 6 1 4 3	4 6 3 3 1
6 2 4 4 4	5 4 4 4 1	5 4 5 2 6	1 2 6 3 4
2 4 1 3 3	1 6 5 6 1	5 5 2 3 1	5 1 1 6 4
5 3 2 6 2	1 6 2 4 4	4 3 3 3 2	4 5 1 4 5
4 3 2 4 5	5 5 1 5 3	4 2 6 2 2	4 1 1 6 5
3 6 3 2 5	5 5 2 2 2	5 2 2 1 3	4 2 2 1 3
6 1 4 5 2	1 6 3 4 1	2 1 1 2 4	4 3 1 2 1
4 1 1 3 5	4 5 1 6 4	2 2 4 5 6	6 5 5 6 3
2 1 3 5 5	2 2 1 5 2	5 6 2 6 5	1 5 6 4 2
1 5 4 1 2	6 3 4 1 4	6 3 2 3 3	2 4 6 5 4
3 3 6 3 5	1 3 2 1 4	3 5 3 1 2	5 6 6 3 2
4 3 6 3 6	6 5 2 4 1	4 4 6 5 2	6 5 3 5 4
3 6 5 5 1	3 2 6 2 2	3 4 4 4 1	6 2 6 2 3
3 5 5 5 3	4 2 1 5 6	4 4 4 4 4	2 6 3 2 2
3 6 4 2 1	1 6 4 6 5	3 3 2 3 2	1 4 4 4 1

Source: *Statistics For Decision Making* by Ronald Gulezian (QAI 1990)

Exhibit 11 presents the results of the first 50 die rolls presented in Exhibit 10 plotted against time. The order of observation begins with the first column of five results, moves horizontally to the next column of five, up to the first five. The sixth set of five begins with the group vertically under the first, and so on. Although, not identical, the remaining points, if plotted, would appear similarly.

Exhibit 12 presents five sequences that are based on the first 50 samples of five results of the die rolls presented in Exhibit 10. Each sequence is based on a different statistic computed from the 50 samples:

A. Sample arithmetic mean

B. Sample standard deviation

C. Sample median

D. Sample range

E. Sample fraction 1's and 2's

All five sequences are *random* in the sense that there exists no predictable pattern. The visual impression obtained from each is different, however, since each statistic is based on a different function of the sample observations.

EXHIBIT 11
TIME PLOT OF RESULTS OF DIE ROLLS

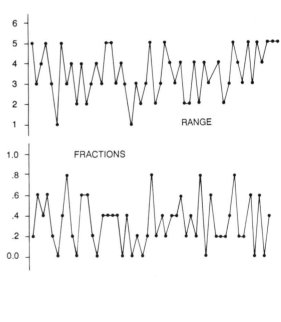

EXHIBIT 12
TIME PLOTS OF SIMULATED SAMPLE STATISTICS BASED ON DIE ROLLS
Examples of Derived Random Sequences

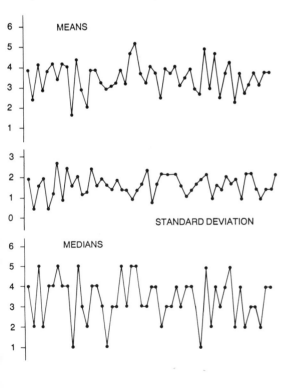

Random Sampling Distributions

Each of the random sequences of the individual statistics introduced above is based on a probability distribution that is derived from
the underlying distribution of die rolls, which is uniform and is the
universe distribution in this example. **Exhibit 13** presents graphs of
the actual (or theoretical) and the simulated distributions associated
with the die rolls.

Exhibit 13
DISTRIBUTION OF INDIVIDUAL DIE ROLLS

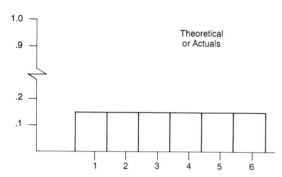

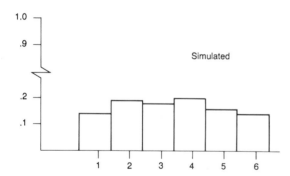

Notice that the simulated distribution is not completely flat like the
theoretical case. This is due to the fact that it is based on a limited
history, or a finite number of rolls. Recalling the basic definition of
probability, the simulated distribution will "approach" the theoretical
as the number of trials is increased.

Presented in **Exhibit 14** are five graphs, or histograms, associated
with each of the statistics plotted as a time sequence in Exhibit 12. Since
each of these distributions is simulated and is based on a limited history,
each represents an approximation of the actual distribution. They do
serve to convey an important idea, however. That is, we can form a
probability distribution of a sample statistic, and, as a consequence,
sample statistics can be viewed as random variables. In other words,
associated with every statistic calculated on the basis of sample observations is an underlying probability distribution describing its behavior.
Such a derived distribution is given a special name and is referred to as

a *random sampling distribution*. By examining Exhibit 14, it is immediately obvious that sample statistics behave like other random variables. Also, how they behave varies from statistic to statistic. They do, however, exhibit properties of central tendency, dispersion, and skewness. Merely for purposes of illustration at this point, the theoretical and simulated means of each distribution are presented in **Exhibit 15**. The simulated quantities are based on measures introduced earlier. Although not equal, the two sets of results appear close.

EXHIBIT 14
SIMULATED SAMPLING DISTRIBUTIONS OF VARIOUS
SAMPLE STATISTICS BASED ON DIE ROLLS

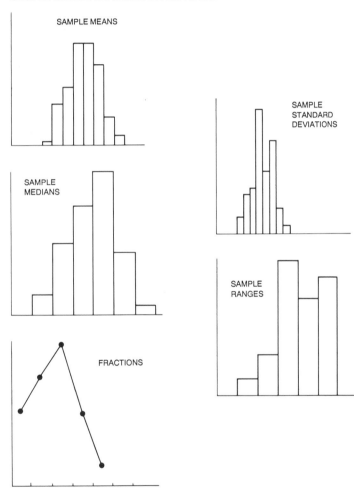

Overall, it should be noted that simulated approximations of various sampling distributions have been illustrated in order to begin to emphasize the idea that sample characteristics are subject to random variation, and that these characteristics behave in recognizable ways. In practical applications, one does *not* sample endlessly in order to establish histograms similar to those presented in Exhibit 14; the histograms were introduced in order to present a concept.

Notice that in this final portion of the chapter, we have taken the results of simple die rolls and used them to relate concepts from all chapters of Module A to those of probability, probability distributions, randomness, and sampling. Actually, the resulting concept of a random sampling distribution and its parameters is fundamental when dealing with problems of inference.

EXHIBIT 15
THEORETICAL OR ACTUAL AND SIMULATED
PARAMETER VALUES

	Actual\Universe	Simulated
Mean	$\mu = 3.5$	$\bar{\bar{X}} = 3.4$
Standard Deviation	$\sigma = 1.71$	$\bar{S} = 1.59$
Median	$X_{50} = 3.5$	$\bar{\bar{X}} = 3.44$
Range	5	$\bar{R} = 3.64$
Proportion 1's & 2's	.333...	$\bar{p} = .33$

Much work done in theoretical statistics is concerned with the distributional properties of sample statistics. In practical applications, we are able to use the results in order to measure or account for the uncertainty associated with sample results and apply them to inferences made about universe or process characteristics. Although the underlying distributions of the statistics considered in Exhibit 14 all are different, the normal plays a vital role as an approximation to some of these.

STATISTICAL INFERENCE

Chapter 4 briefly addressed the sampling problem in terms of simulation in order to provide some insight into the concept of randomness in general and how it relates to the problem of sampling. Use of sample information is basic to all of statistical quality control, and it is the understanding of the nature of randomness and random samples that is of key importance. The underlying concepts are seemingly simple, yet they really are subtle and difficult to fully appreciate. Although many generalizations are available, inconsistencies emerge when dealing at a practical rather than at a theoretical level. The difficulty really lies in terms of separating overall concepts from particular methods that apply in particular circumstances.

In this chapter we address three fundamental and related concepts:

1. Random sampling distributions
2. Estimation
3. Hypothesis testing

Overall, we use sample information to make estimates about universes or to test hypotheses. Since sample information is incomplete in the sense that it does not constitute the entire universe about which a conclusion is to be made, uncertainty is present. This uncertainty is captured or measured in terms of the random sampling distribution. Together, the three concepts form the basis for constructing and using control charts, which are introduced briefly in the next section and discussed more fully in the subsequent module.

The symbols used in this section are defined as follows:

μ = the universe or process mean

$\mu_{\bar{x}}$ = the mean of the sampling distribution of the sample mean

STATISTICAL
INFERENCE

μ_0 = a target or specified process mean, or a hypothesized value of the universe mean (read as "Mu-sub-zero")

$\hat{\mu}$ = an estimate of the universe or process mean

$\overline{X}$ = the sample arithmetic mean

$\overline{\overline{X}}$ = the mean of a collection of sample means

σ^2 = the universe or process variance

σ = the universe or process standard deviation

$\hat{\sigma}$ = the estimated universe or process standard deviation

S^2 = the sample variance

S = the sample standard deviation

$\overline{S}$ = the mean of a collection of sample standard deviations

$\sigma_{\overline{x}}^2$ = the variance of the sampling distribution of the sample mean

$\sigma_{\overline{x}}$ = the standard error of the mean, or the standard deviation of sampling distribution of the sample mean

$\hat{\sigma}_{\overline{x}}$ = the estimated standard error of the sample mean

R = the sample range

$\overline{R}$ = the mean of a collection of sample ranges

r = the relative range

d_2 = the mean of the relative range, or a correction factor for estimating σ using the sample range

d_3 = the standard deviation of the relative range, or a correction factor for estimating the standard error of the range

c_4 = a bias correction factor for estimating σ using the sample standard deviation

μ_{s^2} = the mean of the sampling distribution of the sample variance

$\sigma_{s^2}^2$ = the variance of the sampling distribution of the sample variance

σ_{s^2} = the standard error of the sample variance

μ_s = the mean of the sampling distribution of the sample standard deviation

σ_s = the standard error of the sample standard deviation

$\hat{\sigma}_s$ = the estimated standard error of the sample standard deviation

μ_R = the mean of the sampling distribution of the sample range

σ_R = the standard error of the sample range

$\hat{\sigma}_R$ = the estimated standard error of the sample range

n = the sample size

H_0 = the null hypothesis, or the hypothesis tested

H_1 = the alternate hypothesis

α = the level of significance, or the probability of committing a Type I error (read as "alpha")

CV = the critical value, which defines the regions of acceptance and rejection

z_α = the standardized normal variate from the table of Areas of the Normal Curve used in a one-tailed test

$z_{\alpha/2}$ = the standardized normal variate from the table of Areas of the Normal Curve used in a two-tailed test

β = the probability of committing a Type II error

P_w = the power of the test

Sampling and Statistical Inference

When dealing with the problem of statistical process control, we are confronted with a two-fold problem. On the one hand, basic statistical concepts are used that are taught in a particular way. On the other hand, the resulting methods have been adapted or modified when applied to process control. Part of this problem is due to convention and part is due to convenience and simplicity. For us, it means that we must align the language from the two areas as we build the basic concepts.

In general, when problems are solved statistically, they are defined in terms of a universe, or synonymously a population, of items. A **universe**, or population, represents the totality of elements about which a conclusion is drawn or a decision is to be made. With respect to quality control, these items could be manufactured parts or assemblies produced in batches, or continuously, liquid products in a pipe, a shipment of items, or the output of clerical or service workers.

Characteristics of populations or universes are referred to as **parameters**. When a problem is dealt with statistically, the problem is quantified either in terms of the universe distribution and/or its parameters.

Although it is used in many ways, statistics becomes important when samples are used to draw conclusions about a population or universe in terms of its distribution or parameters. Since **samples** are subsets or portions of a universe, they are incomplete and the conclusions based on them are subject to uncertainty. Methods of statistical inference are used to account for this uncertainty.

Statistical inference can be divided into two related parts, estimation and hypothesis testing. Estimation deals with ascertaining the magnitude of one or more universe parameters or the shape of the universe distribution. Hypothesis testing is concerned with choosing among alternatives or establishing whether or not a specified parameter value or distribution shape is tenable based on sample evidence. Principles of estimation are needed in hypothesis testing. Hypothesis testing underlies the concept of a control chart, whereas establishing existing process capability is more of a problem of estimation.

Whenever we employ methods of statistical inference, we must assume that samples are generated at random. A random sample is one for which there exists a calculable probability of occurrence associated with the sample or its elements. If random samples are not used, we have no objective way of assessing the meaning of the results we obtain. The basic tool for assessing sample results is the random sampling distribution.

Sampling With and Without Replacement

Typically when a sample is drawn from a population, items are successively removed from the population and are not replaced prior to the next draw. That is, in practice samples generally are drawn without replacement. However, much of fundamental statistical theory is based on sampling from infinite universes, which is equivalent to sampling with replacement from a finite population; in either case, draws can be made indefinitely without changing the composition of the universe.

When sampling without replacement where the sample size is small relative to the size of the universe, the composition of the universe changes so little with each draw that "with replacement methods" may be applied. We shall assume sampling with replacement throughout the remainder of this book since we are to deal with the output of processes that presumably are unending. By way of contrast, methods of acceptance sampling, which are not considered, oftentimes assume sampling without replacement, since samples are drawn from individual shipments of a definite size.

SAMPLING DISTRIBUTIONS

When a sample is used to make an inference about a population, a sample characteristic corresponding to a universe parameter is used. This sample characteristic is referred to as a *sample statistic*, or briefly, as a statistic. Sample statistics are frequently, but not always, defined similar to the corresponding universe parameter. For example, if a problem is expressed in terms of the universe mean, the sample arithmetic mean often is the statistic used.

If, say, the sample arithmetic mean is used to estimate the universe mean, in general, the result of a single sample is used. Obviously, we want our estimate to be "close" to the true value. There is no way, however, to know whether the particular result is close or not because the universe mean is unknown; that is the reason for estimating it. What we can do, however, is to develop properties of the sample statistic based on long run considerations that give us a basis for assessing the methodology used to obtain the particular result. It should be noted that in process control applications more than one sample result is used in particular instances; however, the basic concepts to be discussed still apply.

Since different samples from the same population have different values of a particular statistic, we can describe the variation of these values in the form of a distribution. When random sampling is used, we can determine the probability of occurrence of the values of a sample statistic. Hence, under random sampling the value of a statistic varies according to chance and is a *random variable*. The probability distribution of all possible values of a sample statistic is defined as the *random sampling distribution*, or simply the sampling distribution, of that statistic.

The concept of a sampling distribution is fundamental but somewhat difficult to understand. In order to understand the nature of a sampling distribution, imagine that you have a universe that is a large lot of items, and interest lies in some critical dimension, say, diameters. Based on material discussed earlier, we can construct a frequency distribution of the diameters and graph the results as a histogram. We also can compute the arithmetic mean of all part diameters; we shall call this the universe mean.

Suppose we select a random sample of items from the population of parts and record their diameters. We could, of course, construct a frequency distribution for this group of part diameters. The frequency distribution of this sample should reflect the nature of the universe distribution, but obviously it would not be exactly the same. An arithmetic mean of this sample also can be computed.

Envision a process where samples of the same size repeatedly are drawn from the universe until all possible combinations of items are selected. In the case of a continuous process, one could think of this as sampling endlessly. Recognize that there are many unique samples of parts and that an arithmetic mean diameter can be computed for each sample. Since the values of the sample mean will differ, we can construct *another* frequency distribution that is associated with these sample arithmetic means. Hence, from the original distribution of universe measurements we can derive a distribution of a sample statistic. In the situation described it is the sample arithmetic mean; however, a similar distribution could be developed for *any* other sample statistic. The frequency distribution of a sample statistic developed in this way is defined as a sampling distribution.

Probability distributions that describe the relative frequency of occurrence of numerous sample statistics are available. These probability distributions provide the relative frequency of occurrence of values of various statistics used to estimate universe parameters. Consequently, it is not necessary to undertake the repeated sampling process outlined above in order to describe the random behavior of many sample statistics used in practice. With the aid of theoretical probability distributions we can determine the frequency with which values of a sample statistic occur and describe the variation, or extent of error, associated with various statistics used as estimates of universe parameters.

A sampling distribution describes the way the values of a statistic vary from sample to sample. Regardless of the statistic, virtually every sampling distribution has a mean, a variance, and a standard deviation. Because of its importance, the standard deviation of a sampling distribution is distinguished from the population standard deviation by a special name; it is referred to as a standard error. The population standard deviation describes the variation among elements of the universe, whereas the standard error measures the variability in a statistic due to sampling; it is a measure of sampling error.

In the case of the distribution of a sample mean, its mean represents the average of all possible sample means, or the "mean of the means" so to speak; the corresponding variance and standard deviation measure the variability among all possible values of the sample mean. In this case, the standard deviation is called the *standard error* of the mean. It is instructive to examine the table of symbols appearing in **Exhibit 1** in order to understand the meaning of these characteristics.

The symbols in the universe column of the exhibit are familiar from previous sections, and those in the second column are symbols of descriptive measures used in Chapters 2 and 3 of Module A. The universe characteristics – μ, σ^2, and σ – correspond to single numbers since they represent parameters, whereas different values of $\overline{X}$, S^2, and S exist that correspond to each of the many possible samples that can be selected from a universe.

EXHIBIT 1
CHARACTERISTICS OF DISTRIBUTIONS ASSOCIATED WITH
THE DISTRIBUTION OF THE SAMPLE ARITHMETIC MEAN

	Universe	Any Sample of the Many Possible	Sampling Distribution
Mean	μ	$\overline{X}$	$\mu_{\overline{x}}$
Variance	σ^2	S^2	$\sigma_{\overline{x}}^2$
Standard Deviation	σ	S	$\sigma_{\overline{x}}$

The symbols appearing in the last column are new and are introduced here for the first time. The subscript attached to each symbol denotes the variable described by each measure and is used to distinguish these measures from the universe parameters. Hence, the symbol $\mu_{\overline{x}}$ (read "Mu-sub-x-bar") represents the mean of the distribution of all possible sample arithmetic means where the sample mean is the variable described. Similarly, $\sigma_{\overline{x}}^2$ and $\sigma_{\overline{x}}$ are the variance and standard deviation of the distribution of the sample mean. Each of these quantities represents a single number and is a parameter of the sampling distribution of the sample arithmetic mean. $\sigma_{\overline{x}}$ is called the standard error of the mean and measures the error owing to sampling the mean. The diagram appearing in **Exhibit 2** summarizes the material presented above. Histograms indicate that we are dealing with an empirical or observable phenomenon; smoothed frequency curves drawn through the universe distribution and the sampling distribution signify that these frequency distributions can be described by probability distributions. The diagrams are associated with the development of the distribution of the sample mean; however, the general concept applies to any statistic.

Based on the above discussion and a careful examination of the diagram, you should be aware of five general points: (1) one universe distribution exists with fixed parameters, (2) for any universe many possible samples exist, (3) the value of any sample statistic varies from sample to sample, (4) one sampling distribution with a particular set of parameters corresponds to a given statistic, and (5) the sampling distribution describes the way the values of a sample statistic vary from sample to sample.

The preceding material in this section has introduced the ideas of sampling, inference, estimation, and sampling distributions. A good bit has been said, yet it probably still is a mystery regarding where this material leads. The material presented is conceptual and is important if one is interested in understanding why certain measures and methods are used in process control.

The remainder of this portion of the chapter is directed toward sampling properties of the sample mean and standard deviation and to calculating probabilities associated with a sample statistic, which is illustrated in terms of the sample mean. This is of importance when we get to charting principles at a later point, and also relates to some considerations presented at the end of the chapter.

EXHIBIT 2
DEVELOPMENT OF THE RANDOM SAMPLING DISTRIBUTION OF THE SAMPLE MEAN

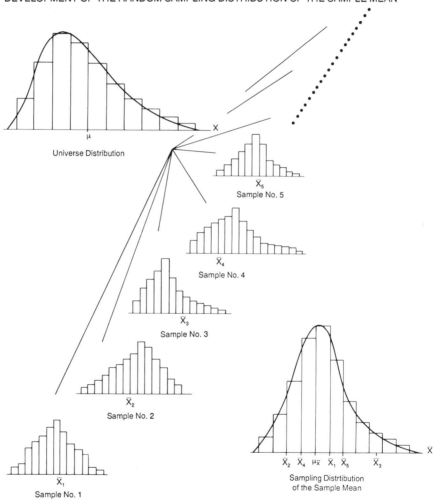

Source: *Statistics For Decision Making* by Ronald Gulezian (QualityAlert Institute 1990)

Facts About Sample Means

We already have mentioned that it is desirable that our sample estimates be "close," in some sense, to the parameters we estimate. Since this must be done in a long-run sense, rather than in one particular estimating situation, it is necessary to build a knowledge base regarding the long-run behavior of relevant statistics under various circumstances or conditions. This knowledge comes from the field of theoretical statistics and is based on conditions relating to known universe or proven parameter values. Strange as it may sound, we develop our knowledge about the long-run behavior of sample statistics based on what would happen with known values, which in turn is applied to the real problem of using particular sample results to estimate unknown parameter values.

Four fundamental results about the behavior of sample arithmetic means are important:

1. The arithmetic mean of the sampling distribution of the sample mean, $\mu_{\bar{x}}$, equals the mean of the population, μ, regardless of the form of the population distribution. That is,

$$\mu_{\bar{x}} = \mu$$

2. The variance, or the square of the standard deviation, of the sampling distribution of the sample mean, $\sigma_{\bar{x}}^2$, equals the population variance, σ^2, divided by the sample size, n. That is,

$$\sigma_{\bar{x}}^2 = \frac{\sigma^2}{n}$$

Hence, the standard deviation of the distribution of sample means equals the square root of the variance.

$$\sigma_{\bar{x}} = \sqrt{\sigma_{\bar{x}}^2}$$
$$or$$
$$\sigma_{\bar{x}} = \frac{\sigma}{\sqrt{n}}$$

3. Sample means from normal populations are normally distributed, regardless of the size of the sample.

4. Sample means from populations that are not normal are not normally distributed, but the distribution approaches the normal as the sample size approaches infinity. This is known as the Central Limit Theorem. Consequently, the normal distribution approximates the distribution of the sample mean for large samples regardless of the shape of the universe distribution.

Although we shall not need the various facts directly in our later work, they are being introduced so that we may clarify some possible inconsistencies associated with measures actually used and, potentially, provide a more complete understanding of the basic principles. Also, the Central Limit Theorem has been introduced for completeness but without explanation. For practical purposes, we shall assume that the sampling distribution of the mean is normal in applications we encounter.

Calculating Probabilities and Limits Using the Four Facts

Once we know the form of the sampling distribution of a sample statistic, we are able to calculate probabilities associated with the occurrence of values of that statistic. Such probabilities are useful when assessing results displayed on process control charts. In addition, we are able to calculate limits associated with the sample result based on given probabilities of occurrence. These are useful for understanding how to construct such charts.

The four facts given above enable us to find the probability of occurrence of values of the sample mean for all sample sizes when the population is normal and approximations when the population is not normal; simply use the appropriate parameter values and proceed according to the steps provided in Chapter 4 for the normal curve. The formula for the standardized variate remains the same except different symbols are used to reflect the difference in the variable involved. That is,

$$z = \frac{\overline{X} - \mu_{\overline{x}}}{\sigma_{\overline{x}}}$$

or

$$z = \frac{\overline{X} - \mu}{\sigma_{\overline{x}}}$$

In words, z is found by subtracting the mean of the sampling distribution from the value of the sample mean and dividing by the standard error. Since the mean of the sampling distribution of sample means equals the mean of the universe, the second of the two expressions can be used to compute z.

Example 1 A continuous manufacturing process produces items whose weights are normally distributed with a mean weight of 8 pounds and a standard deviation of 3 pounds. A random sample of 16 items is to be drawn from the process. What is the probability that the arithmetic mean of the sample exceeds 9 pounds? Interpret the result.

Given: Population normally distributed

$$\mu = 8$$
$$\sigma = 3$$
$$n = 16$$

Find: $P(\overline{x} > 9)$

Compute: $\mu_{\overline{x}} = \mu = 8$

$$\sigma_{\overline{x}} = \frac{\sigma}{\sqrt{n}}$$

$$= \frac{3}{\sqrt{16}}$$

$$= \frac{3}{4} \text{ or } 0.75$$

By way of contrast, the population distribution and the sampling distribution of the mean are shown in the following diagrams:

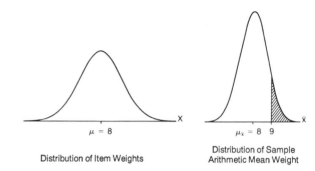

Distribution of Item Weights

Distribution of Sample
Arithmetic Mean Weight

The required probability is represented as the shaded region in the second diagram. Hence,

$$z = \frac{\overline{x} - \mu_{\overline{x}}}{\sigma_{\overline{x}}}$$

$$= \frac{9 - 8}{0.75}$$

$$= \frac{1}{0.75} = 1.33$$

A z-value of 1.33 from the Table of Areas of the Normal Curve yields 0.4082. The required probability is obtained by subtracting this number from 0.5, which yields

$$P(\overline{x} > 9) = 0.5 - 0.4082$$

$$= 0.0918 \text{ or } 0.09 \text{ (rounded)}$$

Recall, the tabulated probability is subtracted from 0.5 owing to the nature of the Table of Areas. Using the rounded result, the calculated probability can be interpreted to mean that 0.09 or 9 percent of all possible samples of 16 items drawn from the population will possess a sample mean value greater than 9 pounds.

Example 2 A random sample of 100 is to be selected from a large group of invoices whose mean amount is $9500, with a standard deviation equal to $800. Find and interpret the probability that the sample arithmetic amount falls between $9404 and $9612.

The result is based on the Central Limit Theorem since we are not told how individual accounts are distributed. Once again, by way of contrast, consider the following diagrams of the possible universe distribution and the sampling distribution.

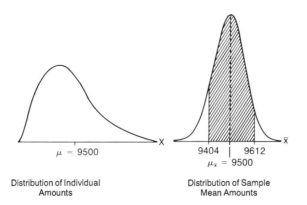

Distribution of Individual
Amounts

Distribution of Sample
Mean Amounts

The required probability is given as the shaded region in the diagram of the normal curve appearing on the right. Hence,

$$\mu_{\bar{x}} = \mu = 9500$$

$$\sigma_{\bar{x}} = \frac{\sigma}{\sqrt{n}}$$

$$= \frac{800}{\sqrt{100}} = 80$$

and

$$z = \frac{\bar{x} - \mu_{\bar{x}}}{\sigma_{\bar{x}}}$$

$$z_1 = \frac{9404 - 9500}{80}$$

$$= -1.2$$

$$z_2 = \frac{9612 - 9500}{80}$$

$$= 1.4$$

Based on the Table of Areas of the Normal Curve, a z-value of 1.2 yields 0.3849 and a z-value of 1.4 yields 0.4192. The answer is obtained by *adding* these two probabilities. Hence,

$$P(9404 \leq \bar{x} \leq 9612) = 0.3849 + 0.4192$$

$$= 0.8041 \text{ or } 0.8 \text{ (rounded)}$$

Using the rounded figure, we can interpret this result as approximately 0.8 or 80 percent of all possible samples of 100 invoices have mean amounts between \$9404 and \$9612. We use the term "approximate" since the result is based on the Central Limit Theorem.

Note. The probability computed in this example reveals nothing about the proportion of invoices possessing various amounts; this only can be determined directly from the universe distribution, provided its form is known.

Example 3 **A Related Problem** Using the information provided in Example 1, find the values of the sample arithmetic mean within which the middle 95 percent of all sample means will fall.

In this problem we are given a probability and must determine values of the sample mean, indicated as $\bar{x}_1$ and $\bar{x}_2$ in the diagram below.

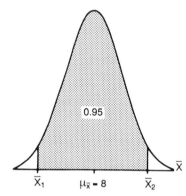

Since we are to find the values of the sample mean corresponding to an interval centered on the mean of the distribution, we can use the same value of z for both values of $\bar{x}$. This z-value is obtained by using half the specified probability, or 0.4750 (i.e., 95/2), and locating z in the Table of Areas of the Normal Curve. Hence, z = 1.96. Based on the formula for z, we must solve for the values of $\bar{x}$ in terms of the known values which yields:

$$\bar{x}_1 = \mu_{\bar{x}} - z\sigma_{\bar{x}}$$
$$= 8 - 1.96(0.75)$$
$$= 8 - 1.47$$
$$= 6.53$$

$$\bar{x}_2 = \mu_{\bar{x}} + z\sigma_{\bar{x}}$$
$$= 8 + 1.96(0.75)$$
$$= 8 + 1.47$$
$$= 9.47$$

Facts About Sample Variances and Standard Deviations (Optional)

Earlier, attention was focused on the concept of a standard deviation since this measure will be used at a later point. In a number of other statistical applications, attention is directed toward the variance, which can be defined as the square of the standard deviation. As descriptive

measures of variation, or dispersion, they both impart the same information about the spread of a set of data or a distribution; the difference
lies in the fact that they are expressed in different units.

When dealing with sample results, however, the properties of
sample variances and sample standard deviations differ. Consequently,
it is difficult for some to understand why simply switching from one to
the other is not completely straightforward. Quite often, statisticians
prefer variances while quality engineers are accustomed to working
with standard deviations. Both measures are considered here to provide
some perspective. We shall discuss the variance first and then the
standard deviation.

In keeping with current conventions, we redefine the sample variance by the formula

$$S^2 = \frac{\Sigma(X - \overline{X})^2}{n - 1}$$

where n is the sample size and $\overline{x}$ is the mean of a sample. Based on this
form of the sample variance, we know the following:

1. The arithmetic mean of the sampling distribution of the sample variance, μ_{s^2}, equals the population variance, σ^2, regardless
   of the form of the population distribution. That is,

   $$\mu_{s^2} = \sigma^2$$

2. The variance of the distribution of the sample variance, $\sigma_{s^2}^2$, is
   found to be

   $$\sigma_{s^2}^2 = \frac{1}{n}\left(\mu_4' - \frac{n - 3}{n - 1}\mu_2'^2\right)$$

   where μ_2' and μ_4' are referred to as the second and fourth
   moments about the origin. Moments about the origin, in general, are means of various powers of a variate. For example,
   the arithmetic mean is the first moment about the origin.

3. When the population is normal, the above formula for the
   variance of the sample variance simplifies to

   $$\sigma_{s^2}^2 = \frac{2}{n - 1}\sigma^4$$

   where σ is the population standard deviation. The standard
   deviation, or standard error, σ_{s^2}, of the sample variance from
   normal populations is given as the square root of the above
   formula, or

   $$\sigma_{s^2} = \sqrt{\frac{2}{n - 1}}\sigma^2$$

   where σ^2 is the population variance.

4. No general formula for the distribution of the sample variance
   exists. Sample variances from normal populations have a

known form that is not normal. The form of the distribution is referred to as a Chi-square distribution. This distribution is considered in many basic and also more advanced texts on statistics.

5. The Chi-square distribution is a continuous distribution which is skewed to the right. A function of Chi-square approaches a normal as the sample size is increased. Convergence is not as rapid as in the case of sample means.

The formulas given above for the mean and variance of the sample variance cannot be generalized to the sample standard deviation from any population by merely taking the square root. In other words, no exact general formula exists for the characteristics of the distribution of the standard deviation similar to the variance. Results are available for normal populations as follows:

1. The arithmetic mean of the sampling distribution of the sample standard deviation, μ_s, is given by the formula

$$\mu_s = \sqrt{\frac{2}{n-1}} \frac{\left[\frac{n-2}{2}\right]!}{\left[\frac{n-3}{2}\right]!} \sigma$$

where n is the sample size, σ is the population standard deviation, and the exclamation mark represents a factorial. Recall, the factorial of an integer is the product of that integer and all integers less than it. When needed, the factorial of $1/2$ is given as $\sqrt{\pi/2}$. Fortunately, the multiple of σ given in the above formula is tabulated in the Appendix in **Table C-1**. The quantity has been designated as c_4 in process control work. Hence, the above formula reduces to

$$\mu_s = c_4\sigma$$

and can be calculated easily in terms of the values provided in the table.

2. The exact formula for the standard deviation, or standard error, of the sample standard deviation, σ_s, for normal populations is given as

$$\sigma_s = \sqrt{1 - c_4^2}\ \sigma$$

For convenience, this is expressed in terms of c_4 rather than the cumbersome expression that it represents.

3. Exact probabilities associated with sample variances from normal populations can be obtained from the Chi-square distribution for variances mentioned above. The Chi-square distribution is tabulated in the Appendix as **Table A-5**.

Although we shall not use all of the facts provided about the sample variance and sample standard deviation directly, they have been presented for the record, so to speak, in order to provide an appreciation for the complexity of the ideas underlying the tools used later. They also allow us to better appreciate the relationship between methods used in statistical process control and ones possibly encountered in basic statistics courses.

Biasedness and Unbiasedness (Optional)

Repeatedly, we have indicated that the problem of estimating parameters using sample data lies in establishing the "closeness" of an estimate to the quantity to be estimated. By now, you should have an appreciation for the fact that "closeness" only can be considered in terms of what occurs in the long run. Moreover, it is the sampling distribution of a sample statistic that provides us with information about the long run behavior of a sample statistic. Since different statistics have different sampling distributions the properties of statistics vary.

Statistical theory provides a number of criteria for determining how good a particular statistic is as an estimator of a universe parameter. We have already introduced the concept of *reliability* or sampling error, which is measured in terms of the standard error of a statistic. That is, the standard deviation of the sampling distribution, or the standard error, is a measure of how close together results from repeated samples are to one another. The smaller the standard error, the closer the values from sample to sample and the greater the repeatability, so to speak, from trial to trial.

Of the remaining criteria available, biasedness is important to consider with respect to methods used in process control. This defines "close" in the long run in terms of being correct "on the average". More precisely, we say that a statistic is an **unbiased** estimator of a universe parameter if the arithmetic mean of its sampling distribution equals the parameter to be estimated. In other words, if one were to sample repeatedly until exhausting all possible samples and calculate the mean of the resulting statistic, if the mean of the results equals the parameter to be estimated, the statistic is said to be unbiased; this applies to any statistic in general. Recognize that one does not actually undertake the repeated sampling process in practice. We are interested in the result, or the concept, of what "would" happen if we did. Also, it is conventional to use the arithmetic mean of a statistic: there is no reason why another measure of central tendency could not be used.

By recalling the results presented earlier in the section, we can assess whether the sample mean, variance, and standard deviation are biased or unbiased:

$$\mu_{\bar{x}} = \mu \text{ (Any population form)}$$
$$\mu_{s^2} = \sigma^2 \text{ (Any population form)}$$
$$\mu_s = c_4\sigma \text{ (Normal universe only)}$$

Immediately we can see that the sample mean, $\bar{x}$, is an unbiased estimator of the universe mean, μ; the sample variance, S^2, is an unbiased estimator of the population variance, σ^2; and the sample standard deviation, S, is a biased estimator of the standard deviation, σ, of a normal universe. In general, there is a tendency for the sample standard deviation to understate the population standard deviation, on the average. There is no general result, however, that expresses the mean of the sample standard deviation as a function of the population standard deviation for any population. The above result for the normal is commonly accepted as an approximation in most cases in practical quality control work.

If we also recall the formula for the sample variance

$$S^2 = \frac{\Sigma(X - \overline{X})^2}{n - 1}$$

we now have a basis for understanding why this is used rather than one based on a divisor equal to the sample size "n".

When "n" rather than "n-1" is used, the corresponding sample variance is a biased estimator of the population variance. By subtracting one from the sample size, the result is an unbiased estimator. As descriptive measures of variation, both impart the same basic information. When estimating a population variance, the unbiased version of the sample variance compensates for the persistent understatement.

Although taking the square root does not automatically yield an unbiased estimate of the population standard deviation, sometimes it is implied that the square root results in an unbiased estimator also. Consequently, the version with "n-1" is becoming the adopted version of the "sample standard deviation" when a single sample is used, both as a descriptive measure for a sample and as an estimator of the universe standard deviation. When a standard deviation is estimated in process control, the normal correction, c_4 , is used to adjust for bias almost universally.

ESTIMATING THE MEAN AND STANDARD DEVIATION

Based on the ideas presented above, we can specify the formulas used to estimate the mean and standard deviation of a process characteristic. Although some of the formulas have been given in Chapter 2 and Chapter 3, we shall repeat them here.

The mean and standard deviation of a single sample are computed using the formulas

$$\overline{X} = \frac{\Sigma X}{n}$$

$$S = \sqrt{\frac{\Sigma(X - \overline{X})^2}{n - 1}}$$

$$= \sqrt{\frac{\Sigma X^2 - (\Sigma X)^2/n}{n - 1}}$$

The first of the two formulas for the sample standard deviation is the definitional form and the second is a shortcut. Both provide the same answer, except for possible rounding errors.

When estimating the mean and standard deviation of a process, typically estimates are based on results from a number of samples selected at equally spaced points in time. The process mean and standard are estimated as

$$\bar{\bar{X}} = \frac{\Sigma \bar{X}}{k}$$

$$\hat{\sigma} = \frac{\bar{S}}{c_4}$$

where

$$\bar{S} = \frac{\Sigma S}{k}$$

Since the sample mean is an unbiased estimator, the average of the means of k samples also is unbiased. The symbol $\hat{\sigma}$ represents the estimated process standard deviation, which equals the mean of the sample standard deviations, $\bar{S}$, divided by the correction for bias, c_4. This could, of course, be used to adjust a single sample result for bias, although it typically is not applicable in process control.

Example 4   Consider a sample of five observations

   12.04   14.61   5.97   9.48   9.49

The mean of the sample is found as

$$\bar{X} = \frac{12.04 + 14.61 + 5.97 + 9.48 + 9.49}{5}$$

$$= \frac{51.59}{5}$$

$$= 10.32$$

Preliminary results to be used in the shortcut formula for the sample standard deviation are found

$$\Sigma X = 12.04 + 14.61 + 5.97 + 9.48 + 9.49$$

$$= 51.59$$

$$\Sigma X^2 = 12.04^2 + 14.61^2 + 5.97^2 + 9.48^2 + 9.49^2$$

$$= 144.962 + 213.452 + 35.641 + 89.870 + 90.060$$

$$= 573.985$$

The sample standard deviation is found as

$$S = \sqrt{\frac{573.985 - (51.59)^2/5}{5 - 1}}$$

$$= \sqrt{3.228}$$

These results could be used to estimate a universe or process mean and standard deviation. Alternatively, S could be adjusted for bias to yield

$$\hat{\sigma} = \frac{S}{c_4}$$

$$= \frac{3.228}{.9400}$$

$$= 3.434$$

where c_4 is obtained from **Table C-1** in the Appendix. Since $\bar{\bar{X}}$ already is unbiased, we can relabel it as an estimate of the universe or process mean as

$$\hat{\mu} = \bar{\bar{X}}$$

$$= 10.318$$

Example 5   Estimate the process mean and standard deviation based on 10 successively drawn samples of five measurements provided below.

Sample Number

Observation Number	1	2	3	4	5	6	7	8	9	10
1	12.04	9.54	11.27	11.88	10.95	14.90	10.91	7.42	12.63	9.95
2	14.61	10.72	12.52	7.69	9.50	8.53	7.46	11.66	10.39	8.40
3	5.97	12.44	6.25	11.99	11.56	9.48	11.34	11.35	3.90	10.97
4	9.48	11.37	5.85	7.87	10.42	9.62	8.28	8.58	11.10	9.50
5	9.49	12.57	7.86	8.69	14.22	10.67	9.67	11.46	10.72	9.81
$\bar{X}$	10.32	11.33	8.75	9.62	11.33	10.64	9.53	10.09	9.75	9.73
S	3.23	1.26	3.00	2.14	1.78	2.50	1.66	1.96	3.38	.92

The process mean and standard deviation are estimated as follows:

$$\bar{\bar{X}} = \frac{10.32 + 11.33 + 8.5 + \cdots\cdots + 9.75 + 9.73}{10}$$

$$= \frac{101.09}{10}$$

$$= 10.11$$

$$\bar{S} = \frac{3.23 + 1.26 + \cdots\cdots + 3.38 + .92}{10}$$

$$= \frac{21.83}{10}$$

$$= 2.183$$

Using **Table C-1** in the Appendix to find c_4 for samples of size 5, we have

$$\hat{\sigma} = \frac{2.183}{.9400}$$

$$= 2.322$$

and

$$\hat{\mu} = \bar{\bar{X}}$$

$$= 10.11$$

Estimating Standard Errors (Optional)

A distinction has been made in this chapter between a universe distribution and the sampling distribution of a sample statistic that is used as an estimator of a universe parameter. For our purposes, the important distinction is between the distribution of a process characteristic and the corresponding sampling distribution of the statistics used to estimate or monitor its performance. Our interest here lies in the process mean and the process standard deviation, for which we have introduced methods of estimation.

The methods of estimation used have been justified in terms of the arithmetic mean of the corresponding sampling distributions, namely, in terms of biasedness or unbiasedness. We also have addressed sampling error, which is measured in terms of the amount of variation in the sampling distribution of a sample statistic. We indicated that the measures are referred to as standard errors, for which we have provided formulas. In general, when dealing with sampled data exclusively rather than targeted process parameters, we need to estimate the standard errors also. When using the results for control chart purposes, the resulting adjustments automatically are made in terms of factors tabulated in the Appendix and need not be calculated.

The two that we need to estimate are the standard error of the mean $\sigma_{\bar{x}}$, and the standard error of the sample standard deviation, σ_s. Using the symbols $\hat{\sigma}_{\bar{x}}$ and $\hat{\sigma}_s$ for the estimated standard errors, these are obtained by substituting the estimated standard deviations into the formulas provided earlier. That is,

$$\hat{\sigma}_{\bar{x}} = \frac{\hat{\sigma}}{\sqrt{n}}$$

$$= \frac{\bar{S}}{c_4\sqrt{n}}$$

$$\hat{\sigma}_s = \sqrt{1 - c_4^2}\,\hat{\sigma}$$

$$= \sqrt{1 - c_4^2}\left[\frac{\bar{S}}{c_4}\right]$$

Example 6 Use the results based on the 10 samples of 5 given in Example 5 to find the estimated standard errors, $\hat{\sigma}_{\bar{x}}$ and $\hat{\sigma}_s$.

$$\hat{\sigma}_{\bar{x}} = \frac{2.322}{\sqrt{5}}$$

$$= 1.0384$$

$$\hat{\sigma}_s = \sqrt{1 - (.9400)^2}\,(2.322)$$

$$= 0.7922$$

Alternatives Using the Sample Range

We indicated in Chapter 3 that both the range and the standard deviation can be used as measures of dispersion. Further, the range is easier to calculate but does not consider the behavior of the observations between the extremes. Although the range and standard deviation of a sample typically will fluctuate together, the range based on small samples will not directly reflect the process range, or the natural tolerance, very well.

The sample range, however, is used extensively in process control. This stems from the fact that it is very easy to understand and to calculate. Also, for small samples it can be used to estimate the process standard deviation reasonably well. In this regard, the sample range can be used in the following ways:

1. To estimate the process standard deviation both for determining process capability and for determining whether a process mean is in control.

2. To determine whether process variation is in control.

The applications are selected and rest on convention and convenience owing to the ease of calculation when computation is done by hand rather than through automated means. Again, recognize that we are interested in using the sample range without reference to the population range.

As in the case of other sample statistics, we can say something about the sampling properties of the range. Our attention is focussed on sampling from normal populations, again owing to the usefulness of the normal in process control work. Due to the nature of the applications involved, we shall introduce the properties of the sample range, R, as it relates to the universe standard deviation, σ. More formally, we shall focus on the relative range, r, which is given as

$$r = \frac{R}{\sigma}$$

That is, the relative range equals the sample range divided by the universe standard deviation, . Properties of the relative range have been derived for normal populations, just as they have been for the sample mean and sample standard deviation. The explicit results are rather complicated and as a consequence we shall present them in terms of tabulated adjustment factors commonly in use.

The mean and standard error of the relative range are given as

$$\mu_r = d_2$$

$$\sigma_r = d_3$$

where d_2 and d_3 are tabulated constants based on sample size, a selection of which are presented in **Table C-2** of the Appendix.

Since the mean of the sample ranges, R, is used in quality control work, we are able to use this value together with the above results to establish estimates of σ and σ_R. That is,

$$\hat{\sigma} = \frac{\overline{R}}{d_2}$$

$$\hat{\sigma}_R = d_3 \left[\frac{\overline{R}}{d_2} \right]$$

Recognize that $\hat{\sigma}$ is an estimate of σ, the process standard deviation; the numerical value for which will not be the same as the one based on $\overline{S}$. Also, $\hat{\sigma}_R$ is an estimate of σ_R, the standard error of the sampling distribution of the sample range.

Example 7   Use the information presented in Example 5 to find the sample ranges for the 10 samples presented and estimate the population standard deviation.

Sample Number	Sample Ranges, R
1	8.64
2	3.03
3	6.67
4	4.30
5	4.72
6	6.37
7	3.88
8	4.24
9	8.73
10	2.57

$$\overline{R} = \frac{\Sigma R}{k}$$

$$= \frac{8.64 + 3.03 + \cdots\cdots + 8.73 + 2.57}{10}$$

$$= \frac{53.15}{10}$$

$$= 5.315$$

$$\hat{\sigma} = \frac{\overline{R}}{d_2}$$

$$= \frac{5.315}{2.326}$$

$$= 2.285$$

**Example 8
(Optional)** Obtain an estimate of the standard error of the sample range, σ_R, based on the results associated with Example 5.

$$\hat{\sigma}_R = d_3 \left[\frac{\overline{R}}{d_2} \right]$$

$$= (.8641) \left[\frac{5.315}{2.326} \right]$$

$$= 1.975$$

Note: Factors such as d_2 and d_3 that are useful when working with the range can be found in Table C-2 of the Appendix.

3σ vs. $3\sigma_{\overline{x}}$ Limits

No matter how much is said, there appears to be confusion regarding limits based on the process standard deviation, σ, and those based on the standard error of the mean, $\sigma_{\overline{x}}$. Although constructed similarly, they relate to different things. Actually, more will be said about this later; however, a preview should help to reinforce the ideas. Since "3-sigma" limits are widely accepted in process control, we shall use only multiples of three here. Any others can be used in their place.

The briefest way to make the distinction between the two is to think of σ relating to process capability, or specifications and tolerances, and $\sigma_{\overline{x}}$ relating to control. If we refer back to limits calculated on the basis of a normal curve, it was indicated that 99.73 percent lie within three standard deviations of the mean. The real question is 99.73 percent of what?

If we consider the interval

$$\mu \pm 3\sigma$$

we are referring to 99.73 percent of individual items produced falling within three standard deviations. Assuming normality, the resulting limits can be considered as natural tolerances for which less than 0.3 percent fall outside the limits. In other words, virtually all of the items produced fall within three standard deviations of the mean. If these natural tolerances coincide with actual specifications, assuming the process is centered on μ, roughly 0.3 percent are defective. Obviously, the further the "natural tolerances" are within the specification limits the smaller the defective rate, which is evident from Exhibit 9 in

Chapter 4 and provides various ppm values corresponding to varying multiples of σ.

By way of contrast, when considering the interval,

$$\mu \pm 3\sigma_{\bar{x}}$$

which has the same structure, there is a distinct difference due to the difference in the error term, $\sigma_{\bar{x}}$. In this case, 99.73 percent of all possible *samples* have sample means between the designated limits. Moreover, the designated limits fall within the 3σ limits. Later, we shall see that sample means falling outside the $3\sigma_{\bar{x}}$ limits provide an indication that the central value, or the process mean, has shifted. Although this may have a bearing on the defective rate, direct interest is not on specifications, but one of control.

Example 9 Assume that the distribution of a critical process dimension is normally distributed with mean, μ, equal to 20 and standard deviation, σ, equal to 1. Further, specifications are given as 16 – 24.

The so-called natural tolerances can be calculated as

$$\mu \pm 3\sigma$$

$$20 \pm 3(1)$$

$$20 \pm 3$$

$$17 - 23$$

Therefore 99.73 percent of the items produced fall between 17 and 23 units, and less than 0.3 percent fall outside.

Since actual specifications are wider than the natural tolerance, 0.06 percent fall outside of the specification limits. The results are displayed in the following diagram.

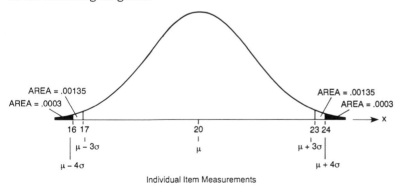

Individual Item Measurements

Now consider the following limits based on samples of 9 observations.

$$\mu \pm 3\sigma_{\bar{x}}$$

$$20 \pm 3 \left[\frac{1}{\sqrt{9}} \right]$$

$$20 \pm 3 \left[\frac{1}{3} \right]$$

$$20 \pm 1$$

$$19 - 21$$

In this case, we can say that 99.73 percent of all possible samples will have a mean, $\bar{x}$, that lies between 19 and 21. Notice that the probability is the same, but the limits are narrower. In terms of individual items, only 26.11 percent fall within and 73.89 percent fall outside. The following diagram portrays the distribution from which the above limits were calculated.

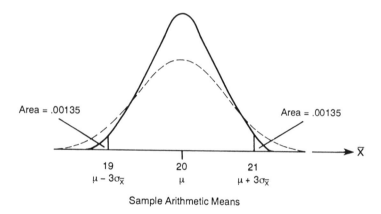

Area = .00135   Area = .00135

19   20   21
$\mu - 3\sigma_{\bar{x}}$ μ $\mu + 3\sigma_{\bar{x}}$

Sample Arithmetic Means

Added Remarks

A lot of ideas have been presented in this section so far. Actually, process control procedures can be implemented without presenting the material in the way in which it is done here. The intent, however, is to provide a stronger basis for understanding the origin of the results applied at a later point.

Basically, we have introduced the concept of a sampling distribution of a statistic which is fundamental to all problems of statistical inference and underlies methods of statistical process control. Although the concept is universal, we have introduced the idea in terms of means, since these are the most commonly used and understood. The sampling distribution forms the basis for establishing properties of estimators and the methods of estimation that we use in practice.

Different statistics behave differently in a sampling sense. Consequently, although general concepts are applicable, specific results differ. Some are more difficult to deal with mathematically and oftentimes are not complete regarding our knowledge about their behavior. Sometimes this results either in confusion or inconsistencies when coupling statistical theory with quality engineering. Moreover, the long history of quality control without the use of computational aids has led to conventions that are simple expedients. Some liberties are taken in quality control work that otherwise are not considered. In many cases, the methods work effectively; however, it is beneficial to question the applicability of the methods used in particular circumstances.

HYPOTHESIS TESTING

Hypothesis testing is a statistical approach for drawing inferences about universe parameters, which is oriented toward choosing among alternatives. The methodology also can be used to establish the distributional form of a population and the magnitude of universe parameters. The emphasis in this section is placed on the concept of hypothesis testing as a statistical method rather than as a complete process control tool. Consequently, the approach leans more toward ones used in basic statistics texts or courses. Although the underlying concepts are the same, quality control procedures depart, to varying degrees, in terms of language and in terms of specific computational procedures.

The goal is to provide a basis for a deeper understanding of the principles underlying statistical process control than usually provided in quality control manuals and workbooks. Although some reference is made to quality control problems, the next chapter bridges the gap between the concepts presented here and control charting procedures. Neither this section nor the next chapter need to be covered if computational procedures of process control are the only things of interest. These are presented in detail in Module C. Emphasis here is on the concept of testing and is illustrated in terms of arithmetic means and normal sampling distributions.

Background

In general, there are two types of statistical tests of hypotheses: one-tailed and two-tailed. While two-tailed tests are more commonly used in process control, one-tailed tests are more common in acceptance sampling, although this depends on the application or standard on which it is based. Here, we shall begin with an illustration that leads to a one-tailed test and then use a second example that is two-tailed.

Consider a problem involving the manufacture of flashlight batteries. A shipment is received by a distributor who must determine if it is of acceptable quality before it is sold. A defective shipment will be returned to the manufacturer. Since it is impossible to produce batteries that have identical lengths of life, quality must be considered in terms of one or more parameters of the shipment; a high mean life with low variance is desirable.

Assuming the variance exhibits a satisfactory amount of stability, let us concentrate on the mean life of the shipment. Assume that the distributor requires that the mean life should be at least 1000 hours. The problem can be expressed in terms of two quantified statements or *hypotheses* about the shipment quality, or about the universe of batteries:

1. The mean length of life is at least 1000 hours: $\mu \geq 1000$

2. The mean length of life is less than 1000 hours: $\mu < 1000$

Corresponding to each hypothesis is a course of action that depends on the actual state of the shipment:

1. Accept the shipment; hold for sale

2. Do not accept the shipment; return to manufacturer

As the problem is stated, emphasis is placed on making a decision regarding the acceptability of the shipment rather than knowing the magnitude of the mean life of the batteries exactly.

Whenever sampling is employed to make a decision, we need a procedure to account for errors arising because the entire universe is not observed. Hypothesis testing is one such procedure.

In order to perform a test we must distinguish between the two hypotheses or statements about the universe. The one "tested" is referred to as the **null hypothesis**, or the hypothesis tested. The other represents the negation of the null hypothesis and is called the **alternate hypothesis**. The alternate hypothesis must be defined in such a way that rejection of the null hypothesis automatically leads to the acceptance of the alternate. No overlap in the values of the two hypotheses can exist.

Special attention must be given to the way a null hypothesis is chosen since the entire testing procedure is based on this choice. The null hypothesis *always* is chosen as the hypothesis that contains the equality. For example, in the battery shipment problem presented above, the null hypothesis is "the mean length of life is at least 1000 hours," or $\mu \geq 1000$. Of the two possible alternatives this one contains the "equal-sign" in addition to the "greater-than" inequality. The value specified in the null hypothesis is referred to as the hypothesized value or the **hypothesized parameter**. Hence, "1000" is the hypothesized value of the mean length of life in our example.

Special symbols are used to distinguish between the null and alternate hypotheses. The null hypothesis is designated by the symbol H_0 (read as "H-sub-zero") whereas the symbol H_1 (read as "H-sub-one") is used for the alternate hypothesis. Therefore, the two hypotheses in the battery example can be written in terms of these symbols as

Null hypothesis $H_0 : \mu \geq 1000$

Alternate hypothesis $H_1 : \mu < 1000$

The value specified at the equality in the null hypothesis, referred to as the hypothesized parameter, also can be designated as μ_0.

So far, we have introduced the idea that the manner in which hypotheses are formulated depends on the statement of the original problem to be solved. Of the designated hypotheses, the one containing the equality is the one tested.

Once a problem is formulated in terms of hypotheses, a decision rule is developed that tells us which values of a sample statistic lead to acceptance or rejection of the null hypothesis. The statistic that is used to perform a test of a hypothesis is referred to as the **test statistic**. Since the value of a sample statistic exhibits variability from sample to sample, a test of a hypothesis must account for this variability. For example, suppose we hypothesize the mean life of the batteries is 1000 hours and decide to test whether this is true on the basis of the mean of a sample of items drawn from the shipment. Even if this hypothesis is correct, it is unlikely that the mean of the sample will be exactly equal to the shipment mean. This results from the fact that the sample mean may differ from the universe mean owing to chance or sampling variability. Hence, it is necessary to decide how much of a difference between the sample result and the hypothesized value is tolerable before being able to conclude that the shipment mean does not equal 1000. In other words, it is necessary to develop a way of deciding how much variability between the sample result and the hypothesized value can be attributed to chance and how much cannot. If we conclude that the variability is due to chance, we accept the null hypothesis, and if not we reject the null hypothesis.

Aided by the sampling distribution, we are able to establish the maximum difference between the value hypothesized and the sample statistic that is consistent with the null hypothesis; the set of values of the statistic corresponding to this difference that leads to the acceptance of the null hypothesis is called the **region of acceptance**. Conversely, the set of values of the statistic leading to rejection of the null hypothesis is referred to as the **region of rejection**, or **critical region**. The value of the sample statistic that defines the regions of acceptance and rejection is referred to as the **critical value**.

In order to understand these points, consider the null hypothesis in the battery example, $H_0 : \mu \geq 1000$, against the alternate hypothesis, $H_1 : \mu < 1000$. Suppose we intend to select a sample from the shipment, compute the arithmetic mean of this particular sample, and use the result to decide which of the two hypotheses to select. Hence, we are to test a hypothesis where the sample mean is the test statistic.

In order to perform the test we first must decide on the values of the sample arithmetic mean that will lead to the conclusion that the null hypothesis is true and those values that will lead us to conclude that it is false. If we conclude the null hypothesis is false, we reject it and automatically accept the alternate hypothesis. Obviously, if the value of the sample mean happens to be greater than or equal to 1000, we would accept the null hypothesis since the sample result would be consistent with this hypothesis: the sample mean has a value that falls in the same range as the universe mean described in the null hypothesis.

For the example we are considering, the real problem is to decide what values of the sample mean *below* the hypothesized value of 1000 would lead to the rejection of H_0. This results from the fact that the alternate hypothesis specifies values of the universe mean that are "less than" 1000, or $\mu < 1000$. In other words, values of the sample mean that are below the hypothesized mean are *not* consistent with the null hypothesis and become "candidates" that may lead to the rejection of H_0.

Since the sample arithmetic mean is subject to variability due to chance, values of the sample mean close to the hypothesized mean are more likely to occur than others when the null hypothesis is true. Consequently, it is reasonable to reject the null hypothesis when the sample mean is "much less" than the hypothesized mean since these values of the sample mean are unlikely to occur when the null hypothesis is true. Hence, "low values" of the sample mean below 1000 that are unlikely lead to rejection of H_0; which can be seen in the following diagram:

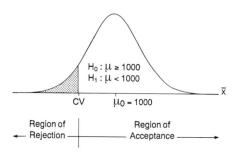

The problem, therefore, is to decide on the point within the extremes beyond which we feel that it is not likely that the sample result comes from a universe possessing the hypothesized value. This point is the *critical value*, designated as CV in the above diagram, which provides the *decision rule* used to choose between the two hypotheses in terms of a region of rejection and a region of acceptance. Values of the sample mean falling in the region of rejection lead to the rejection of the null hypothesis, and values of the sample mean falling in the region of acceptance lead to the acceptance of the null hypothesis.

Probabilities can be attached to values of the sample mean using the sampling distribution of the mean. In the example we are considering, we can use the sampling distribution of the mean to identify values of the sample mean that are "much lower" than 1000 and are unlikely to occur if the universe mean actually is 1000. Hence, we use the sampling distribution to find critical values that define the region of rejection.

There are two types of errors associated with any test. These always are related to the null hypothesis and are defined as follows:

Type I error: Rejection of a true null hypothesis

Type II error: Acceptance of a false null hypothesis

Whenever a hypothesis is formulated, we do not know whether it is true or false. However, we can distinguish between the types of errors that can be made *conditional* on the possible alternative. For example, consider the battery shipment illustration where the null hypothesis states that the mean life is at least 1000 hours, or $\mu \geq 1000$. If this represents the actual state of the shipment and we conclude the mean life is less than 1000 hours, we have committed a Type I error; in practical terms, an acceptable shipment is rejected. A Type II error would be committed if an unacceptable shipment were accepted by concluding that the mean life is at least 1000 hours when it really is less.

The types of errors that can be made are summarized in **Exhibit 3** in terms of the example. We can see from the exhibit that no error is committed if the hypothesis chosen is the same as the true state of the shipment. If, however, we reject the null hypothesis and conclude that the alternate is true when it is not, a Type I error is committed. On the other hand, if we conclude that the null hypothesis is true and accept it when the alternate is true, a Type II error is committed.

EXHIBIT 3
SUMMARY OF TYPES OF ERRORS
IN BATTERY SHIPMENT EXAMPLE

| | Conclusion | |
True State of Shipment	$H_0 : \mu \geq 1000$	$H_1 : \mu < 1000$
$\mu \geq 1000$	No Error	Type I Error
$\mu < 1000$	Type II Error	No Error

By employing the sampling distribution of a statistic, we can measure in advance the probabilities of committing the two types of errors. The probability of committing a Type I error is essential when constructing a test of a hypothesis. The probability of a Type I error is given a special name; it is called the **level of significance**. The probability of a Type II error also can be used to establish a test and is useful in evaluating a test. It is considered at the end of the section.

Like estimation, different computational procedures are required in particular cases, although there is a standard format for conducting a test. Consequently, we shall introduce the general steps for conducting a test first and then present details for means assuming a normal sampling distribution in order to illustrate the concepts.

General Procedure for Testing Hypotheses

The following general steps apply when performing a test of any hypothesis:

Step 1. State the problem in terms of two hypotheses about the universe.

Step 2. Quantify the two hypotheses in terms of a relevant universe parameter. Identify the null and alternate hypothesis, such that the null hypothesis always contains an "equality."

Step 3. When appropriate, specify courses of action corresponding to the null and alternate hypotheses.

Step 4. Identify the test statistic.

Step 5. Determine the sampling distribution of the test statistic under the assumption that the "equality" in the null hypothesis is true.

Step 6. Select the value of the level of significance, α, representing the probability you personally are willing to risk in committing a Type I error. The guide to follow is that this probability must be "small." Conventional values used in most statistical problems are 0.01, 0.02, 0.025, 0.05, and 0.10.

Step 7. Based on the chosen level of significance, compute the critical value(s) of the sample statistic that provides a decision rule leading to the acceptance or rejection of the null hypothesis. The region of rejection always appears in one or both tails of the sampling distribution of the test statistic. The area or probability in the tails equals the level of significance, α. For a one-tailed test α appears in one tail, and for a two-tailed test half α appears in each tail of the distribution.

Step 8. Select a random sample and compute the value(s) of the relevant statistic(s).

Step 9. Based on the calculated value of the test statistic, apply the decision rule established in Step 7.

Step 10. Draw a conclusion or make a decision in terms of the original problem.

Procedure for Means with Normal Sampling Distribution and Known σ

Once a null hypothesis is specified and the sampling distribution of the corresponding test statistic is identified, the problem is reduced to determining the critical value. The procedure or method of calculation for doing this depends on the form of the sampling distribution.

Here, we are only treating tests for means where we assume the sampling distribution of the sample mean is normal and the population standard deviation is known. Calculation of a critical value in these cases reduces to the computation of the value of a normal variate corresponding to a specified probability in the tail of the distribution. The area in the tails is α, the probability of committing a Type I error.

Diagrams summarizing all cases of tests of hypotheses for means based on the given assumptions is presented in **Exhibit 4**. It can be seen that the region of rejection found by calculating the critical values(s) represents the extreme values of the sample mean associated with the tails of the random sampling distribution of the mean. Critical values are computed in the same way as any other values of a normally distributed variable corresponding to a given probability level; the probability level is specially chosen so that there is a small probability of a Type I error.

Two examples are given below. The first is based on the battery shipment problem and delineates all ten steps in order to provide a complete view of the testing procedure. The second example, while not as detailed, introduces a two-tailed test and relates to material in Module C.

Example 10 Shipments of flashlight batteries meet distributor specifications provided the mean length of life is at least 1000 hours. If a shipment does not meet this requirement, it is returned to the manufacturer. Shipments received in the past have consistently had a standard deviation of 50 hours, and length of life can be assumed to be normally distributed. A random sample of 49 batteries is drawn from an incoming shipment in order to determine whether specifications are met. The sample arithmetic mean is 995 hours. Based on the information given, determine whether the distributor should keep the shipment or return it to the manufacturer.

EXHIBIT 4
SUMMARY OF TESTS OF HYPOTHESES FOR THE UNIVERSE
MEAN USING A NORMAL SAMPLING DISTRIBUTION AND KNOWN σ

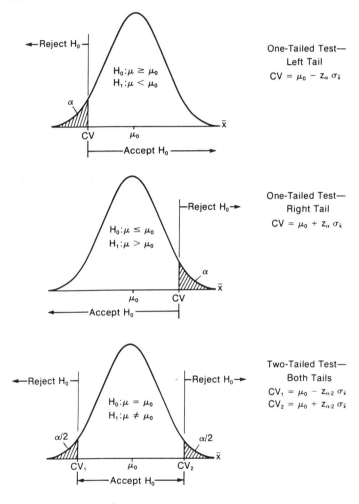

One-Tailed Test—
Left Tail
$CV = \mu_0 - z_\alpha \sigma_{\bar{x}}$

One-Tailed Test—
Right Tail
$CV = \mu_0 + z_\alpha \sigma_{\bar{x}}$

Two-Tailed Test—
Both Tails
$CV_1 = \mu_0 - z_{\alpha/2} \sigma_{\bar{x}}$
$CV_2 = \mu_0 + z_{\alpha/2} \sigma_{\bar{x}}$

Step 1. Problem Goal Stated in Terms of Hypotheses

Based on the statement regarding the distributor's specification, one hypothesis is

The shipment mean life is at least
1000 hours, or $\mu \geq 1000$

The other possible hypothesis must consider shipment states not consistent with the stated specification, resulting in the hypothesis

> The shipment mean life is less than
> 1000 hours, or $\mu < 1000$

Step 2. Choosing The Null and Alternate Hypotheses

One of the two hypotheses specified in Step 1, the first ($\mu \geq 1000$) contains an equality in addition to the "greater-than" inequality. The equality must appear in the null hypothesis. Hence,

$$H_0: \mu \geq 1000$$
$$H_1: \mu < 1000$$

Step 3. Specifying Courses of Action

This follows directly from the problem statement:

H_0: Shipment meets specifications — Keep the shipment
H_1: Shipment does not meet specifications — Return shipment to manufacturer

Step 4. Test Statistic

Since the hypotheses are stated in terms of the universe mean, the test statistic is the sample arithmetic mean, $\bar{x}$.

Step 5. Identify the Sampling Distribution of the Test Statistic

Since shipment standard deviations exhibit stability over time, we shall treat the universe standard deviation as known, equal to 50 hours (i.e., $\sigma = 50$). Based on this assumption and the fact that we are told that the distribution of battery life is normal, the distribution of the sample mean can be assumed to be normal.

Step 6. The Level of Significance

Since a level of significance is not provided in the problem, we must select one. Assume we are willing to take a small risk of returning an acceptable shipment and choose a 1 percent level, or $\alpha = 0.01$.

Step 7. Critical Value and Decision Rule

Based on the alternate hypothesis ($H_1: \mu < 1000$), "low values" of the sample mean lead to rejection of the null hypothesis. Consequently, the region of rejection is in the left tail of the sampling distribution. This is defined in terms of the critical value

$$CV = \mu_0 - z_\alpha \sigma_{\bar{x}}$$

$$= \mu_0 - z_{.01}\left[\frac{\sigma}{\sqrt{n}}\right]$$

$$= 1000 - 2.33\left[\frac{50}{\sqrt{49}}\right]$$

$$= 1000 - 2.33(7.14)$$

$$= 983.36$$

The value of z substituted above corresponds to an area in the tail of a normal curve equal to 0.01, the value of α. The value of the standard error is obtained by substituting into the formula for $\sigma_{\bar{x}}$ that is included as part of the calculations. The value of μ_0, the hypothesized mean, is found from H_0 in Step 2. Based on the critical value, the decision rule is

Decision Rule: Accept H_0 if $\bar{x} > 983.36$
            Reject H_0 if $\bar{x} \leq 983.36$

Step 8. Select Sample and Compute Value of Test Statistic

The mean of the random sample is 995 hours. If this were not given it would have to be computed from individual sample observations.

Step 9. Applying the Decision Rule

$\bar{x} = 995 > CV = 983.36$

The sample arithmetic mean equal to 995 is greater than the critical value of 983.36. Therefore, based on the decision rule found in Step 7 we accept the null hypothesis.

Step 10. Conclusion in Terms of Original Problem

Based on the courses of action specified in Step 3, we conclude that the shipment meets specifications since the null hypothesis was accepted; hence, the shipment should be kept for sale.

The results of the testing procedure are illustrated in the following diagram:

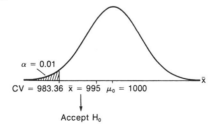

Example 11 A continuous filling process is said to be in the "state of control" and produces acceptably filled containers if the mean weight of the process equals 431 grams. Although

the process standard deviation appears to exhibit stability over time at $\sigma = 1.5$ grams, the process mean may vary owing to operator error or problems of process adjustment.

Periodically, random samples of five containers are selected to determine whether the process is producing acceptably filled containers. Although it is important not to underfill containers below the label claim limit, interest here lies in maintaining a stable process mean that has been established to be at a level that considers the label claim, but also considers overfill. Consequently, a two-tailed test is used since departures of the process mean from 431 grams in either direction is important to detect. If the result of a test indicates the process is out of control, the source of trouble is sought. It has been decided that a 0.01 level of significance should be used.

A random sample of 5 containers is selected, resulting in a mean of 428 grams and a standard deviation equal to 1.4 grams. Test an hypothesis to determine whether trouble should be sought or the process should be left alone.

To test: $H_0: \mu = 431$ (Leave the process alone)
$\qquad\quad H_1: \mu \neq 431$ (Look for trouble)

Given $\mu_0 = 431$
$\qquad \sigma = 1.5$
$\qquad \underline{n = 5}$
$\qquad \overline{X} = 428$
$\qquad S = 1.4$
$\qquad \alpha = .01$
$\qquad \alpha/2 = .005$

Since a two-tailed test is required, two critical values, CV_1 and CV_2, are required and can be obtained using the formulas given at the bottom of Exhibit 4. Unusually low and unusually high values of the sample mean lead to rejection of the null hypothesis, resulting in a region of rejection in *both* tails.

Compute $CV_1 = \mu_0 - z_{\alpha/2}\,\sigma_{\overline{x}}$

$$= \mu - z_{.005}\left[\frac{\sigma}{\sqrt{n}}\right]$$

$$= 431 - 2.57\left[\frac{1.5}{\sqrt{5}}\right]$$

$$= 429.276$$

Compute $CV_2 = \mu_0 + z_{\alpha/2}\,\sigma_{\overline{x}}$

$$= 431 + 2.57\left[\frac{1.5}{\sqrt{5}}\right]$$

$$= 432.724$$

Since $\overline{X} = 428$ falls below the lower critical value, the null hypothesis is rejected and the decision is to look for trouble in the process. The example is summarized in the following diagram:

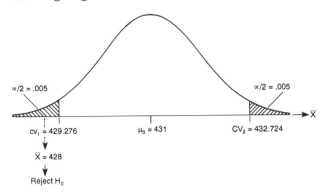

Note. In this example, the values of the universe standard deviation, σ, and the sample standard deviation, S, were provided; the universe value was used to compute the critical values, CV_1 and CV_2. When interested only in a test of the universe mean, the "actual" standard deviation or a target value generally would be used when it is available. However, if there is a question about the stability of the process standard deviation, it would be necessary to test whether the standard deviation is at an acceptable level before using it to test the mean, or the sample estimate may be used. Tests about standard deviations are presented in Module C using a control chart. Also, in the case of a two-tailed test, as illustrated in this example, the level of significance, α, is split such that each half appears in each tail of the sampling distribution. Hence, the value of z is designated as $z_{\alpha/2}$ and is found in the Table of Areas of the Normal Curve corresponding to half the specified level of significance (i.e., $\alpha/2 = .005$).

ADDITIONAL IDEAS

The basis for testing hypotheses lies in the fact that the value of a sample statistic varies from sample to sample. Consequently, the value from a single sample will, in all likelihood, be different from the parameter of the universe sampled. By employing the procedures for testing a hypothesis, it is possible to determine whether a sample result differs from a hypothesized value as a result of chance or whether the difference reflects a real difference between the true universe value, which is unknown, and the value hypothesized.

When a value of a sample statistic falls in the region of rejection, the difference between the sample statistic and the hypothesized parameter is referred to as a **significant difference:** the difference is greater than can be attributed to chance alone. Hence, we reason that if chance did not produce the difference, it can be attributed to a "non-chance" factor and results from the fact that the sample came from a universe with a parameter value specified in the alternate hypothesis.

By specifying the level of significance, α, as "small", we act as if it is too unlikely to reject a true null hypothesis. Therefore, when a value of a statistic does fall in the region of rejection, we act as if the result did not occur due to chance. Since α is subject to our control, the risk of rejecting a true null hypothesis can be made as small as desired.

When a sample statistic falls in the region of acceptance, although its value differs from the hypothesized value, the difference is attributed to chance. This is due to the fact that we consider such a difference likely to occur if the null hypothesis is true. A difference between a sample statistic falling in the region of acceptance and the hypothesized parameter is referred to as a **non-significant difference**.

Whenever a hypothesis is accepted, it is not "proved" to be true but is held to be tenable or reasonably defensible. In other words, there is insufficient evidence to reject it. Of course, an accepted hypothesis can be false, constituting a Type II error.

The Level of Significance

The level of significance, α, has been defined as the probability of committing a Type I error and is chosen to be small enough that such an error is unlikely to occur. Since α is a probability, we can interpret it in terms of a frequency of occurrence in a large number of repeated trials. Hence, α provides the proportion of the time that a true null hypothesis will be rejected if we repeated the sampling and testing procedure a large number of times under similar conditions. For example, consider the battery shipment illustration in which we used a level of significance of 0.01 to test the hypothesis

$H_0:\mu \geq 100$ (Shipment meets specifications)

against the alternative

$H_1:\mu < 1000$ (Shipment does not meet specifications)

We can interpret the value 0.01 to mean that 0.01 or 1 percent of a large number of tests of the same hypothesis based on different random samples of the same size would lead us to conclude incorrectly that the shipment does not meet specifications when in reality it does. Consequently, 1 percent of the time we would incorrectly consider shipments as unacceptable if we continually applied the same testing procedure. When performing a test on a one-time basis, this is considered to be of no real practical importance. In cases where sampling is done repetitively, however, these occasional occurrences should be accounted for in one's methodology.

Testing with the z-Statistic

Alternative procedures for performing a test of a hypotheses exist. In general, these are used either for convenience or because of the nature of a particular test statistic. Whenever a normal sampling distribution applies, as in the case of means presented earlier, a test can be performed in terms of the standardized variate, z. Briefly stated, a comparison is made between the z-value computed on the basis of the observed sample mean and the hypothesized mean, and a z-value corresponding to a chosen level of significance.

The test statistic is calculated using the formula

$$z = \frac{\overline{x} - \mu_0}{\sigma_{\overline{x}}}$$

This value of z is referred to as the "z-statistic" where $\overline{x}$ is the value of the sample mean of a particular sample, μ_0 is the value hypothesized in the null hypothesis, and $\sigma_{\overline{x}}$, is the standard error of the mean.

In order to perform a test of a hypothesis, the value of z calculated on the basis of the above formula is compared with a critical value of z based on α obtained from the table of Areas of the Normal Curve. The decision rules are similar to the ones presented earlier except that values of z are used instead of values of the sample mean directly.

In order to understand how to perform a test using the z-statistic, consider Example 10 where we test $H_0{:}\mu \geq 1000$ against the alternative $H_1{:}\mu < 1000$, where n = 100, α = 0.01, $\overline{X}$ = 995, and σ = 50. The value of z corresponding to X under the null hypothesis is computed using the formula for the z-statistic as

$$z\text{-statistic} = \frac{\overline{X} - \mu_0}{\sigma_{\overline{x}}}$$

$$= \frac{995 - 1000}{50/\sqrt{49}}$$

$$= -0.7$$

The value of z corresponding to a level of significance, α, equal to 0.01 is obtained from the Table of Areas and equals 2.33. Since the region of rejection is in the left tail, the critical value of z is negative and equals

$$CV = z_\alpha$$

$$= -2.33$$

Based on this type of test, we find that the calculated value of z equal to –0.7 is greater than the critical value equal to –2.33, which leads to the acceptance of the null hypothesis.

By comparing these results with those of Example 10, you will find that the observed and critical values of z are located in the same positions with respect to the region of acceptance and rejection. In other words, the same conclusion is drawn, regardless of the statistic used to perform the test. When a value of the z-statistic falls in the regions of acceptance or rejection so does the value of the sample mean. Both procedures lead to the same conclusion.

Testing with p-Values

Another way of performing a statistical test is in terms of a probability value, or **p-value**, associated with a particular sample result. A p-value is the probability of obtaining a difference equal to or greater than the one actually observed between the value of the test statistic and the hypothesized value.

When a p-value is available, it can be compared directly to the level of significance in order to perform a test. In the case of a one-tailed test, the null hypothesis is rejected if the p-value is less than or equal to α. For a two-tailed test, the null hypothesis is rejected if the p-value is less than or equal to half the level of significance, or $\alpha/2$. The conclusion using this procedure always will be the same as the one based on the other procedures discussed.

In order to illustrate the use of a p-value, consider the battery shipment problem again. In order to find the appropriate probability in this case, we need to calculate the value of the z-statistic corresponding to the sample result based on the hypothesized value. That is,

$$z = \frac{\overline{x} - \mu_0}{\sigma_{\overline{x}}}$$

$$= \frac{995 - 1000}{50/\sqrt{49}}$$

$$= -0.7$$

This yields a probability of .2580 from the Table of Areas of the Normal Curve. Hence, the p-value is found as

$$p\text{–value} = P(\overline{x} \leq 995 \,|\, \mu = 1000)$$

$$= .5 - .2580$$

$$= 0.2420$$

Since the p-value exceeds α equal to 0.01, the conclusion is to accept the null hypothesis and conclude that the mean shipment life is at least 1000 hours. This is the same conclusion reached in Example 10. Diagrammatically, the result is shown as the shaded region as follows:

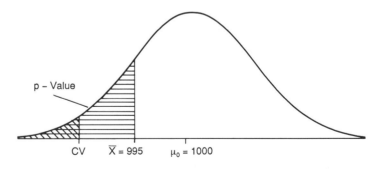

In this case, we can see that the probability of obtaining the observed result or less given μ_0 equals 0.2420. Essentially, this corresponds to a "likely" event since it exceeds α. If, on the other hand, the sample mean equalled 981, the p-value can be calculated similarly and equals 0.0039.

In such a case, one would reject the null hypothesis since this result is less than α equal to 0.01.

Type II Errors and the Power of the Test (Optional)

The hypothesis tests of this section are constructed on the basis of α, the level of significance. Once the null and alternate hypotheses are specified, the critical values, regions of acceptance and rejection, and decision rule can be determined using α. If the null hypothesis is true, we automatically know the probabilities of accepting and rejecting the null hypothesis. Moreover, we have control over these probabilities. By focusing on the probabilities of accepting and rejecting false null hypotheses, we can fully appreciate how a test actually works.

The null hypothesis is false whenever a universe possesses a parameter value consistent with the alternate hypothesis. In many applications it is important to know how well this is detected by a specific test. In other words, it is important to know the extent to which a test can discriminate between true universe values and false null hypotheses. This is accomplished either in terms of the probability of a Type II error or in terms of the power of the test. Recall, a Type II error is the acceptance of a false null hypothesis. The **power of a test** is defined as the probability of rejecting a false null hypothesis. Power, therefore, is the probability of drawing a correct conclusion when the null hypothesis is *false*.

The probability of committing a Type II error and the power of the test assume different values depending on the value of the universe parameter consistent with the alternate hypothesis. In other words, many different values are related to an alternate hypothesis and a different probability of a Type II error and a different power correspond to each value.

In general, the probability of a Type II error, ß, is determined by finding the probability associated with values of the test statistic in the region of acceptance, conditional on a parameter value specified in the alternate hypothesis. The sampling distribution of the test statistic, which may have a different form under the alternate hypothesis, is used to find the required probability.

The power of the test, P_w, or the probability of rejecting a false null hypothesis can be found as the complement of β by the formula

$$P_w = 1 - \beta$$

In other words, P_w can be found by subtracting the probability of a Type II error from one. Power also can be found directly as the area under the sampling distribution in the region of rejection. For a given value of μ_1, or a value of the mean consistent with the alternate hypothesis, the sum of β and power must equal one.

In order to understand these ideas reconsider the battery shipment example. Assume the shipment mean actually equals 980 hours instead of the hypothesized value 1000. Since the null hypothesis would be false under this assumption, we can compute the probability of committing a Type II error. The results are described in **Exhibit 5**. The first distribution shown in the exhibit is the sampling distribution of the mean under the null hypothesis. This is the distribution used to compute the critical value that forms the basis for the test.

EXHIBIT 5
ILLUSTRATION OF THE PROBABILITY OF A
TYPE II ERROR IN THE BATTERY SHIPMENT EXAMPLE

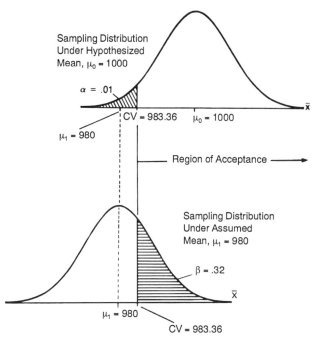

The second distribution shown in the exhibit is the sampling distribution of the mean under the assumption that the universe mean equals 980. The probability of the Type II error equal to 0.32 is the area under this distribution associated with the region of acceptance. Assuming normality, this value is found by computing z corresponding to the region between the assumed mean μ_1, and the critical value, finding the probability from Table A-1 and subtracting this value from 0.5.

The value of z is computed as

$$z = \frac{CV - \mu_1}{\sigma_{\bar{x}}}$$

$$= \frac{983.36 - 980}{7.14}$$

$$= \frac{3.36}{7.14}$$

$$= 0.47$$

Based on Table A-1, this value of z yields a probability of 0.1808. The probability of a Type II error in this case is

$$\beta = 0.5 - 0.1808$$

= 0.3192 or 0.32 (rounded)

This value can be interpreted to mean that 0.32 or 32 percent of all possible samples, based on repeated tests, would result in the conclusion that the shipment mean is at least 1000 hours when it really is 980 hours. Stated another way, we can say that 32 percent of the time unacceptable shipments with a mean of 980 hours would be kept for sale when they should be returned to the manufacturer.

The corresponding power of the test can be found by subtracting β from 1.

$$P_w = 1 - \beta$$

$$= 1 - 0.32$$

$$= 0.68$$

This value represents the probability of reaching a correct conclusion when the null hypothesis is false and the true universe mean equals 980 hours. Hence, 0.68 or 68 percent of the time the testing procedure would correctly lead us to reject unacceptable shipments with a mean of 980 hours.

The interpretations of β and P_w in the above example are conditional on a specific mean of 980 hours. In practice, we do not know the magnitude of the universe mean; therefore, β and P_w are related to the value of a universe mean that we assume to be true. By doing this for many universe values consistent with the alternate hypothesis, we are able to determine the ability of a test to discriminate between true universe values and a false null hypothesis.

Exhibit 6 presents the probability of committing a Type II error and the power of the test for selected values of the universe mean in the battery example. β and P_w are given for values of the universe mean that are consistent with the alternate hypothesis: all values of the universe mean are less than 1000. The larger the difference between the value of the universe mean shown in the table and the hypothesized value of 1000, the smaller the probability of committing a Type II error. Conversely, the power of the test increases as the difference becomes greater.

EXHIBIT 6
β AND P_w FOR SELECT VALUES OF μ CORRESPONDING TO THE BATTERY SHIPMENT EXAMPLE

Universe Mean	Probability of a Type II Error (β)	Power of The Test (P_w)
970	0.03	0.97
975	0.12	0.88
980	0.32	0.68
985	0.59	0.41
990	0.82	0.18
995	0.95	0.05

The probability of committing a Type II error generally is "high" when the difference between the universe mean and the hypothesized value is small. In other words, small differences between a true value and the one hypothesized are not detected very often. The corresponding power is low so that a correct conclusion using sample data is infrequent when the true value is "close" to the value hypothesized. As the difference between the actual value and the one hypothesized increases, ß becomes smaller and the power increases. In other words, larger differences will lead to correct decisions more frequently when using sample information. The ability of a test to detect false null hypotheses becomes greater when the difference between the hypothesized value and the true value is larger.

By graphing the probability of a Type I and a Type II error on a single graph, we can visualize how well a test draws correct and incorrect conclusions in terms of the possible values that can be assumed by the universe mean. This is accomplished with a power curve, which is a graph that depicts the probability of rejecting the null hypothesis for all possible values of the universe parameter. In order to illustrate the concept, consider **Exhibit 7**, which presents the power curve for the test used in the battery example.

EXHIBIT 7
POWER CURVE CORRESPONDING TO
BATTERY SHIPMENT EXAMPLE

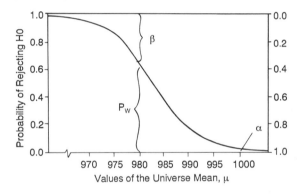

The first point to note about the exhibit is that all values of μ to the left of 1000 correspond to the alternate hypothesis, and the remaining values correspond to the null hypothesis. Hence, values of μ equal to or greater than 1000 correspond to the true null hypothesis, whereas values to the left of 1000 correspond to a false null hypothesis. Although the power of the test is associated with false null hypotheses, the power curve considers the probability of rejecting H_0 in cases where it can be true or false.

The height of the curve corresponding to values of μ consistent with the alternate hypothesis (or a false null hypothesis) represents the power of the test, P_w. This can be read on the left-hand scale. The distance from the top axis to the curve gives the probability of a Type II error, β, which is read on the right-hand scale. P_w. increases and β decreases as the difference between the actual and hypothesized mean increases.

The power curve can be used to evaluate how well a test performs. Every test of a hypothesis has a power curve that is based on the value of α and the sample size, n, and is independent of a specific sample result. In other words, the power curve is determined by the test, not the results of a sample; the sample results are used to perform the test and reach a conclusion.

In order to understand how to relate the power curve to a particular test, it is useful to introduce the concept of a power curve under ideal conditions. This is the curve that corresponds to a test of a hypothesis where it is not possible to commit an error of either kind: false null hypotheses are always rejected ($P_w = 1$), and true null hypotheses are never rejected ($\alpha = 0$). When sampling, the ideal cannot be realized; however, a test whose power curve is closest to the ideal when compared with other tests is the better test.

The power of a test can be increased in two ways: (1) by increasing the sample size, n or (2) by increasing the level of significance, α. By holding α fixed and increasing n, the power of the test increases, which means that the power curve is closer to the ideal. Similarly, when n is held constant and α is increased, the power curve is closer to the ideal. In the first case, protection against a Type I error remains the same, and increased power is obtained at a higher cost of sampling. In the latter case, the cost of sampling is held constant, but the risk of committing a Type I error increases.

Consider the battery shipment problem where, on the one hand, we increase α from 0.01 to 0.05 and, on the other hand, we increase n from 49 to 100. The resulting power curves are shown in **Exhibit 8**. In both cases we can see that power is increased and the curve is closer to the ideal. Obviously, as the power increases the probability of a Type II error diminishes.

EXHIBIT 8
INCREASING POWER IN THE BATTERY SHIPMENT EXAMPLE

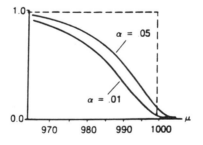 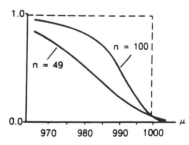

Other Methods and Tests

The computational procedures presented in this chapter apply to tests of hypotheses for means and, strictly, assuming a normal sampling distribution and known σ. The overall approach or concept, however, applies to tests of any parameter. What differs is the form of the sampling distribution of the test statistic and the way critical values are calculated.

Many other tests are available that either apply to similar problems under varying assumptions or to other problems involving different test statistics. For example, when testing means based on samples drawn from normal universes with an unknown standard deviation, the common statistical test used is based on Student's t-distribution and is simply referred to as the t-test. This is more applicable when single samples are used to estimate error and when samples are small. Owing to the nature of process control problems, the normal results are generally used.

Means generally are used in process control when process characteristics of interest are measurable. When binary outcomes are of interest, the test statistic generally used is the proportion or percent of occurrences of one of the outcomes. Obviously, when only acceptability is of interest, one may use the percent defective, or the number defective, in a sample. Exact tests in such cases are available and are more commonly used as a basis for acceptance sampling and inspection. With respect to process control applications, normal approximations are more common and are presented in Module D.

Although it has not been made completely clear in quality control applications regarding what to do when measurements are not normal, some recognition is being given to tests for normality of underlying process distributions. The most commonly used, which is not the most powerful, is the chi-square test of goodness of fit. The chi-square test and other such tests can be found in many basic statistics texts and those on nonparametric methods.

Another class of tests that date back to the origins of statistical inference fall under the umbrella of analysis of variance or experimental design methods. Although useful in selected plant floor applications, these also can have a widespread impact on product and production design. Primarily, they involve experimental control of various relevant factors and the effect of these factors on critical product variables or measurements. Taguchi methods, which may sound familiar, represent a subset of the vast body of experimental design methodology available today.

Statistics and Statistical Quality Control

The emphasis in this chapter has been placed on hypothesis testing as it would be presented if it were purely given as a statistical concept. This also applies to the material at the end regarding the power of the test, which is the way statisticians view the problem as opposed to quality engineers. The purpose in doing this is to provide some appreciation for the fundamental statistical principles. Briefly, before closing the chapter, we shall attempt an overview of the role of the ideas presented in a quality control setting.

Methods of hypothesis testing form the basis for process control techniques and for acceptance sampling procedures; however, they are used in different ways in each case. The basic process control tool is the control chart that is composed of a center line and control limits. The center line is equivalent to a hypothesized parameter, and the control limits can be viewed as critical values corresponding to a two-tailed test (although one-sided charts also are used). Since the chart is used periodically at fixed time points, use of the control chart in its simplest form is equivalent to performing a series of tests of hypotheses for a relevant process characteristic over time.

Although a level of significance can be set at any value, traditionally control limits are set at a fixed point that are three standard errors from the hypothesized parameter, regardless of the form of the sampling distribution. For normal sampling distributions, this results in an α equal to 0.0027, which by conventional standards is very small. When the sampling distribution is not normal, 3-sigma limits will generally lead to small unknown values that are still effective.

Control charts for measurements, which are the most common, typically are based on small samples. Consequently, with a very small α and small n, power is very low. Therefore, the opportunity for Type II errors is large. The underlying philosophy is to detect important process changes.

By way of contrast, with respect to acceptance sampling more emphasis is placed on establishing percent defective levels. Larger α's are employed, and attention is directed toward controlling the probability of Type II errors at reasonably smaller values. Further, sample sizes are subject to greater choices and are based on indexed values of ß. The power of the test is an important component of the test selection process. Most of the decision making related to acceptance sampling is based on standardized tables and graphs and is not subject to widespread problems of implementation found in process control. The most common standards which may sound familiar are ML-STD 105D and ML-STD 414, [40] and [41].

Again, part of the goal associated with this chapter is to present underlying principles that can be used to bridge the gap between the field of statistics and quality engineering. It should be noted in this regard that the concepts of power and the power curve are not dealt with directly by quality control professionals. Although it imparts the same information, quality control standards are provided in terms of an operating characteristic (OC) or operating characteristic curve.

The **operating characteristic** simply can be defined as the probability of committing a Type II error, or ß. Hence, an operating characteristic curve provides the probability of *accepting* null hypotheses conditional on various values of the universe parameter. When used in acceptance sampling, the curve provides probabilities of accepting shipments for, say, various shipment percent defective levels associated with a particular sampling plan.

Actually, the OC curve simply is the power curve turned upside down. Hence, power provides the probability of correct decisions whereas the operating characteristic emphasizes the probability of committing an error. Both provide the same information, but viewed differently.

Comments on Estimation

When we illustrated the procedure for hypothesis testing, we introduced a specific value for the hypothesized mean and assumed that α was known. In a process control setting, this would be equivalent to use of specified target values. In many cases it is desirable to establish how a process naturally performs before either making adjustments or establishing targeted values.

In such cases, it is necessary to estimate the process mean and the process standard deviation. Although variations exist, traditional statistical approaches employ the results of a single sample. With respect to process control applications, however, the results of numerous samples are used in conjunction with the procedures presented earlier in this chapter. These will be employed in the material presented in the chapters that follow.

Calculators and Computers

Since this chapter serves as a general introduction to hypothesis testing in addition to considering specific tests for means, it is of interest to mention the role of computing devices in hypothesis testing at this point. When performing individual tests like the ones presented in this chapter, there is no real problem doing the calculations by hand or with a simple "4-function" calculator, especially for small samples. In situations where sample statistics must be computed directly from large samples or when elaborate procedures involve many kinds of calculations, and in situations where more than one test is performed on the same data, there can be savings in time and effort by using a more sophisticated computing device such as a computer or a calculator with a stored program feature. Programs exist for performing many kinds of tests in addition to the ones presented here. These are available in the form of computer programs as well as programs that are part of optional libraries that can be purchased with various hand-held and desk-top calculators. Many of the programs represent subroutines that are part of larger integrated programs used to accomplish multiple tasks.

The output for a particular test of a hypothesis generally is the numerical value of the test statistic based on the sample input data. The resulting test statistic then can be compared with a tabulated critical value to determine whether the result is significant or not. Many computer programs and some calculators with a stored program feature can provide a "p-value" along with the value of the test statistic presented. When a p-value is made available, it can be compared directly to the level of significance in order to perform a test, so that it is not necessary to refer to tables such as those in the Appendix.

BASIC
CHARTING

The previous chapter presented the basic sampling problem, the concept of estimation and related estimates, and principles of hypothesis testing. Since differences exist between estimates used in general statistical problems and in quality control, more attention was given to estimates used in process control. In the case of hypothesis testing, a more traditional approach was used. The ideas presented serve as a backdrop for the material that follows in this and the remaining chapters. Although not essential, they do provide a basis for a deeper understanding of the process control tools presented.

This chapter attempts to introduce the basic statistical process control tool, the control chart, in order to provide a fundamental understanding of the concept. In a sense, it forms a bridge between basic statistical principles and specific process control tools presented in later chapters.

The symbols used in this chapter are defined as follows:

μ	= a universe or process mean
μ_0	= a standard or target process mean
$\overline{X}$	= a sample mean
$\overline{\overline{X}}$	= the mean of a series of sample means
σ	= a universe or process standard deviation
$\hat{\sigma}$	= an estimate of the process standard deviation
S	= a sample standard deviation
$\overline{S}$	= the mean of a series of sample standard deviations
CL	= center line
UCL	= upper control limit

LCL   = lower control limit
X   = a value of a particular measurement
n   = sample size
k   = the number of samples

Statistical Process Control

Statistical process control can be defined as a body of methods and activities that can be used as an aid to establish, achieve, and maintain process stability. In addition, it can be used to establish the capability of a process and to measure the extent to which a process is improved. All of this is accomplished in terms of measurement and monitoring of variation with respect to relevant process characteristics.

Recall from material presented earlier that all recorded measurements are subject to variation, no matter how small this variation may be. This being the case, all measurements relating to a process characteristic can be thought of as generated from an underlying distribution. A process is stable or under control when this underlying distribution remains the same over time.

A control chart is used to monitor a process to determine whether or not it is stable. After selecting the characteristics of importance to be measured, the appropriate sample statistics are identified. Stability or control, therefore, is established in terms of the behavior of these statistics. In other words, instead of looking at an entire distribution, one monitors parameters of the distribution in terms of their corresponding estimators or statistics in order to assess the stability of the underlying distribution.

This approach is based on the concept that there are *two* types of variation with respect to the statistics on which to focus: *natural* or random variation, and non-random variation due to some *assignable cause* or to a change in the process. Control charts can provide evidence that a change has occurred, but do not indicate what it is. This is based on specific knowledge of the process itself.

Once assignable causes have been eliminated and a process is deemed stable, it is possible to determine the *capability* of the process in meeting specifications. Although process improvement should always be a goal, if a process is not capable, measures must be taken toward improvement unless unacceptable product can be tolerated. Though not always the case, these measures usually involve more fundamental changes either in the underlying system or in design. Improvement is directed toward meeting established targets and the reduction of the inherent or random variation in the process.

Basic Control Chart Structure

Although differences exist in special cases, control charts have a certain structure shown in **Exhibit 1**. The three horizontal lines shown correspond to the upper control limit (UCL), the center line (CL), and the lower control limit (LCL), respectively. Values of an appropriate statistic or sample estimator are plotted on the vertical scale. The lower horizontal scale corresponds to the sampling interval, or to points in time, which usually but not always are equidistant. A sample is selected at each time point, on which a value or values of sample statistics are calculated and plotted.

EXHIBIT 1
BASIC CONTROL CHART STRUCTURE

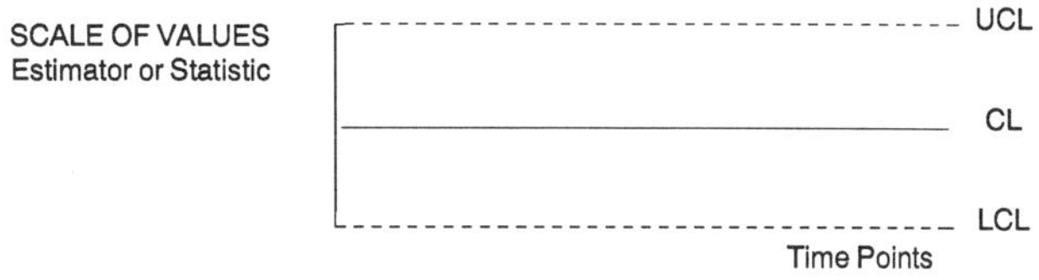

SCALE OF VALUES
Estimator or Statistic

UCL

CL

LCL

Time Points

The center line of a control chart represents the value of the parameter of the process requiring control. This value, for example, could be the process mean, the process standard deviation, or, say, the fraction or percent defective associated with the process. Depending on the situation at hand, the center line can represent a standard, or a target, or it can be estimated from the process.

Traditionally, control limits are established as three standard errors from the center line. That is,

Control Limits:
Center Line ± 3 × Standard Error

The standard error corresponds to the standard deviation of the distribution of the sample statistic chosen to monitor the relevant process parameter. The standard error can be based on a standard or a target value or can be estimated from process data.

In this chapter, we shall illustrate the basic concepts in terms of means only. The goal here is to demonstrate how charts are constructed and the basic reasoning underlying their use. Later chapters provide more detail regarding actual applications of control charts. First, we shall illustrate the calculation of control limits and then present the corresponding charts with plotted data.

Example 1   The objective in this example is to construct a control chart associated with the Pennsylvania Super-7 Lottery that became famous in the recent past when the pot reached $110,000,000. Results for a one year period are presented in **Exhibit 2**.

Essentially, lottery holders draw tickets containing seven numbers between 1 and 80. Each week 11 numbers are drawn at random. A ticket holder that has 7 of the 11 numbers wins the main prize.

Consequently, we can think of the numbers generated each week as a random sample of 11 observations from a process that is distributed uniformly over the integers 1-80. In this case, we are working backwards since we supposedly know the inputs. In the case of an actual manufacturing process, we would not have this informa-

EXHIBIT 2
RESULTS OF THE PENNSYLVANIA SUPER–7 LOTTERY
FOR A ONE YEAR PERIOD

DATE					SUPER 7 WINNING NUMBERS						
10/1/87	06	07	10	13	19	35	55	64	65	74	79
10/21/87	03	04	10	24	28	44	48	60	69	70	79
10/28/87	07	11	14	27	40	43	46	65	68	75	76
11/4/87	02	04	19	23	26	41	43	51	54	73	74
11/11/87	10	11	32	36	46	51	53	85	64	71	72
11/18/87	14	20	32	33	37	47	60	64	68	71	77
11/25/87	01	03	13	14	29	31	39	45	52	71	78
12/2/87	03	05	22	30	32	37	38	46	47	48	56
12/9/87	06	12	20	24	28	34	47	64	68	77	79
12/16/87	02	07	21	24	26	36	41	55	58	62	74
12/23/87	04	05	06	30	33	37	45	51	52	53	59
12/30/87	08	14	18	32	51	52	56	62	65	75	78
1/6/88	02	11	20	22	26	41	43	57	61	63	80
1/13/88	04	05	08	12	32	46	48	50	59	65	74
1/20/88	01	17	24	36	39	43	48	49	54	63	71
1/27/88	04	10	22	23	52	55	58	64	66	76	78
2/3/88	05	15	18	22	25	40	41	46	49	60	76
2/10/88	08	16	20	31	32	38	40	46	52	55	67
2/17/88	04	30	32	42	50	55	58	59	74	76	80
2/24/88	14	32	43	47	51	56	60	64	74	78	80
3/2/88	02	03	05	17	29	23	25	40	41	66	70
3/9/88	01	02	20	30	34	47	48	57	58	62	77
3/16/88	07	26	30	38	42	45	46	50	59	62	80
3/23/88	11	12	16	33	40	43	49	55	63	67	77
3/30/88	17	18	24	36	38	43	47	58	64	71	74
4/6/88	03	10	22	25	27	50	55	56	64	65	80
4/13/88	02	16	28	29	31	35	53	57	61	69	72
4/20/88	02	08	19	28	31	39	43	52	55	57	60
4/27/88	01	05	23	28	29	30	33	39	41	56	72
5/4/88	02	12	19	25	49	51	55	58	72	79	80
5/11/88	11	12	21	23	31	44	46	53	64	73	77
5/18/88	06	07	12	16	31	40	44	60	62	66	77
5/25/88	10	14	16	20	43	50	57	58	61	64	77
6/1/88	10	14	16	31	35	46	54	56	75	76	77
6/8/88	05	18	32	40	49	55	67	68	75	78	80
6/15/88	09	14	15	17	23	37	49	61	68	69	75
6/22/88	11	21	30	37	39	40	41	53	57	70	80
6/29/88	02	25	31	40	41	63	64	66	69	73	76
7/6/88	07	08	14	28	32	38	40	41	48	59	74
7/13/88	04	07	10	11	32	44	46	54	55	56	74
7/20/88	08	25	28	31	33	35	40	44	49	73	80
7/27/88	10	20	25	36	41	42	49	55	60	74	77
8/3/88	04	08	20	25	32	41	51	53	69	74	75
8/10/88	02	04	07	14	27	50	55	59	69	70	73
8/17/88	02	10	12	25	30	35	40	41	46	59	73
8/24/88	18	23	29	35	50	57	59	62	70	73	77
8/31/88	25	28	40	41	43	50	51	58	59	60	71
9/7/88	11	15	26	34	47	53	58	67	68	74	75
9/14/88	01	10	12	15	28	34	37	42	45	57	61
9/21/88	11	14	22	30	31	41	52	56	60	61	78
9/28/88	09	10	21	27	43	51	64	65	68	70	77
10/5/88	04	12	14	21	26	58	61	62	68	69	71

tion and would either introduce target values or make estimates directly from the process.

Since we can assume an underlying uniform distribution, we know the mean and the standard deviation theoretically based on the concepts in Chapter 4, and we can use the results to calculate the corresponding values*. The results are summarized as follows:

Uniform Distribution:

$$f(x) = \frac{1}{80}; x = 1, 2, ..., 80$$

Mean: $\mu = \dfrac{80 + 1}{2}$

$= 40.5$

Standard Deviation:

$$\sigma = \sqrt{\frac{80^2 - 1}{12}}$$

$= 23.09221$

Assuming our interest is to determine whether the process is in control with respect to the mean, we can construct a control chart for means based on the above parameter values as follows:

Center line, CL $= \mu$
$= 40.5$

Standard error of the sample mean, $\sigma_{\bar{x}} = \dfrac{\sigma}{\sqrt{n}}$

$= \dfrac{23.09221}{\sqrt{11}}$

$= 6.963$

Upper control limit, $UCL_{\bar{x}} = \mu + 3\sigma_{\bar{x}}$
$= 40.5 + 3(6.963)$
$= 40.5 + 20.889$
$= 61.389$

Lower control limit, $LCL_{\bar{x}} = \mu - 3\sigma_{\bar{x}}$
$= 40.5 - 3(6.963)$
$= 40.5 - 20.889$
$= 19.611$

The basic control chart based on known inputs appears as follows:

---

* Completely general expressions for the properties of the uniform distribution can be found in Hastings, N.A.J. and J.B. Peacock, *Statistical Distributions*, Butterworths, London 1975.

$$UCL = 61.389$$

$$CL = 40.5$$

$$LCL = 19.611$$

1 2 3 4 52

Week Number

Example 2 In order to complete the control chart constructed in Example 1, it is necessary to calculate the sample means for the 52 samples and plot them on the chart.

The means of the 52 samples are presented in **Exhibit 3**. Recall, sample means are calculated using the formula

$$\overline{X} = \frac{\Sigma X}{n}$$

EXHIBIT 3
MEANS AND STANDARD DEVIATIONS OF SAMPLES OF ELEVEN BASED ON
SUPER–7 LOTTERY

Week No.	Sample Mean, $\overline{X}$	Sample Standard Deviation, S	Week No.	Sample Mean, $\overline{X}$	Sample Standard Deviation, S
1	38.8	29.03	27	41.2	22.68
2	39.9	27.76	28	35.8	20.02
3	42.9	25.80	29	32.5	20.26
4	37.3	24.81	30	46.7	26.36
5	45.8	21.58	31	41.4	23.64
6	47.5	21.70	32	38.3	25.69
7	34.2	25.88	33	42.7	23.62
8	33.1	17.25	34	44.5	25.32
9	43.5	27.14	35	51.5	25.29
10	36.7	23.34	36	39.7	25.41
11	34.1	20.66	37	43.5	20.33
12	46.5	24.68	38	49.8	24.20
13	38.7	24.61	39	35.4	20.85
14	36.6	25.71	40	35.7	24.24
15	40.5	20.40	41	40.5	20.81
16	46.2	26.60	42	44.0	22.34
17	36.1	21.28	43	41.1	25.52
18	36.3	17.81	44	39.1	28.60
19	50.91	22.84	45	35.7	19.09
20	54.45	20.13	46	50.3	20.86
21	28.3	23.64	47	47.3	14.04
22	39.6	24.66	48	48.0	23.30
23	44.1	19.53	49	31.1	19.76
24	42.4	22.65	50	41.5	20.65
25	44.5	20.34	51	45.9	25.28
26	41.5	25.18	52	42.4	26.62

Using information obtained from Exhibit 2, the necessary calculations are illustrated for the first and last samples.

Sample 1:

$$\overline{X} = \frac{6 + 7 + 10 + 13 + 19 + 35 + 55 + 64 + 65 + 74 + 79}{11}$$

$$= 38.8$$

Sample 52:

$$\overline{X} = \frac{4 + 12 + 14 + 21 + 26 + 58 + 61 + 62 + 68 + 69 + 71}{11}$$

$$= 42.4$$

The completed control chart is presented in **Exhibit 4**. A discussion of the chart appears at the end of this chapter.

Note. The position taken in this book is that an understanding of the statistical concepts underlying basic process control tools are to be provided. Much of the work done in process control is based on established methods for which many intermediate calculations are provided in the form of tables. These are provided in terms of various so-called *factors*. Due to the use of these factors, oftentimes the underlying structure of various calculations is not fully understood.

In order to understand this we can simply demonstrate the use of a factor associated with Example 1 to make the point. The control limits for a mean chart with known inputs are given as

$$\mu \pm 3\sigma_{\overline{x}}$$

Since $3\sigma_{\overline{x}}$ can be written as $3\left[\dfrac{\sigma}{\sqrt{n}}\right]$, $3/\sqrt{n}$ is independent of the parameter values and can be tabulated, and is commonly given as a factor, A, where

$$A = \frac{3}{\sqrt{n}}$$

In the case of Example 1, this is given as

$$A = \frac{3}{\sqrt{11}}$$
$$= 0.90453$$

When multiplied by σ, we get

$$A\sigma = .90453(23.09221)$$

$$= 20.888$$

and the control limits are then calculated as

$$\mu \pm A\sigma$$
$$40.5 \pm 20.888$$
$$19.612 - 61.388$$

Other factors in addition to A are tabulated in the Appendix and will be introduced in the more procedural sections appearing later.

EXHIBIT 4
CONTROL CHART
Means of Eleven Weekly Numbers for the Pennsylvania Super-7 Lottery

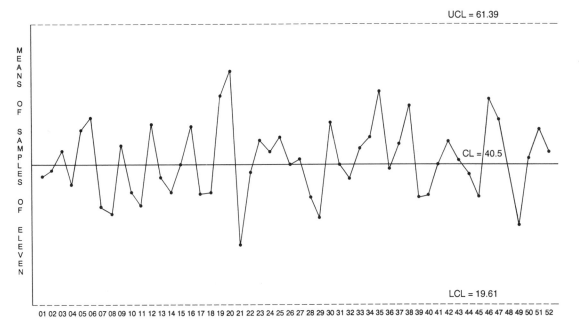

WEEK NUMBER WITHIN YEAR

Example 3 We mentioned that a control chart can be constructed on the basis of a standard or target or known parameter values, or it can be based on accumulated process data. Here, by way of contrast, we shall illustrate the latter procedure for a mean chart using the lottery data.

Before constructing the chart, we need to estimate μ and σ. In addition to the sample means, values of the sample standard deviations are provided in Exhibit 3. Recall, these are calculated using the formula

$$S = \sqrt{\frac{\Sigma(X - \overline{X})^2}{n - 1}}$$

or the shortcut

$$S = \sqrt{\frac{\Sigma X^2 - (\Sigma X)^2/n}{n - 1}}$$

Using the shortcut, we illustrate the calculation for the first sample.

$$S = \sqrt{\frac{(6^2 + 7^2 + \ldots + 74^2 + 79^2) - (6 + 7 + \ldots + 74 + 79)^2/11}{11 - 1}}$$

$$= 29.03$$

The estimate of μ based on the material presented in Chapter 2 or 5 is given as

$$\bar{\bar{X}} = \frac{\Sigma \bar{X}}{k}$$

$$= \frac{38.8 + 39.9 + \cdots + 45.9 + 42.4}{52}$$

$$= \frac{2157.3}{52}$$

$$= 41.486$$

In order to estimate σ we first must calculate $\bar{S}$ based on the material presented in Chapter 3 or 5.

$$\bar{S} = \frac{\Sigma S}{k}$$

$$= \frac{29.03 + 27.76 + \cdots + 25.28 + 26.62}{52}$$

$$= \frac{1200.77076}{52}$$

$$= 23.09175$$

Then using the principles presented in Chapter 5, our estimate of σ becomes

$$\hat{\sigma} = \frac{\bar{S}}{c_4}$$

$$= \frac{23.09175}{.9754}$$

$$= 23.67413$$

where c_4 is obtained from **Table C-1** in the Appendix.

Control limits then are found as

$$\bar{\bar{X}} \pm 3 \left[\frac{\hat{\sigma}}{\sqrt{n}} \right]$$

where $\bar{\bar{X}}$ is the center line and $\hat{\sigma} / \sqrt{n}$ is an estimate of $\sigma_{\bar{x}}$. We have,

$$CL = \bar{\bar{X}}$$

$$= 41.486$$

$$UCL = \bar{\bar{X}} + 3\frac{\hat{\sigma}}{\sqrt{n}}$$

$$= 41.486 + 3\left[\frac{23.67413}{\sqrt{11}}\right]$$

$$= 62.90$$

$$LCL = 41.486 - 3\left[\frac{23.67413}{\sqrt{11}}\right]$$

$$= 20.07$$

Due to the nature of the lottery data, these results are very close to the ones presented earlier. Consequently, there is no real need to plot the corresponding chart. When dealing with actual process data *differences* can arise, and the corresponding charts can provide different information. Ultimately, one seeks results similar to the lottery data. More will be said about this shortly.

Note. Again, we can illustrate the use of tabulated factors. In this case, there are two possibilities:

$$1. \bar{\bar{X}} \pm A\,\hat{\sigma}$$
$$2. \bar{\bar{X}} \pm A_3\bar{S}$$

where

$$A = \frac{3}{\sqrt{11}}$$

$$= \frac{3}{\sqrt{n}}$$
$$= .90453$$

and

$$A_3 = \frac{3}{c_4\sqrt{n}}$$

$$= \frac{3}{.975411\sqrt{11}}$$

$$= 0.927347$$

The result of using either (1) or (2) in terms of A or A_3, respectively, are the control limits 20.07 and 62.90 already calculated.

ADDITIONAL IDEAS

Previous portions of this chapter simply presented the basic structure of control charts and illustrated how to construct a chart in terms of means. Two cases were considered: (1) targeted parameter values based on known inputs, and (2) a chart based on observed data. Although additional considerations are involved when constructing actual charts, these are to be considered in later chapters. At this point, we shall discuss some of the reasoning underlying chart construction and attempt to tie it together with the fundamental statistical principles presented in earlier chapters and the material that follows concerning specific types of control charts.

Reasoning Behind Construction of Control Charts

If you have read Chapter 5 that deals with statistical tests of hypotheses, you will have a fuller appreciation for the nature of a control chart and how it is used. When constructing a test of a statistical hypothesis, we hypothesize a specific value of a universe parameter and then, based on the sampling distribution of the corresponding test statistic, calculate a critical value or critical values. The critical values provide a basis for making a decision whether a specific value of a sample statistic is different from the hypothesized value due to chance or due to non-chance factors; namely, that the sample result was generated from a universe with a parameter value other than the one hypothesized.

Regardless of the method of calculation, the center line of a control chart is equivalent to an hypothesized process parameter, and the control limits correspond to critical values (for a two-tailed test) corresponding to the sample characteristic monitored. Hence, at any particular point in time, plotting a single point on the control chart is equivalent to testing a hypothesis. Consequently, when a sample point falls within the control limits we conclude that there is no change in the process.

If, on the other hand, a sample point falls outside a control limit, we conclude that a change in the process has occurred: something other than chance is affecting the process. Commonly, it is said that an "assignable cause of variation" (non-chance factor) is present in the process. In such cases, one must search for the cause and eliminate or correct it, or take action, so to speak. When a point falls within the control limits, no action is to be taken since the variation between the sample result and the center line is what one "usually" would expect to occur if chance were the only force operating on the process. In such cases, we conclude that the process is subject to natural, or inherent, variation.

When we introduced α, or the probability of committing a Type I error in Chapter 5, it was noted that α should be small but that its value was to be designated by the one performing the test. Although a similar approach can be taken in the case of a control chart, traditionally the control limits are set at three standard errors from the center line. Implied, therefore, in the case of a normal sampling distribution (which is generally the case for means) is the fact that α automatically is fixed at a value equal to 0.0027, or approximately equal to 0.003 or 3-in-one-thousand.

A level of significance equal to three-in-one-thousand is considered to be very small. Consequently, at this level, processes that do not change will not incorrectly be considered to change very often. Correspondingly, processes that do change will not be detected very often. More precisely, large changes in a process will be detected more than small ones. This results from the fact that the power of the underlying test is not very high, in part, due to the small α.

A second reason for low power is that in the more frequently used control charts, namely those for measurement type data, small samples typically are used. Aside from the fact that small samples are easily collected, more importantly, small samples allow for chance variation to be measured within samples while providing the opportunity for non-chance factors, or process changes, to be observed among samples. More is said about this at the end of Chapter 7 in terms of the concept of subgrouping.

The ideas presented in the previous two paragraphs relate to the most fundamental application of control charts, which were espoused by their founder, Walter Shewhart [14]. Overall, the intent is to use control charts to pick up "more serious" or important changes in a process, with the recognition that changes always occur and that one should not constantly adjust the process in order to accommodate changes that are unimportant. Actually, extensions and modifications to the basic control chart concept are available, Grant and Leavenworth [11] and Lucas and Saccucci [39]; however,here we are interested only in the fundamentals.

The preceding discussion provides some understanding of the relationship between a control chart and a basic test of a statistical hypothesis. The explanation given was based on the use of a chart at any particular point in time. Another important notion that will be discussed in detail in Chapter 8 is that control charts are considered in terms of a stretch of time. Consequently, one way to view the basic use of a control chart is in terms of a series of isolated tests of hypotheses at successive points in time. On the other hand, observing samples over time provides an opportunity for the concept of randomness to play itself out, so to speak. Consequently, different types of information are provided by a control chart in terms of changes occurring within a process by observing patterns over time.

The Lottery Example

One of the first reactions one might have when reading this chapter very possibly could be, why is a lottery example used? It would be more relevant to use actual process data! Such a reaction is natural, but misses an important point.

The lottery example, obviously, was chosen intentionally. One reason for this choice was to introduce the basic concept of a control chart and how to construct it. The other reason lies at the heart of all existing process control techniques and is one that is rarely understood. That is, the lottery can be assumed to be a completely random process with which we all have some familiarity. The point that is difficult to understand is that a process is stable or in control when it behaves like a lottery. In other words, all points should fall within the control limits, and no explicable pattern of variation of the sample points exists anywhere within the chart.

When no explicable patterns exist, the variation among the samples can be attributed to chance. Pure chance phenomena cannot be decomposed or explained. This means that there is no basis for predicting individual sample results. One can, of course, predict long-run or overall values of the mean, the standard deviation, and the underlying distribution shape.

Two final points are worth noting. First, in addition to individual points falling outside control limits, a process can be concluded to be out of control or unstable if non-random or systematic patterns are present in the chart. A discussion of such patterns appears in Chapter 8. Second, it should be noted that it is now common to refer to patterns of control as "predictable" patterns of variation. The sense in which this is correct is in terms of the stability of the underlying distribution or its mean and standard deviation. Strictly speaking, random patterns are not predictable, by definition.

The Sampling Process and Randomness

In the case of the lottery, individual values are generated randomly, and, therefore, samples based on these values also can be assumed to be random samples. In order to use methods of statistical inference effectively—and therefore control charts—samples must be assumed to be generated at random.

When dealing with many statistical problems, one has the opportunity to select items at random. In the case of process control, typically one selects consecutive items constituting the sample at a particular point in time. In such a case, the items are not selected at random. It is assumed, however, that owing to the inherent or natural variation that is part of every process, this variation is due to chance. Consequently, measurements of a process characteristic can be assumed to be generated at random, and, therefore, samples of these measurements constitute random samples. Much, however, depends on the method of subgrouping and the nature of a particular process.

Three-Sigma Limits and Normality

In the case of the lottery example the underlying parent or universe distribution is uniform. Consequently, the sampling distribution of the sample mean really is not normal. For reasonable sized samples (generally 5 or more for uniform populations), the sampling distribution of the mean can be approximated by a normal curve based on the Central Limit Theorem. Hence, the probability of three in a thousand outside the 3-sigma limits is close to correct.

Typically, 3-sigma limits are used by convention regardless of the nature of the type of control chart, or the test statistic used. Although 3-sigma limits are a normal concept, it can be demonstrated that the approach does work in practice as a rational basis for correcting a process or leaving it alone. The probabilities in the tails, or beyond 3-sigma, however, are not the same and are not known exactly. Modifications can be introduced in order to establish exact probability limits. Also, other multiples beside 3-sigma can be used. Obviously, in the latter instance, the probability of a Type I error and power will change even in the case of a normal distribution.

Control, Capability and Acceptance

Ultimately all quality efforts should be directed toward satisfying customer requirements associated with a given product at minimum cost, whether these requirements directly correspond to engineering specifications or not. Ideally, many other activities and techniques beside control charting should be involved in a total quality effort. This being the case, at a very basic level continual confusion exists regarding the concepts of process control, process capability, and acceptance.

Every product or service possesses characteristics that can be identified or measured. The concept of acceptance strictly relates to the relevant characteristics of the product. Numerous procedures and techniques have been developed to accept shipments of a given product: see Grant and Leavenworth [11] for a description. Product acceptance, directly, however, is not the subject matter of this book.

This book is concerned mainly with procedures for monitoring any characteristic that enables a process to be kept in a state of statistical control. Consequently, characteristics that are monitored can be associated with a final product, unassembled parts, machines, workers, materials, gages or measuring instruments, or any combination of these.

Process capability refers to the ability of a process to meet requirements or specifications with respect to any characteristic, but mainly with respect to product characteristics. Therefore, a process should be stable, or in control, with respect to the characteristics considered before one can assess its capability. Moreover, it should be understood clearly that control limits do not directly apply to capability. Control limits generally are based on the sampling distribution of a relevant sample statistic used to monitor a process. Limits relating to capability are based solely on specifications associated with individual items or units of observation. More will be said about this in Chapter 9.

Although we have isolated the concepts of acceptance, control, and capability, they can be related and treated within the framework of process control. Basically, if done properly, control charting can provide sufficient information about the process and the behavior of product characteristics so that acceptance procedures and capability studies can be minimized. In other words, if a process is properly controlled and doing what it should do, the product should be made properly. The emphasis rests with preventing conditions that produce unacceptable products rather than inspecting out bad product after it is produced.

Types of Control Charts

In this chapter, we have introduced the basic procedure for constructing a control chart and illustrated this in terms of means. Many types of control charts exist that apply to different types of problems and data. Moreover, more sophisticated forms and variations of the basic chart also are available that either are superior or possess features that can be used in special applications.

A broad way of classifying control charts is in terms of the type of data ultimately used. The broadest distinction can be made in terms of variable as opposed to attribute data. Variable data are based on measured quantities whereas attribute data are associated with qualitative characteristics. Obviously, the length of a steel rod can be considered as a variable, whereas characterization of a part as defective is qualitative even though measurements may be used to establish the part's acceptability.

When variable data are employed, a control chart to monitor both central tendency and variability is used. Different types of variable charts are available. In combination, we shall consider mean and standard deviation charts ($\overline{X}$ and S), mean and range charts ($\overline{X}$ and R), median and range charts ($\tilde{X}$ and R), and what are referred to as moving average and moving range charts (MA and MR). Each has certain advantages and features that can be applied under different circumstances and is covered in Chapters 7 and 12.

Attribute charts are of two basic types, namely, in terms of defectives or defects. A defect represents a flaw or imperfection whereas a defective represents an unacceptable item. Consequently, more than one defect can result in a defective. Control charts for defectives either are in the form of the proportion, or percent defective (p-charts) or in terms of the number of defective items (np-charts). Control charts for defects are in terms of the number of defects (C or U-charts). The two types of attribute charts are presented in Chapters 10 and 11.

VARIABLES DATA

Module Summary

The preceding modules provide a framework for this module, which directly deals with the main process control tools currently in use. Module A introduces the fundamental notion of variation in measurements, the properties of data, and basic methods of description. A link is made between frequency distributions, which are cross-sectional descriptions, and time plots. Both concepts are important in the material that follows.

Module B introduces the concepts of probability and probability distributions, which represents the underlying statistical model for dealing with variation. Probability distributions form the basis for methods of estimation and hypothesis testing, which are sampling methods underlying the process control tools introduced in this module. The last chapter of Module B does provide a basic preview of control charts as a link between the methods of inference and the concepts presented in this module.

139

VARIABLES
DATA

This module is concerned with the most common process control tools, control charts for measured data. Both mean and standard deviation charts and mean and range charts are considered. Although range charts are more common on the plant floor, both are emphasized equally here for two reasons. First, it will become apparent that similar information, in general, is imparted by both. Second, if standard deviation charts are encountered or used, familiarity with these is provided.

In order to provide some motivation for the material presented, a filling process example with a sufficient number of illustrative problems is used as a unifying theme throughout all chapters of the module. Although a number of different ideas are introduced, Chapter 7 is concerned primarily with construction of variables charts and the reason for doing so.

Chapter 8 provides an underlying basis for interpreting patterns in control charts and presents a number of patterns commonly encountered. Some commonly accepted rules and guidelines also are included along with the statistical basis for these rules.

Chapter 9 introduces four basic procedures for describing the capability of a process, once control is established. The filling process example then is extended in order to demonstrate, in statistical terms, the meaning of improvement. This also serves to unify the essential components of the quality control problem.

MEAN, STANDARD DEVIATION, AND RANGE CHARTS

The last chapter titled "Basic Charting" introduced the concept of a control chart, which is the simplest of process control tools. Although some ideas of control chart construction and interpretation were provided, by restricting focus on the Super-7 Lottery data the intent was to reinforce the preceding concepts of probability and statistics in terms of their relationship to what is to come.

In this chapter we are concerned with a more formal treatment of the *construction of control charts*. The two chapters that follow deal with interpreting various out-of-control patterns beyond the simplest case presented here and methods for determining the capability of a process once it has been determined that it is in control. Together, the three chapters provide a complete view of process control.

Basic Reasoning

Repeatedly, we have stated that no matter how well a process is managed, a certain amount of natural variation exists and is inherent in the process. Again, no matter how small the differences among individual items may be, differences do exist. These differences result from a set of uncontrollable forces, or *common causes*, in the process.

When these chance causes are the only ones operating on a system or process, we say that the process is in a state of statistical control; the process is stable. On the contrary, other sources of variation that may be controllable also can appear within a process. These could result from such things as changes in raw materials or operator performance, or from variations in machine settings or wearing parts. When such causes of variation, termed *special causes*, are present, we say that a process is out of control; the process is unstable.

A control chart, therefore, is a basic device or tool used to monitor a process in terms of one or more key characteristics in order to determine which source(s) of variation are operating on the process. Evidence of a special cause of variation leads to actions to remove it.

Although control charts can indirectly provide information about the capability of a process to meet specifications, it should be recognized that at this point our interest lies solely with process stability. Process capability is to be treated later. In general, it can be assumed that by controlling or correcting special causes of variation, productivity and quality can be improved, and costs can be reduced. Reducing natural variation, which is more difficult and can require fundamental changes to the process, can result in further improvements.

Variable vs. Attribute Data

Relevant product and process characteristics may be one of two types. Either they are *measurable* on a continuous scale or they are *qualitative*, or categorical. Obviously, such things as length, diameters, and resistance can be classed as variables. Clearly when we simply classify items as acceptable or unacceptable, we are dealing with two qualitative states, or categories, such that the resulting data are considered as an attribute.

In some cases it is inconvenient or impossible to use variable type data, whereas in others both classification types may be used. For example, typically, color is treated qualitatively since ultimately it is the visual perception of acceptability that is important. However, devices such as a spectrophotometer or a densitometer can be used to measure various aspects of color, the output of which is measurable and can be used in the control of a printing process. Variable data have certain advantages over attribute data with respect to process control, whereas attribute data are more usefully applied, though not exclusively, to acceptance problems.

This and the next two chapters deal with process control charts and process capability exclusively in terms of *variable data*. Chapters 10 and 11 of Module D concentrate on control charts for attributes.

Types of Variable Control Charts

It has been stated that control charts are used to determine whether processes are stable or not. Strictly, a process is stable when there exists a single system of chance causes such that the process distribution remains the same over time. Typically, however, we do not monitor an entire frequency distribution over time but use measures which reflect a distribution's essential features. Based on selected product or process characteristics, it is sufficient to monitor central tendency and dispersion. When working with variable data, generally *two* charts are used for each relevant characteristic, one for central tendency and the other or process variation. One first looks for stability with respect to process variation *before* being concerned with process centering.

Since different measures of central tendency and dispersion are available, obviously alternative charts can be developed based on the measures selected. For example, different charts can be based on the properties of the sample estimators, types of applications, convenience, and the means of recording data. The most commonly used combination is the mean and the range chart. We use the sample mean because of its sampling properties and its common acceptance. The sample range is used because of the ease with which it is understood and calculated by hand. Generally speaking, the standard deviation chart is better than the range chart, especially for larger samples.

Further ease-of-hand computation can be achieved by using the sample median instead of the mean in order to monitor central tendency. For special applications, such as a chemical process, other forms referred to as a moving average and moving range chart can be used. These also possess greater sensitivity in detecting process shifts.

Mean, standard deviation, and range charts are addressed in this and the subsequent chapter. Considerations applying to most all charts are provided also. Median, moving average, and moving range charts are deferred until Chapter 12. Additional special purpose charts beyond the simpler ones presented here also are cited.

The symbols used in this chapter are defined as follows:

μ	=	the universe or process mean
μ_0	=	a target or specified process mean
X	=	a value of a particular measurement
$\overline{X}$	=	a sample mean
$\overline{\overline{X}}$	=	the mean of a series of sample means
n	=	sample size
k	=	the number of samples
S	=	the sample standard deviation
S_0	=	the targeted sample standard deviation corresponding to the process target σ_0
$\overline{S}$	=	the mean of a series of sample standard deviations
R	=	the sample range
R_0	=	the value of the range corresponding to the process target σ_0
$\overline{R}$	=	the mean of a series of sample ranges
σ	=	the universe or process standard deviation
σ_0	=	a target or specified process standard deviation
$\sigma_{\overline{x}}$	=	the standard error of the mean, or the standard deviation of the sampling distribution of the mean
$\hat{\sigma}_{\overline{x}}$	=	the estimated standard error of the mean
σ_s	=	the standard error of the sample standard deviation
$\hat{\sigma}_s$	=	the estimated standard error of the sample standard deviation

σ_R	=	the standard error of the sample range
$\hat{\sigma}_R$	=	the estimated standard error of the sample range
$CL_{\bar{x}}$	=	the center line on a mean chart
$UCL_{\bar{x}}$	=	the upper control limit on a mean chart
$LCL_{\bar{x}}$	=	the lower control limit on a mean chart
CL_s	=	the center line on a standard deviation chart
UCL_s	=	the upper control limit on a standard deviation chart
LCL_s	=	the lower control limit on a standard deviation chart
CL_R	=	the center line on a range chart
UCL_R	=	the upper control limit on a range chart
LCL_R	=	the lower control limit on a range chart
A	=	tabulated factor used to calculate limits on mean charts based on targets μ_0 and σ_0
A_2	=	tabulated factor used to calculate control limits on a mean chart based on process ranges
A_3	=	tabulated factor used to calculate control limits on a mean chart based on process data
B_3	=	tabulated factor used to calculate the lower control limit on an S chart based on process data
B_4	=	tabulated factor used to calculate the upper control limit on an S chart based on process data
B_5	=	tabulated factor used to calculate the lower control limit on an S chart based on the target σ_0
B_6	=	tabulated factor used to calculate the upper control limit on an S chart based on the target σ_0
c_4	=	tabulated bias adjustment factor for the sample standard deviation as an estimate of σ
d_2	=	tabulated bias adjustment factor for the sample range as an estimate of σ
D_1	=	tabulated factor used to calculate the lower control limit for a range chart based on the target σ_0
D_2	=	tabulated factor used to calculate the upper control limit for a range chart based on the target σ_0
D_3	=	tabulated factor used to calculate the lower control limit on a range chart based on process data
D_4	=	tabulated factor used to calculate the upper control limit on a range chart based on process data

CONSTRUCTING VARIABLES CONTROL CHARTS

We have indicated that a control chart is a basic tool that is used to separate or identify chance and non-chance causes with respect to a process. When used properly, it can be helpful in detecting problems and maintaining a stable process.

Control charts are graphs of a key process characteristic plotted on the vertical axis and equidistant time intervals (or the sample number) plotted on the horizontal axis. Generally, two limits appear on the chart, an upper and a lower control limit. A center line is placed between the limits, which corresponds to a standard, or target, or a calculated average.

Although other ways exist for calculating control limits, it is customary in the United States to place them at three standard errors from the center line. Within this framework, there are two basic ways to establish a center line and calculate control limits. The one is based on standard values, or targets, expressed in terms of the process mean and standard deviation. The other involves using estimates based on actual process data. Frequently, estimates used together with control charts can be employed to establish the standards. More will be said about the choices at the end of this chapter.

When process data are employed in order to estimate the process standard deviation, alternatives exist. With respect to quality control, either the sample standard deviation or the sample range is used. The basis for using these measures is presented in some detail in Chapter 5. Here, we shall use the results in order to demonstrate how the measures are used to construct control charts for the mean, standard deviation, and the range.

Control Charts for Mean and Standard Deviation; Standard Values μ_0 and σ_0 Given

When it is of interest to construct control charts for the mean and the standard deviation based on established standards, or targets, the following formulas for the center line and control limits for each chart can be calculated using the following formulas:

Mean or $\overline{X}$-chart:

$$\text{Center line, } CL_{\overline{x}} = \mu_0$$

$$\text{Upper control limit,} \quad UCL_{\overline{x}} = \mu_0 + 3\sigma_{\overline{x}}$$

$$= \mu_0 + A\sigma_0$$

$$\text{Lower control limit,} \quad LCL_{\overline{x}} = \mu_0 - 3\sigma_{\overline{x}}$$

$$= \mu_0 - A\sigma_0$$

Standard Deviation or S-chart:

$$\text{Center line, } CL_s = S_0$$

$$= c_4\sigma_0$$

$$\text{Upper control limit, } UCL_s = S_0 + 3\sigma_s$$

$$= B_6\sigma_0$$

$$\text{Lower control limit, LCL}_s = S_0 - 3\sigma_s$$

$$= B_5\sigma_0$$

The *values* plotted on the corresponding control charts are the values of the sample mean, $\overline{X}$, and the sample standard deviation, S, based on samples successively drawn over time. Recall, the formulas used to calculate these are given as

$$\text{Sample mean, } \overline{X} = \frac{\Sigma X}{n}$$

$$\text{Sample standard deviation, } S = \sqrt{\frac{\Sigma(X - \overline{X})^2}{n - 1}}$$

The symbols μ_0 and σ_0 represent established or targeted values of the process mean and standard deviation. $\sigma_{\overline{x}}$ and σ_s are the standard errors of the sample mean and sample standard deviation, or the standard deviations of the respective sampling distributions. Although a targeted value of the process standard deviation, σ_0, is specified, the center line of the standard deviation chart is S_0 which is to accommodate sample standard deviations based on an "n-1" divisor and the fact that S is a biased estimator of σ. Details regarding the estimation problems and sampling are discussed in detail in Chapter 5.

Note. The terms A, c_4, B_5 and B_6 are adjustment factors based on the sample size, n, which can be obtained from **Table C-1** in the Appendix. Although the more direct formulas provided initially can be used, the ones involving the tabulated adjustment factors are simpler.

Control Charts for Mean and Range; Standard Values of μ_0 and σ_0 Given

When the sample range is used to monitor variation instead of the standard deviation based on a targeted value of σ_0, the mean chart is the *same* as the one given above, but different formulas are needed for the range chart. The formulas are given as

Range or R-chart:

$$\text{Center line, CL}_R = R_0$$

$$= d_2\sigma_0$$

$$\text{Upper control limit, } \text{UCL}_R = R_0 + 3\sigma_R$$

$$= D_2\sigma_0$$

$$\text{Lower control limit, } \text{LCL}_R = R_0 - 3\sigma_R$$

$$= D_1\sigma_0$$

In this case, the center line is an average value of the sample range and corresponds to the specified value of the process standard deviation. Again, d_2, D_1, and D_2 are tabulated factors and can be found in **Table C-2** in the Appendix.

Control Charts for the Mean and Standard Deviation Based on Estimates from Process Data

When established standards are not used to establish control limits, these can be based on estimates of the process mean and standard deviation from a series of samples selected at fixed time intervals. The center line and control limits for an $\overline{X}$ and S chart based on process data are calculated using the following formulas:

Mean or $\overline{X}$-chart:

$$\text{Center line, } CL_{\overline{x}} = \overline{\overline{X}}$$

$$\text{Upper control limit, } UCL_{\overline{x}} = \overline{\overline{X}} + 3\hat{\sigma}_{\overline{x}}$$

$$= \overline{\overline{X}} + A_3\overline{S}$$

$$\text{Lower control limit, } LCL_{\overline{x}} = \overline{\overline{X}} - 3\sigma_{\overline{x}}$$

$$= \overline{\overline{X}} - A_3\overline{S}$$

Standard Deviation or S-chart:

$$\text{Center line, } CL_s = \overline{S}$$

$$\text{Upper control limit, } UCL_s = \overline{S} + 3\hat{\sigma}_s$$

$$= B_4\overline{S}$$

$$\text{Lower control limit, } LCL_s = \overline{S} - 3\hat{\sigma}_s$$

$$= B_3\overline{S}$$

Recall from Chapters 2, 3, or 5 that $\overline{\overline{X}}$ is the mean of the individual sample means and $\overline{S}$ is the mean of the sample standard deviations, each of which is given by the formulas

$$\overline{\overline{X}} = \frac{\Sigma\overline{X}}{k}$$

$$\overline{S} = \frac{\Sigma S}{k}$$

where k is the number of individual samples. The formulas for $\overline{X}$ and S are given above as well as in Chapters 2, 3, and 5.

Note. Again, recognize from the formulas that the control limits in each case are three *standard errors* from the corresponding center line. In this case the standard errors are estimates, which are denoted as $\hat{\sigma}_{\bar{x}}$ and $\hat{\sigma}_s$. For ease of calculation, the alternatives in terms of factors A_3, B_3, and B_4 are used. These are tabulated in the Appendix in **Table C-1**.

Control Charts for the Mean and Range Based on Estimates from Process Data

Unlike the case where standards or targeted values are used, when a range chart is employed to monitor variability, *both* the mean chart and the range chart are based on sample range estimates, and the mean chart, therefore, is *not* based on the formulas given immediately above. This results from the fact that if the range is used, due to its ease of calculation to measure variability, it is unreasonable to calculate sample standard deviations in order to estimate σ and $\sigma_{\bar{x}}$. Since estimates of these are necessary to obtain limits for the mean, an alternate procedure is used based on the sample range.

Center lines and control limits for the mean and range charts based on process data using the sample range are then given as

Mean or $\bar{X}$-chart:

$$\text{Center line, } CL_{\bar{x}} = \bar{\bar{X}}$$

$$\text{Upper control limit, } UCL_{\bar{x}} = \bar{\bar{X}} + 3\hat{\sigma}_{\bar{x}}$$

$$= \bar{\bar{X}} + A_2\bar{R}$$

$$\text{Lower control limit, } LCL_{\bar{x}} = \bar{\bar{X}} - 3\hat{\sigma}_{\bar{x}}$$

$$= \bar{\bar{X}} - A_2\bar{R}$$

Range or R-chart:

$$\text{Center line, } CL_R = \bar{R}$$

$$\text{Upper control limit, } UCL_R = \bar{R} + 3\hat{\sigma}_R$$

$$= D_4\bar{R}$$

$$\text{Lower control limit, } LCL_R = \bar{R} - 3\hat{\sigma}_R$$

$$= D_3\bar{R}$$

The center line for the mean chart, $\bar{\bar{X}}$, is calculated similarly; however, it should be noted that although the symbol for the estimated standard error of the mean, $\hat{\sigma}_{\bar{x}}$, re-appears, its numerical value will *not* be the same since it is estimated in terms of the range instead of the sample standard deviation. Details regarding the differences between the estimates appear in Chapter 5. Here, we can avoid the computational burden by using the factors A_2, D_3, and D_4 which are tabulated in the Appendix in **Table C-2**.

Example 1 A particular filling process involves filling small containers with a powdery substance by means of filling machines, each with ten nozzles. The label claim on each container specifies the weight of the contents to be 425 grams.

Based on independent considerations management has set targeted values for the process mean and standard deviation equal to 442 grams and 4.5 grams, respectively. The purpose of the targets is to insure against falling below the label claim limit and, therefore, was not based on properly collected data nor on considerations of process control.

For purposes of illustration, let us assume that weights are recorded for containers from the first five nozzles at one-half hour intervals over a nine-hour period. The results are presented in **Exhibit 1**.

EXHIBIT 1
FILLING PROCESS WEIGHTS IN GRAMS

Time	Sample Number	Nozzle 1	Nozzle 2	Nozzle 3	Nozzle 4	Nozzle 5	$\overline{X}$	S	R
7:00	1	448.3	449.1	447.9	445.0	448.1	447.7	1.57	4.1
7:30	2	449.6	446.6	448.4	449.5	445.4	447.9	1.85	4.2
8:00	3	440.5	439.9	435.8	433.9	439.1	437.8	2.85	6.6
8:30	4	435.4	440.6	437.7	438.1	438.7	438.1	1.87	5.2
9:00	5	440.5	438.6	439.7	432.9	438.3	438.0	2.98	7.6
9:30	6	435.2	438.1	436.8	435.6	437.9	436.7	1.31	2.9
10:00	7	445.4	445.7	445.0	439.9	444.8	444.2	2.41	5.5
10:30	8	441.8	445.9	443.3	442.6	443.9	443.5	1.55	4.1
11:00	9	439.5	439.7	440.4	434.0	438.7	438.5	2.57	6.4
11:30	10	435.8	437.5	437.2	436.6	437.0	438.8	.66	1.7
12:00	11	437.8	438.4	438.0	432.0	437.4	438.5	2.64	6.4
12:30	12	433.6	435.3	436.1	434.8	437.8	438.3	1.23	3.2
1:00	13	439.2	440.1	440.6	434.0	438.9	438.6	2.64	6.6
1:30	14	435.5	440.0	437.5	437.1	438.6	437.5	1.83	4.5
2:00	15	439.9	439.6	440.7	434.5	439.2	438.8	2.45	6.2
2:30	16	435.8	439.6	437.5	437.4	438.4	437.5	1.53	3.8
3:00	17	440.9	441.5	442.2	434.1	439.9	439.7	3.25	8.1
3:30	18	437.2	439.7	439.3	437.6	440.0	438.8	1.27	2.8
						Totals	7918.9	36.46	89.9

Based on the *specified targets*, calculate control limits for the mean and standard deviation and construct the corresponding control charts based on the data.

Given: $\mu_0 = 442$

$\sigma_0 = 4.5$

$$CL_{\bar{x}} = \mu_0$$

$$= 442$$

$$UCL_{\bar{x}} = \mu_0 + A\sigma_0$$

$$= 442 + 1.34(4.5)$$

$$= 448.0$$

$$LCL_{\bar{x}} = \mu - A\sigma_0$$

$$= 442 - 1.34(4.5)$$

$$= 436.0$$

$$CL_s = S_0$$

$$= C_4\sigma_0$$

$$= .94(4.5)$$

$$= 4.23$$

$$UCL_s = B_6\sigma_0$$

$$= 1.96(4.5)$$

$$= 8.82$$

$$LCL_s = B_5\sigma_0$$

$$= 0(4.5)$$

$$= 0.00$$

Based on the values calculated above and the data appearing in Exhibit 1, the resulting $\overline{X}$ and S charts are presented in **Exhibit 2**.

Example 2 In this example we shall use the same problem setting presented in Example 1 in order to illustrate a *range chart* based on a targeted process standard deviation. The mean chart is the *same*. In an actual problem, *either* the S chart or the R chart is displayed, depending on the circumstances.

The extra calculations to establish the range chart, which are not considered in Example 1 based on the targeted value of σ, are performed as follows:

$$\text{Given: } \sigma_0 = 4.5$$

$$CL_R = d_2\sigma_0$$

$$= 2.326(4.5)$$

$$= 10.47$$

$$UCL_R = D_2\sigma_0$$

$$= 4.92(4.5)$$

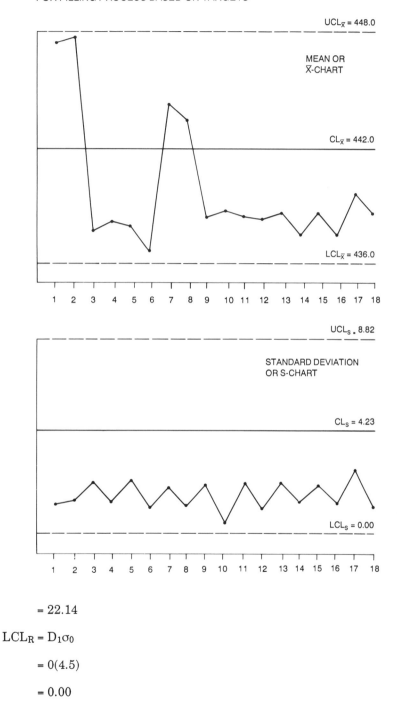

EXHIBIT 2
MEAN AND STANDARD DEVIATION CHARTS
FOR FILLING PROCESS BASED ON TARGETS

$= 22.14$

$$LCL_R = D_1\sigma_0$$

$$= 0(4.5)$$

$$= 0.00$$

The mean and range charts based on the targeted values
of $\mu_0 = 442$ and $\sigma_0 = 4.5$ with data from Exhibit 1 are
presented in **Exhibit 3**.

EXHIBIT 3
MEAN AND RANGE CHARTS FOR FILLING
PROCESS BASED ON TARGETS

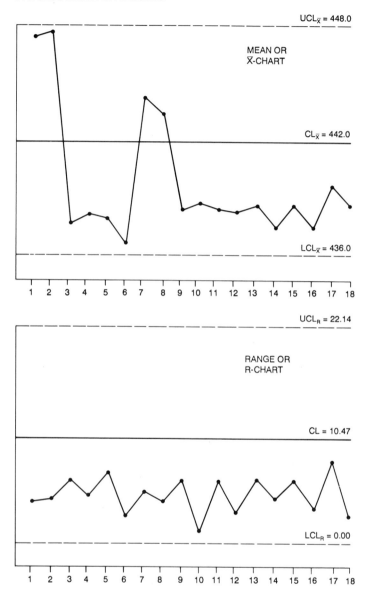

MEAN OR
X̄-CHART

UCL$_{\bar{X}}$ = 448.0

CL$_{\bar{X}}$ = 442.0

LCL$_{\bar{X}}$ = 436.0

RANGE OR
R-CHART

UCL$_R$ = 22.14

CL = 10.47

LCL$_R$ = 0.00

Example 3 Construct mean and standard deviation control charts for the filling process problem presented in Example 1 using control limits calculated from *data* provided in Exhibit 1.

Since the values of $\bar{X}$ and S are provided for each sample along with the corresponding totals, we obtain the values of $\bar{\bar{X}}$ and $\bar{S}$ as

$$\bar{\bar{X}} = \frac{\Sigma\bar{X}}{k}$$

$$= \frac{7918.9}{18}$$

$$= 439.9$$

$$\bar{S} = \frac{\Sigma S}{k}$$

$$= \frac{36.46}{18}$$

$$= 2.03$$

Control limits are then calculated as follows:

$$CL_{\bar{x}} = \bar{\bar{X}}$$

$$= 439.9$$

$$UCL_{\bar{x}} = \bar{\bar{X}} + A_3\bar{S}$$

$$= 439.9 + 1.43(2.03)$$

$$= 442.8$$

$$LCL_{\bar{x}} = \bar{\bar{X}} - A_3\bar{S}$$

$$= 439.9 - 1.43(2.03)$$

$$= 437.0$$

$$CL_s = \bar{S}$$

$$= 2.03$$

$$UCL_s = B_4\bar{S}$$

$$= 2.09(2.3)$$

$$= 4.81$$

$$LCL_s = B_3\bar{S}$$

$$= 0(2.03)$$

$$= 0.00$$

The corresponding mean and standard deviation chart based on limits calculated from the actual process data appear in **Exhibit 4**.

Example 4 Construct a *mean and a range* chart for the filling process problem presented in Exhibit 1 using control limits calculated from data provided in Exhibit 1.

MEAN, STANDARD
DEVIATION, AND
RANGE CHARTS

Recognize in this example, unlike Example 2, that *both* sets of limits must be recalculated because if the range is used to measure variability, it is unreasonable to use the standard deviation to obtain limits for the mean chart.

EXHIBIT 4
MEAN AND STANDARD DEVIATION CHARTS FOR FILLING PROCESS
BASED ON NATURAL LIMITS CALCULATED FROM PROCESS DATA

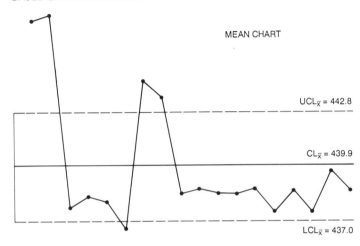

MEAN CHART

$UCL_{\bar{x}} = 442.8$

$CL_{\bar{x}} = 439.9$

$LCL_{\bar{x}} = 437.0$

STANDARD DEVIATION
CHART

$UCL_S = 4.81$

$CL_S = 2.03$

$LCL_S = 0.00$

Based on the totals appearing in Exhibit 1, the values of $\bar{\bar{X}}$ and $\bar{R}$ are found as

$$\bar{\bar{X}} = \frac{\Sigma \bar{X}}{k}$$

$$= \frac{7918.9}{18}$$

$$= 439.9$$

$$\bar{R} = \frac{\Sigma R}{k}$$

$$= \frac{89.9}{18}$$

$$= 5.0$$

Control limits are then found as

$$CL_{\bar{x}} = \bar{\bar{X}}$$

$$= 439.9$$

$$UCL_{\bar{x}} = \bar{\bar{X}} + A_2\bar{R}$$

$$= 439.9 + 0.58(5)$$

$$= 442.8$$

$$LCL_{\bar{x}} = \bar{\bar{X}} - A_2\bar{R}$$

$$= 439.9 - 0.58(5)$$

$$= 437.0$$

$$CL_R = \bar{R}$$

$$= 5$$

$$UCL_R = D_4\bar{R}$$

$$= 2.11(5)$$

$$= 10.55$$

$$LCL_R = D_3\bar{R}$$

$$= 0(5)$$

$$= 0$$

The corresponding mean and range chart based on limits calculated from the actual process data appear in **Exhibit 5**.

Example 5   Based on the results in Example 4, eliminate the subgroups based on points out of control limits on the mean chart, recalculate the limits based on the remaining data, and construct a mean, standard deviation, and range chart for the filling process problem.

Again, recognize that in an actual application either the $\bar{X}$ and S charts or the $\bar{X}$ and R charts, in pairs, would be constructed. Also, in the case where limits are calculated on the basis of actual process data, the limits for the mean chart depend on the measure of variation used, either S or R. For purposes of illustration, the three charts are introduced together here.

Based on Exhibit 5, we can see that samples numbered 1, 2, 6, 7, and 8 fall outside the control limits on the mean

EXHIBIT 5
MEAN AND RANGE CHARTS FOR FILLING PROCESS BASED ON NATURAL
LIMITS CALCULATED FROM PROCESS DATA

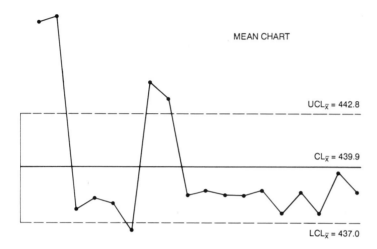

MEAN CHART

$UCL_{\bar{x}} = 442.8$

$CL_{\bar{x}} = 439.9$

$LCL_{\bar{x}} = 437.0$

RANGE CHART

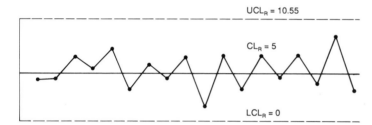

$UCL_R = 10.55$

$CL_R = 5$

$LCL_R = 0$

chart. In terms of the remaining 13 subgroups, we obtain the following results.

$$\bar{\bar{X}} = \frac{\Sigma \bar{X}}{k}$$

$$= 5699.2$$

$$= 441.6$$

$$= 438.4$$

$$\bar{S} = \frac{\Sigma S}{k}$$

$$= \frac{27.82}{13}$$

$$= 2.14$$

$$\bar{R} = \frac{\Sigma R}{k}$$

$$= \frac{69.16}{13}$$

$$= 5.32$$

Based on S:

$$CL_{\bar{x}} = \bar{\bar{X}}$$

$$= 438.4$$

$$UCL_{\bar{x}} = \bar{\bar{X}} + A_3\bar{S}$$

$$= 438.4 + 1.43(2.14)$$

$$= 441.5$$

$$LCL_{\bar{x}} = \bar{\bar{X}} - A_3\bar{S}$$

$$= 438.4 - 1.43(2.14)$$

$$= 435.3$$

$$CL_s = \bar{S}$$

$$= 2.14$$

$$UCL_s = B_4\bar{S}$$

$$= 2.09(2.14)$$

$$= 4.47$$

$$LCL_s = B_3\bar{S}$$

$$= 0(2.14)$$

$$= 0$$

Based on R:

$$CL_{\bar{x}} = \bar{\bar{X}}$$

$$= 438.4$$

$$UCL_{\bar{x}} = \bar{\bar{X}} + A_2\bar{R}$$

$$= 438.4 + 0.58(5.32)$$

$$= 441.5$$

$$LCL_{\bar{x}} = \bar{\bar{X}} - A_2\bar{R}$$

$$= 438.4 - 1.58(5.32)$$

**MEAN, STANDARD
DEVIATION, AND
RANGE CHARTS**

$$= 435.3$$

$$CL_R = \overline{R}$$

$$= 5.32$$

$$UCL_R = D_4\overline{R}$$

$$= 2.11(5.32)$$

$$= 11.23$$

$$LCL_R = D_3\overline{R}$$

$$= 0(5.32)$$

$$= 0$$

The mean, standard deviation, and range charts based on samples within the control limits on the mean chart of the previous example are presented in **Exhibit 6**. Recognize in *this particular case* the numerical values of the limits for the mean chart are the same based on S and R; therefore, only one mean chart is presented.

EXHIBIT 6
MEAN, STANDARD DEVIATION, AND RANGE CHARTS
FOR FILLING PROCESS BASED ON PROCESS DATA
WITH POINTS OUTSIDE NATURAL LIMITS REMOVED

MEAN CHART

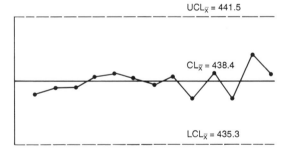

$UCL_{\overline{x}} = 441.5$

$CL_{\overline{x}} = 438.4$

$LCL_{\overline{x}} = 435.3$

RANGE CHART

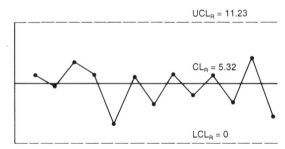

$UCL_R = 11.23$

$CL_R = 5.32$

$LCL_R = 0$

STANDARD DEVIATION
CHART

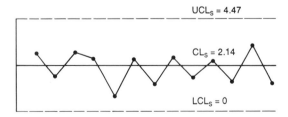

$UCL_S = 4.47$

$CL_S = 2.14$

$LCL_S = 0$

Note. Since this chapter is directed toward construction of variables control charts, we are not directly interested in estimating the process standard deviation, σ. Although this is of fundamental importance and is considered in Chapter 9, we shall obtain an estimate here in order to use it as a basis for the next example. In general, before obtaining such an estimate from process data, control should be established; this is something we have not yet considered.

Recall from Chapter 5 that we can estimate the process standard deviation in two ways, either in terms of S or R. Using both procedures, we have

$$(1) \quad \hat{\sigma} = \frac{\overline{S}}{c_4}$$

$$= \frac{2.14}{.94}$$

$$= 2.28$$

$$(2) \quad \hat{\sigma} = \frac{\overline{R}}{d_2}$$

$$= \frac{5.32}{2.326}$$

$$= 2.29$$

The results of the two alternatives are very close. Rounding both to the first decimal place yields an identical value of 2.3.

Example 6   **Exhibit 7** presents simulated filling process weights from a normally distributed process with a mean, μ, equal to 442 and a standard deviation, σ, equal to 2.3 for thirteen samples, or subgroups, of five nozzles each.

EXHIBIT 7
SIMULATED FILLING PROCESS WEIGHTS BASED ON NORMAL PROCESS WITH $\mu = 442$ AND $\sigma = 2.3$

Sample Number	Nozzle 1	Nozzle 2	Nozzle 3	Nozzle 4	Nozzle 5	$\overline{X}$	S	R
1	444.6	442.6	441.8	441.8	445.1	443.2	1.57	3.3
2	439.9	447.2	442.6	442.9	444.1	443.3	2.65	7.3
3	440.1	441.1	438.9	443.7	438.9	441.0	2.07	4.8
4	442.9	442.6	440.5	442.4	442.2	442.1	.94	2.4
5	446.0	438.1	445.1	442.1	439.8	442.2	3.37	7.9
6	444.2	440.0	439.1	441.0	437.2	440.3	2.59	7.0
7	445.4	443.7	445.8	441.9	444.4	444.2	1.55	3.9
8	442.6	441.9	439.5	441.2	439.3	440.9	1.46	3.3
9	441.6	439.4	441.7	446.2	442.2	442.2	2.47	6.8
10	443.3	443.5	441.6	438.8	446.6	442.8	2.86	7.8
11	441.0	438.7	444.9	444.8	439.1	441.7	3.00	6.2
12	440.7	442.5	447.2	442.7	443.0	443.2	2.40	6.5
13	441.7	440.4	442.1	440.7	441.7	441.3	.73	1.7
					Totals	5748.4	27.66	68.9

The $\overline{X}$ and S and $\overline{X}$ and R charts with limits based on the data are presented in **Exhibit 8** and **Exhibit 9**, respectively.

**MEAN, STANDARD
DEVIATION, AND
RANGE CHARTS**

Since we have illustrated the procedures for calculating control limits, the calculations have been omitted in this case. Use of the inputted values of μ and σ would yield almost identical results since the simulated mean equal to 442.2 and estimated standard deviation equal to 2.27 are very close to the simulated inputs. This example has been introduced in order to provide a basis for comparing the previous results to a completely random process which is in a state of statistical control.

EXHIBIT 8
MEAN AND STANDARD DEVIATION CHARTS FOR FILLING PROCESS
BASED ON SIMULATED RANDOM PROCESS WITH $\mu = 442$ AND $\sigma = 2.3$

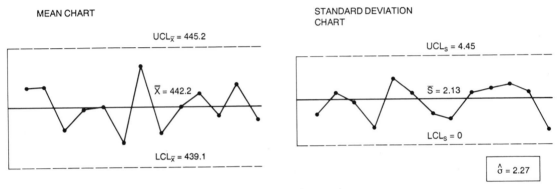

EXHIBIT 9
MEAN AND RANGE CHARTS FOR FILLING PROCESS BASED ON
SIMULATED RANDOM PROCESS WITH $\mu = 442$ AND $\sigma = 2.3$

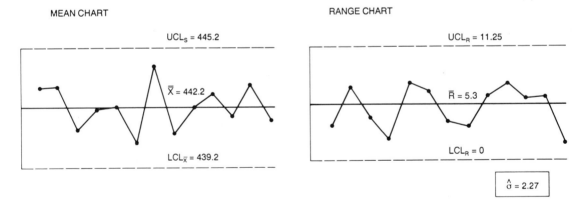

Notes and Observations Based on the Examples

The examples just presented were introduced for two basic reasons. On the one hand, they illustrate the procedures for constructing variables control charts. Also, they were chosen in order to have wider appeal in terms of the type of information provided by such charts. Before entering into a discussion needed to obtain a better understanding of the nature of control charts, it is useful at this point merely to list relevant observations that can be made from the charts; these observations will then be included in the subsequent discussions when appropriate.

1. All charts illustrated are based on sample statistics and do not directly provide information concerning specifications or tolerances on individual containers, or items, in general.

2. Different information can be supplied by charts based on targeted values and ones constructed based on process data.

3. The way in which charts are used based on targets as opposed to those based on process data may differ.

4. Charts originally constructed on the basis of process data eventually may be used as a basis to establish targets.

5. The *basic* way in which control charts are interpreted is that points outside the control limits indicate that the process is not in control. In addition, non-random patterns of points within control limits *also* indicate an out-of-control process.

6. In the filling process problem illustrated, the way in which targeted values were set can be improved on.

7. Samples or subgroups are based on five observations, one from each of the five nozzles. The measure of variation underlying the calculations, therefore, is based on variability among nozzles. If differences among the performance of the nozzles exist or if variation from fill to fill within the nozzles differs, then different methods of sampling, charting, and measuring variation should be considered.

8. By examining Exhibits 2 and 3, and Exhibits 4 and 5 one can see that the standard deviation and range charts basically exhibit similar patterns and, therefore, provide basically the same information, although departures between the two charts will occur.

9. By examining Exhibit 2 or Exhibit 3, one sees that all sample points fall *within* the control limits on the mean, standard deviation, and range charts. Recall that the limits are based on targeted inputs for μ and σ.

10. Although all points fall within the control limits, there is evidence of *lack of control* on all charts. In the case of means, the initially high values are associated with overshooting in order to prevent underfill. Overcompensation was instituted to correct for this, followed by another adjustment upward. The resulting two points above the center line were misinterpreted to be too high, which resulted in another adjustment downward leading to an apparent more stable run of points below the center line. Aside from the adjustment, it appears as if the process is operating about an average of 440 grams rather than 442 grams.

11. By examining Exhibit 4 and Exhibit 5, which provide control charts based on natural limits calculated directly from the process data,

the points made in observation 10 become clearer. The natural limits are tighter than those based on the targeted inputs. Over-and-under adjustment points on the mean chart are outside the control limits. The run of points still is apparent on the mean chart, and the S and R charts are more properly centered. The near-perfect oscillatory movement about the center line on the S and R charts indicates a non-random cause. This results from a fact undisclosed at the outset: the original instructions were not followed such that nozzles 6-10 also were sampled and mislabeled as 1-5. Nozzles 6-10 replaced nozzles 1-5 on an alternating basis.

12. Exhibit 6 provides control charts based on recalculated limits after samples corresponding to points outside the control limits on the previous mean chart are eliminated. Notice that all points fall within the control limits on all charts, and the plotted points vary about the center line. Note, however, that the sample means vary very tightly about the center line, which also is an indication of non-randomness. The standard deviations and ranges continue to have a near perfect oscillatory movement about the respective center lines.

13. The charts appearing in Exhibit 6 really are not precise and a theoretically unjustified liberty was taken. Since intervening sample points tied to particular points in time were removed, the location of the corresponding coordinates should be left open; in the form presented, the charts appear as if the data were generated at equidistant points in time. The points eliminated correspond to over-and-under adjustment. If over-and-under adjustment were the only problem, this should be eliminated and the revised limits should be applied to newly collected data. Actually, since variation appears not to be in control, this should be considered first. Since more than one non-random source of variation appears to be present in the charts, the recalculated results are subject to question. In the form presented, however, the charts in Exhibit 6 without the adjustment problem are useful to illustrate out-of-control patterns within the control limits, especially when compared to the completely random patterns shown in Exhibit 8.

14. Exhibit 8 and Exhibit 9 present $\overline{X}$ and S charts and $\overline{X}$ and R charts for a simulated process based on the targeted mean, $\mu_0 = 442$, and the estimated process standard deviation, $\hat{\sigma} = 2.3$. A number of points are worth noting about these exhibits:

a. The charts are based on idealized conditions of randomness, which provide a *reference frame* for the way the process should appear when charted under the assumption that the targeted mean is controlled and natural or inherent variability holds at equal to 2.3 grams.

b. It has not yet been established that = 2.3 is a true measure of chance variation; this was used for purposes of illustration only.

c. Notice the patterns appear more erratic and spread-out than those in Exhibit 6.

d. Because the patterns are random, if another sequence of thirteen samples are generated under the same conditions, the exact same patterns will *not* appear when charted. They will be *somewhat similar* in the sense that the points typically will fall within the control limits, they will vary about the center line, and they will be about as erratic and spread-out. There is no way to develop a template, of sorts, that will allow one to establish exactly where the points should fall. Ways are available, however, to establish how the points should not fall if the process is not in control.

e. To get a better sense of the nature of the original sequence of observations generated, these are like the Super-7 Lottery results presented in Chapter 6, except the lottery results are based on an underlying uniform distribution whereas in this case the results are based on a normal distribution.

15. It should be emphasized that *no* direct consideration has been given to specifications on the weights of the contents of individual containers. This will briefly be discussed at the end of this chapter and in more detail in Chapter 9 when we consider process capability.

16. All of the examples were presented in terms of numerous sample results plotted on each chart at one time. Briefly, such an approach is useful when solving a particular problem or when attempting to establish targets. The approach was used to illustrate the construction of a completed chart. Once up-and-running, control charts based on established targets can be used in real-time in order to monitor a process, one point at a time.

ADDITIONAL IDEAS

At the very beginning of this chapter we briefly introduced the reasoning underlying the use of control charts. We then provided procedures and formulas for calculating control limits and illustrated them in terms of the filling process problem. Of the many examples from which to choose the filling problem was selected because of its realness, its seeming simplicity, and its ease of understanding in terms of basic concepts; yet, it also possesses features through which we may discuss related ideas and concepts. These have been introduced in the following discussion.

Basic Reasoning

There are two basic premises underlying the use of control charts: (1) random variation is associated with the output of any process, and (2) a process is in a state of control if a stable pattern of random variation exists. By this we mean that the same probability distribution underlies the generation of measurements at differing points in time. Strictly, when any part of the distribution changes the process is not in control.

A control chart, in its most basic form, is a simple graphical device that is used to observe and monitor one or more parameters of the distribution of relevant process characteristics in order to determine whether or not the process parameters have changed. When no change is indicated in terms of no points out of control, the decision rule is to leave the process alone. Only common or natural causes of variation are operating on the process, which are not subject to control. This is equivalent to concluding that the variation present is *random*.

When a point falls outside the control limits, one concludes that a change in the process distribution has occurred that results from a non-random or *assignable cause* of variation. The process is said to be out of control as a result of *special* or controllable causes of variation. On the surface, out-of-control points provide information that the process is not operating properly and lead to the identification of the problem. In some cases, the nature of the problem may be obvious, whereas in others one must hunt for the problem based on varying amounts of effort and knowledge. The results from the chart do not necessarily identify the problem.

Real-Time vs. Batch Applications

The above comments suggest that a control chart can be used on a *point by point* basis at each point in time. Actually,this was the original intention based on strict statistical considerations. Recognize that control charts can be used in a number of ways. It is convenient to borrow some language from computer applications in order to make the point by distinguishing between real-time and so-called "batch" applications.

Originally, control charts were intended to be used in real time on a point by point basis in order to detect problems in a process roughly at the time they are indicated on the chart. In such cases, it is necessary to have established limits either based on set targets or based on data previously obtained from a stable process. In a true sense, here one is testing an hypothesis at each point in time, rejection of which occurs when a point falls outside a control limit. This leads to the conclusion that a change occurs and to the decision to hunt for the problem and correct it.

Also, in *real time*, as sample points *build-up* or accumulate, it is possible to detect patterns that indicate the values are not generated randomly, even though points do not fall outside the control limits. Such patterns also indicate lack of control and lead to corrective action. An *accumulation* of points is necessary to establish non-randomness. Applied in a different way, a probabilistic basis like that underlying hypothesis testing can be established for deciding when to hunt for a problem. A reference guide with further explanation regarding control chart patterns is provided in Chapter 8.

Another way to apply control charts is to collect data from numerous samples before using the charts. In such cases, which we can refer to as *batch* applications, real-time or near real-time responsiveness is not a consideration. One or more points outside a control limit indicates that a change in the process has occurred retrospectively. Rather than pattern-buildup, one can use the chart for pattern identification relating to past changes in the process. Further, accumulating control chart data in a complete batch mode is useful for developing targets for future real-time applications on an on-going basis and for solving specific problems on a one-time basis.

Uses and Benefits

By reviewing the notes and observations accompanying the examples, one can quickly obtain an idea regarding the way in which the charts can be applied. Here, we shall summarize some uses and benefits of control charts that we may refer to when appropriate.

In general, control charts are useful in that they can be used:

1. To establish what the process is doing
2. To determine what the process can do
3. To determine whether the process is doing what you want it to do
4. As an aid in deciding whether additional data or charts should be used
5. As a basis for leaving a process alone or for not making adjustments
6. As a basis for diagnosing a process and deciding when to hunt for problems or make adjustments to the process
7. As a basis for establishing the magnitude of inherent or natural variability and, therefore, as an input into the capability of a process
8. As a basis for deciding whether to continue using a particular process or whether to change the system or product design
9. To varying degrees as a replacement for inspection and product acceptance plans and procedures
10. As a reference point or basis for continued process improvement

Simply stated, the ultimate goal of a process control program should be to maintain a targeted value of key process or product characteristics, on the average, with as little variability about the average as possible. The variation about the average should be attributed to common or uncontrollable sources of variation only. Hence, control charts are used to detect special or controllable causes of variation and eliminate them, while establishing and maintaining desired targeted values on a continued basis. Once implemented and used effectively, control charts lead to reduction of defects, scrap, rework, and cost, and increased capacity or productivity.

Relation of Control Charts to Previous Ideas

Underlying every measurable quality characteristic associated with a process or the product produced by the process is a distribution describing the way in which the measurements vary. This variation can be attributed to natural causes, in which case the variation can be regarded as stable and unalterable unless a change in the system or design underlying the process is altered.

When problems occur in the process, the underlying distribution will be affected either in terms of shape, central tendency, or amount of dispersion. When one is interested in determining whether problems do arise, typically it is sufficient to focus on central tendency and dispersion. As a result, one or more mean and standard deviation or range charts are used to determine when non-chance or assignable causes of variation are present in a process.

Whether established as a target or calculated from a batch of accumulated process data, the value associated with the center line of a control chart is equivalent to an hypothesized process parameter relating to a test of an hypothesis as presented in Chapter 5. The values of the control limits, therefore, are equivalent to critical values. Recall that critical values represent values of a (sample) test statistic that define regions of rejection relating to a test of a population or process parameter. Critical values are calculated by using the sampling distribution of the relevant test statistic, where the sampling distribution is the probability distribution associated with all possible values that a sample statistic may assume, as *contrasted* with the process distribution of individual items or measurements.

The forms or shape of the sampling distribution of the sample mean, standard deviation, and range are different. In the case of sample means, the assumption of normality is reasonable based on the Central Limit Theorem. Although any multiple of the standard error of the sample mean may be used, typically three standard errors are used. Strictly, this choice is based on the assumption of normality and provides a probability or area in each tail equal to 0.00135; this is equivalent to a level of significance α, for a two-tailed test equal to 0.0027, or approximately 0.003.

Even in cases where the process distribution is normal, the sampling distributions of the sample standard deviation and the sample range are not normal. It is conventional, however, also to construct control limits (critical values) that are three standard errors from the center line, or hypothesized value. Consequently, the level of significance is not equal to 0.003, although its value should be reasonably small.

Since the actual shapes of the distribution of the sample standard deviation and the sample range are not symmetric, the upper and lower control limits may not be equidistant from the center line in cases where subtracting three standard errors leads to a negative value. Since measures of variation cannot be negative, a default lower control limit of zero is used. The true lower limit based on a small tail area probability in such cases would be greater than zero if it were used.

Another situation in which the control limits are affected by an approximation is in cases in which the process distribution is not normal. Here, the formulas for the standard error of the sample standard deviation and range provided in Chapter 5 are based on a normal process. Again, it is conventional to use these values that are embedded within the tabulated factors in the Appendix. In general, the approximations cited do work in practice.

Strictly, based on material we have discussed so far, the concept of hypothesis testing presented earlier applies when using a control chart on a point by point basis in the real-time mode. At each point in time, a value of a sample statistic within the control limits leads to the conclusion that the value at the center line has not changed. A value of the statistic outside a control limit leads to the conclusion that the value hypothesized at the center line has changed. A statistic above the upper control limits leads to the conclusion that the process parameter designated at the center line has increased, and a statistic below the lower control limit leads to the conclusion that the process parameter designated at the center line has decreased. On a mean chart an increase or a decrease may signify trouble, whereas on a standard deviation or range chart an increase may constitute trouble and a decrease may indicate an improvement.

When a point falls outside a control limit, essentially one is rejecting the hypothesized process parameter (value at center line) at an approximate level of significance equal to 0.003. Therefore, we can say that the result is beyond what chance alone would produce and, thus something other than chance is operating on the process. An important point to note is that the implied test is related to a specific parameter, which one concludes has changed. Although there may be an effect on the quality of the final product, the out-of-control point should lead to corrective action with respect to the *process*. By using such a small level of significance, one will not be looking for problems constantly since smaller changes in the process will occur that will not be detected frequently.

In addition to a small α, generally a small sample size is used. Together, these result in little power. Recall, the power of the test is the probability of rejecting false null hypotheses. Hence, the probability of committing Type II errors (accepting false null hypotheses) is relatively high. Although processes constantly are changing, small changes generally are not addressed using the control chart approach suggested. Power, or alternatively the operating characteristic, is considered to be more useful when one is concerned about product acceptance.

When dealing with an accumulation of sample points, either in real-time or in our so-called batch mode, the hypothesis testing procedure alluded to above is not strictly applicable. It is possible, however, to apply the concept of hypothesis testing to a collection of points over time in a number of ways. Further, without use of formal testing procedures, observing time-sequenced patterns provides a basis for seeking corrective action.

Probability Limits and Other Multiples

All of the tabulated factors used to calculate control limits appearing in the Appendix incorporate a multiple of three standard errors, regardless of the statistic charted. Other multiples beside three can, of course, be used. Loosely, when smaller multiples are employed, a control chart will be more responsive to changes since the control limits are closer to the hypothesized value, or the center line. Probabilities of committing Type II errors, therefore, are reduced, and the power of the test is enhanced. Incorrect conclusions when the process has not changed will occur more frequently since the probability of a Type I error will be larger.

Actually, control limits at two standard errors have been advocated as *warning limits* to be used to begin to suspect that trouble exists in a process. Use of two standard errors under the assumption of normality provides a level of significance equal to 0.0455, or roughly 5 percent. Later we shall see how use of zones based on limits at one, two, and three standard errors can be used to assess non-random patterns.

We have indicated that use of control limits at three standard errors provides a level of significance roughly equal to 0.003 under the assumption of normality. Aside from convention, there is no reason why control limits cannot be set just as critical values are when testing hypotheses by pre-setting the level of significance at a desired level and then calculating the corresponding limits. In quality control parlance, this is referred to as the use of *exact probability limits*.

The overall procedure for establishing probability limits has been described in Chapter 5, which is concerned with hypothesis testing, and has been illustrated in terms of normal sampling distributions. Consequently, the material presented there is sufficient to establish probability limits for mean charts.

When dealing with charts for controlling variation, one must use the sampling distribution of the statistic used, which is not normal in the case of the sample standard deviation and the range. The distribution of the range is not simple and is not commonly used. Instead of the sample standard deviation, if exact probability limits are desired, a chart is constructed in terms of its square, or the sample variance. Under the assumption of process normality, the sampling distribution of the sample variance conveniently has a known and easily usable form, referred to as the Chi-square distribution. For details regarding this distribution and its applications see Gulezian [57] or Hamburg [58].

Within, Among, and Total Sums of Squares (Optional)

In simplest terms, the goal in using control charts is to determine when natural or common causes of variation are operating on a process and to leave the process alone, or to establish when special causes of variation are present and to search for the reason and correct the problem leading to such assignable causes of variation. In order to do so, it is necessary to properly measure natural variation in a process. In principle, this is relatively easy; however, it is not always simple in practice.

With more understanding of the nature of measures of variation, it is possible to get a better grasp of what one is measuring and how. Although measures are easily calculated, they do not automatically measure the right thing. Further, added understanding provides insight into the way the data should be collected or the samples or subgroups are formed.

Exhibit 10 reproduces the data used in the filling problem presented earlier. The accompanying information has been modified in order to introduce some additional points. In addition to the subgroup means and standard deviations, the subgroup variances are provided. Also provided are the mean and standard deviation for each nozzle separately over time.

The first point to be emphasized by re-examining the data is that different types of variation can be identified. If we look to all of the original observations (within the box) without regard to subgroup or

time, we can see that variation exists among all ninety observations.
When measured, we can refer to this as *total variation*.

Exhibit 10
FILLING PROBLEM DATA WITH ROW AND COLUMN
MEANS, STANDARD DEVIATIONS AND VARIANCES

Sample Number	Nozzle 1	Nozzle 2	Nozzle 3	Nozzle 4	Nozzle 5	$\bar{X}$	S	S^2
1	448.3	449.1	447.9	445.0	448.1	447.7	1.57	2.4649
2	449.6	446.6	448.4	449.5	445.4	447.9	1.85	3.4225
3	440.5	439.9	435.8	433.9	439.1	437.8	2.85	8.1225
4	435.4	440.6	437.7	438.1	438.7	438.1	1.87	3.4969
5	440.5	438.6	439.7	432.9	438.3	438.0	2.98	8.8804
6	435.2	438.1	436.8	435.6	437.9	436.7	1.31	1.7161
7	445.4	445.7	445.0	439.9	444.8	444.2	2.41	5.8081
8	441.8	445.9	443.3	442.6	443.9	443.5	1.55	2.4025
9	439.5	439.7	440.4	434.0	438.7	438.5	2.57	6.6049
10	435.8	437.5	437.2	436.6	437.0	438.8	.66	.4356
11	437.8	438.4	438.0	432.0	437.4	438.5	2.64	6.9696
12	433.6	435.3	436.1	434.8	437.8	438.3	1.23	1.5129
13	439.2	440.1	440.6	434.0	438.9	438.6	2.64	6.9696
14	435.5	440.0	437.5	437.1	438.6	437.5	1.83	3.3489
15	439.9	439.6	440.7	434.5	439.2	438.8	2.45	6.0025
16	435.8	439.6	437.5	437.4	438.4	437.5	1.53	2.3409
17	440.9	441.5	442.2	434.1	439.9	439.7	3.25	10.5625
18	437.2	439.7	439.3	437.6	440.0	438.8	1.27	1.6129
$\bar{X}$	439.6	440.9	440.2	437.2	440.1	($\bar{X} = 439.9$)		
S	4.51	3.59	3.81	4.55	3.17			

Total variation is made up of different types. That is, it is made up of
the variation *among* nozzles and the variation *within* each nozzle. We
already used the standard deviations of the subgroups in order to construct our charts. The overall variation among nozzles can be measured
by pooling the standard deviations of each subgroup, which reflects the
variability of the observations among the values in each row of the table
about the subgroup means. In a summary way, this also is evidenced in
terms of the differences among the column or individual nozzle means.

Another type of variation can be related to the variability associated
with each column, or individual nozzle. That is, the measured contents
corresponding to each nozzle vary over time. This variation is measured
in terms of the individual column standard deviations at the bottom of
the table. Again, another summary way of observing the within nozzle
variation is to observe the differences existing among the row or subgroup means.

We have been using standard deviations as one measure of variation.
Although ranges were used also, they really are used as a computational
convenience. The principles to be discussed are more meaningfully
presented in terms of components related to the standard deviation.

Recall from Chapter 3 that standard deviations are defined as the square root of the average sum of squares of a set of observations. Consequently, the sum of squares, or sum of squared deviations, from the mean, although not adjusted for the number of observations is the essential component that indicates the nature of the variation measured. Consequently, we shall discuss the different types of variation cited in terms of sums of squares.

A fundamental result taken from basic statistics relates the total sum of squares associated with the entire set of observations to the sum of squares pooled within subgroups over time and the sum of squares among subgroups. This can be written in the following form:

$$\text{Total Sum of Squares} = \frac{\text{Within Subgroup}}{\text{Sum of Squares}} + \frac{\text{Among Subgroup}}{\text{Sum of Squares}}$$

or

$$\sum_{}^{k}\sum_{}^{n}(X - \bar{\bar{X}})^2 = \sum_{}^{k}\sum_{}^{n}(X - \bar{X})^2 + n\sum_{}^{k}(\bar{X} - \bar{\bar{X}})^2$$

The symbols used are re-defined as follows:

X = individual measurements of the characteristic measured

$\bar{X}$ = mean of measurements within a singe subgroup, or sample

$\bar{\bar{X}}$ = mean of all measurements for all subgroups or the mean of subgroup means

n = sample or subgroup size, assumed equal for all sub groups

$\sum^{k}$ = summation operator indicating the sum is taken over all subgroups

$\sum^{n}$ = summation operator indicting the sum is taken over all elements in a sample

$\sum^{k}\sum^{n}$ = double summation indicating the sum is taken over all elements within a sample and among all samples, or over all observations combined.

In words, the above expression states that the total sum of squares associated with a set of observations based on repeated samples can be partitioned into two additive components where one represents the variation within the samples and the other among samples. This has been introduced earlier at the very end of Chapter 3, where actual calculations in a simple case are presented in order to illustrate the concept and to demonstrate that the equality holds.

In order to understand the nature of the variation measured by each sum of squares, one may look inside the parentheses associated with each term and examine the difference calculated. With respect to the total sum of squares on the left, the term $(X - \bar{\bar{X}})$ corresponds to a difference between any of the entire set of observations and the grand mean, or mean of all of the observations. The corresponding sum of squares, therefore, corresponds to the total amount of variation present in the data.

With respect to the within group sum of squares immediately to the right of the equality, the term $(X - \overline{X})$ corresponds to a difference between any value within a subgroup, or particular row, and the subgroup or row mean. The nature of the variation measured is the variation within a subgroup. In the case of the filling example, this represents the variation from nozzle to nozzle, or among nozzles. By using a double summation the within subgroup, or nozzle to nozzle variation, is pooled for all subgroups over time.

The among subgroup sum of squares is based on the term $(\overline{X} - \overline{\overline{X}})$ which corresponds to a difference between the mean of any subgroup and the overall, or grand, mean of all of the observations, or the mean of the subgroup means. The nature of the variation measured, therefore, is the variation from subgroup to subgroup. In the case of the filling problem, therefore, the among subgroup sum of squares is a measure of how the samples over all nozzles vary from one another over time.

Before discussing the measures further, let us show how the values are calculated, based on the data presented in Exhibit 10, in order to provide additional understanding of the meaning of the terms.

$$\text{Total Sum of Squares} = \overset{k}{\Sigma}\overset{n}{\Sigma}(X - \overline{\overline{X}})^2$$

$$= (448.3 - 439.9)^2 + \cdots + (448.1 - 439.9)^2$$

$$+ (449.6 - 439.9)^2 + \cdots + (445.4 - 439.9)^2$$

$$+ \cdots\cdots\cdots\cdots\cdots$$

$$+ (437.2 - 439.9)^2 + \cdots + (440.0 - 439.9)^2$$

$$= 1446.3$$

$$\text{Within Sum of Squares} = \overset{k}{\Sigma}\overset{n}{\Sigma}(X - \overline{X})^2$$

$$= (448.3 - 447.7)^2 + \cdots + (448.1 - 447.7)^2$$

$$+ (449.6 - 447.9)^2 + \cdots + (445.4 - 447.9)^2$$

$$+ \cdots\cdots\cdots\cdots\cdots$$

$$+ (437.2 - 438.8)^2 + \cdots + (440.0 - 438.8)^2$$

$$= 330.7$$

An alternative method of calculating the within subgroup sum of squares in terms of the subgroup variances is

$$\overset{k}{\underset{}{\Sigma}}\overset{n}{\underset{}{\Sigma}}(X - \overline{X})^2 = (n - 1)\overset{k}{\underset{}{\Sigma}}S^2$$

$$= (5 - 1)[2.4649 + \cdots + 1.6129]$$

$$= 4[82.6742]$$

$$= 330.7$$

That is, the within subgroup sum of squares is found as the common sample size minus one multiplied by the sum of the sample variances. This more clearly illustrates the nature of the pooling process, or the way the within sum of squares aggregates the variation within the samples into a single measure.

Among Sum of Squares $= n\Sigma(\overline{X} - \overline{\overline{X}})^2$

$$= (447.7 - 439.9)^2 + (447.9 - 439.9)^2 + \cdots$$

$$\cdots + (439.7 - 439.9)^2 + (438.8 - 439.9)^2$$

$$= 1115.6$$

By adding, we get

Total Sum of Squares = Within + Among

$$= 330.7 + 1115.6$$

$$= 1446.3$$

The first important point to note about the concept of sums of squares is that the within subgroup sum of squares is the basis for the value of $\overline{S}$ used to measure inherent process variability underlying the control limits for the $\overline{X}$ and S charts. This means that within sample variation provides the *basis* for measuring chance variation, or inherent variation, which can be neither explained nor decomposed any further. Consequently, the method of subgrouping, which is *not* determined automatically, is the basis for properly measuring the natural variation present in a process. The subgroups should be established in such a way that *only* variation due to *chance* can occur *within a subgroup*, or sample.

By focusing on the among subgroup sum of squares, we can make another important point. That is, if subgroups are established so that chance variation alone can operate within a subgroup, then non-chance or assignable causes of variation, when present, will affect the among subgroup sum of squares from time point to time point. This variation would be reflected in an among sum of squares that is *not* of the same order of magnitude as the within sum of squares. When chance alone is operating on a process, both within and among sum of squares should be of the same order of magnitude, since chance always will affect the among sum of squares even in the absence of special causes.

Although establishing subgroups in our filling problem across nozzles is subject to some questions that will be discussed, assume for the moment that subgrouping was done properly. On the basis of the earlier examples, we concluded that the filling process was not in control. Notice that the among subgroup sum of squares, which is calculated to be 1115.6, is not of the same order of magnitude as the within sum of squares, equal to 330.7. This discrepancy, then, tends to support the conclusions regarding lack of control observed previously from the charts.

Statistical tests are available to determine whether among and within variation are significantly different. The comparison in the aggregate does not go far enough for process control work since the process is ongoing and changes can occur repeatedly over time. The concepts were introduced, however, in order to provide insight into what is being measured when using control charts, the meaning of control and out-of-control, and how variation for control purposes should be measured. Understanding of the different types of variation lies at the heart of the process control problem.

Note. The concept of sums of squares is very powerful because sums of squares possess a property that does not exist with respect to other measures of variation; that is, they can be divided-up or partitioned into additive components attributable to different effects. Typically, however, it is not used as above in books or the literature concerned with process control. Instead, the underlying concept is considered under the umbrella of "rational subgrouping." The sum of squares concept is used here strictly as an aid in exposition in order to provide a clearer understanding of the methodology used in process control.

The concept of sums of squares is, however, the principal measure used in procedures of analysis of variance that underlie most work in experimental design, which currently is becoming popular in the field of quality control. In problems of experimental design one attempts to pre-plan the effects to be isolated and specially designs a study to isolate these effects in terms of sum of squares or components of variance. A key element is to establish a proper measure of chance error. Although similar principles operate in control charts, one uses judgment and knowledge of a process, together with continual re-trial, in order to measure chance error within subgroups properly. It should be emphasized that the way in which the components of variation are presented here is not in an experimental design context, which would involve numerous other considerations not introduced. Experimental design procedures are useful in determining factors affecting a process and product quality, and establishing process capability and reducing process variation. Detailed discussions of experimental design can be found in Box, Hunter, and Hunter [63] and Montgomery [76].

Subgrouping

To a great extent, the success of the control chart approach rests with the way in which samples or subgroups are constructed. In part, the way in which this is done is based on the particular problem and process with which one is confronted. The overriding general rule, however, is that subgroups should be defined in such a way that each is as *homogeneous* as possible *within* a subgroup such that changes in a process occur among subgroups, or from subgroup to subgroup, over time. In quality control parlance, this is referred to as *rational subgrouping*.

In order to conform to the above rule, the guide is to select items within the smallest possible time interval permissible. Further, aside from considerations of convenience, smaller sample sizes do not provide as much of an opportunity for process changes to occur within a sub group. This alone, however, does *not* insure that the variation within a subgroup is solely based on chance causes.

If, for example, we consider our filling problem presented earlier, a single subgroup was made up of one observation from each of five nozzles from one filling machine at a specific point in time. Although the practice is common, this scheme, in order to be useful, assumes that the nozzles perform similarly, or that the amounts from each nozzle are the same on the average with similar patterns of variation about the average.

Another possible basis for subgrouping in the filling problem is initially to chart each nozzle separately by taking samples of the contents of five containers from each nozzle at successive time intervals. Presumably, the variation within a singe nozzle would more reasonably be due to chance than to the variation among nozzles. The variation within nozzles, of course, still suffers from the problem that other sources of variability over time may still be included. Once it is under-stood how to obtain similar and stable output among the nozzles, one might then use a singe chart based on subgroups selected across nozzles.

A key to the subgrouping problem is to identify the potential sources of variation acting on a process and to develop a basis for understanding which must ultimately be placed under surveillance. Separate charts initially may be used for individual machines, operators, shifts within a day, measuring instruments, and possibly different positions on the same piece of product or output. At a later point, some of the different sources of variation may be condensed within a singe subgroup when it becomes clear how their effects can be controlled or identified on an individual chart.

We have stated that small subgroup sizes are desirable in order to obtain subgroups that are as homogeneous as possible and so charts are not sensitive to each and every unimportant change in a process. When hand computation is used to develop control charts, small samples also are desirable and generally are preferable when using the sample range and the sample median. Recommended sample sizes range between 3 and 5.

To some extent, the degree to which the sensitivity of a chart to pick up changes is important depends on the effect of changes on meeting specifications. If small changes affect the capability of a process, larger sample sizes may be required. Other types of charts that will be cited later also may be used instead.

It is difficult to provide a general rule for determining the frequency with which subgroups are observed; this really depends on charac-teristics of an individual process. When beginning a control chart program, more frequent samplings are desirable until an understanding of the process is secured. Once a better understanding of a process and its problems are understood through charting, a wider sampling interval can be considered.

Choosing Characteristics for Charting

One of the reasons for using the filling problem as an example was that it is relatively simple to understand, it contains features that are useful to highlight certain subsequent points, and it is associated with a relatively self-contained process unaffected by other processes associated with the final delivered product. Due to the problem's simplicity, it is readily apparent that the weight of the contents of filled containers is the variable requiring measurement and control.

The problem of choosing characteristics for controlling a process is not always as simple. In some cases, one may try different ones or make changes from time to time as experience with charting a process grows. Note that in the filling problem we addressed, one actually has a choice between the weight of the container and its contents or the weight of the contents alone. The latter choice assumes that the contents of the sampled containers can be removed and weighed separately.

If the weight of filled containers were used, recognize that two sources of inherent variation that are additive are involved: the weights of the containers and the weights of the contents dispensed by the filling machine. In order to manage overfill and also meet the label claim on the contents, it would not be possible to separate the two sources of variation unless the container weights were known within specified limits, or, preferably, known to vary according to a stable distribution. This could be accomplished by inspection or by control applied to the supplier's process.

In more complicated cases, it might be necessary to search for the characteristics needed to control a process. Reasonable candidates are any variables relating to the reduction or elimination of production costs. These could be related to raw materials, equipment, operators, environmental conditions, or measuring instruments. The ones with the most realizable initial control potential are those associated with adjustments to a process that are readily made by individual operators. Characteristics of the final product used as a basis for final inspection also are good candidates as control variables.

Measurement, Control, Capability, and Process Improvement

Up to this point we have introduced procedures for constructing basic control charts for variables, illustrated the procedures, and provided some insight into the way they are interpreted and applied. Before going further, it is beneficial to summarize the essential features of a statistical process control program in terms of what we have done and what we have yet to do.

There are four key features relating to a complete statistical process control program:

1. Measurement

2. Control

3. Capability

4. Improvement

In order to know anything about a process it is necessary to measure it in terms of one or more key characteristics. Without measurement it is not possible to know how the process is performing nor when to correct it, or when it has been corrected. Although the term control is used in a number of ways, process control in terms of control charts involves monitoring key process and product characteristics primarily to identify and eliminate special causes of variation.

Once special causes of variation are managed and controlled, it is possible to establish the capability of a process that loosely can be defined as the extent to which the process is able to meet product specifications. Various ways of demonstrating capability are available, and these are discussed in detail in Chapter 9. At this point, there are two important points to note about capability and control. First, the charting techniques introduced in this chapter intentionally were separated from the capability problem. This was done in order to focus on the idea that the charts primarily are concerned with the process. This does not mean that the product is unaffected. If done properly, the impact of controlling the process will be demonstrated on the product either in terms of quality or cost, or both.

Second, once a charting system has been effectively implemented, it is possible to use control chart information to determine the capability of a process. In many instances, either with a known amount of inherent variation or by reducing process variation, it is possible for a process to go out of control temporarily without affecting the defect status of the final product.

Once targets are properly established and maintained and capability is demonstrated, it is possible to work on improvement of the process by reducing variation. This can be accomplished in terms of small changes to the process or in terms of changes in the underlying system and/or product design.

The Filling Process Problem

The main reason for introducing the filling problem was to illustrate the construction of variables control charts. Actually solutions to two interrelated questions also result from the illustrations. On the one hand, we have developed a basis for answering the question, How do you solve the filling problem? On the other, we have the question, How do you control the filling process?

The filling problem basically is one of filling containers such that the weight does not go below the label limit and that overfill is minimized. So far, we have described the behavior of the process with the control charts but have said nothing about the problem. If you examine all of the individual weights in Exhibit 1, you will find that none fall under the label limit of 425 grams. Once the points relating to over-and-under adjustment were eliminated, we saw that the process tended to operate at an average level below the initial target. Despite this, the label limit was met since the variation in the process happened to be less than the original target.

Once it is recognized that process adjustment is the main problem and that the variation in the process is stable, it is possible to establish new targets that not only are realistic but will lead to a reduction in overfill while meeting the label claim with reasonable certainty. At this point, control charts can be used to determine when and when not to adjust the process, or to control the filling process at the realistic targets on a continual basis.

If it is desired to reduce overfill further, it is necessary to reduce the inherent variation in the process. The extent to which this is done will depend on the benefit in reducing overfill to the cost of making changes to the process. The most drastic change would be to install new equipment; however, the problem might be solved more simply. Experimentation with nozzle design, the size of the opening in the nozzle, and flow rates may lead to reduced variation at a much lower cost.

Basic Steps and Rules

Before closing this chapter it is useful to provide a summary of the main steps and activities relating to implementation of a control chart program. Some of the items listed are rather obvious while others require knowledge of a particular process. The list does serve as a guide that can be followed and that provides some perspective for the material presented in this entire module. It is assumed for this purpose that control limits will be developed on the basis of process data rather than pre-established targets, since in the latter case modifications to the list are obvious.

A. INITIAL CONSIDERATIONS

1. Detail the process to be studied or controlled
2. Delineate objectives to be satisfied by using control charts
3. Select the initial variable(s) to be charted
4. Establish the method of subgrouping
5. Establish sample size and timing of selection
6. Establish data collection forms, which also allow for comments regarding production changes for hunting for assignable causes
7. Establish individual responsibility for data collection, charting, and corrective action

B. INITIAL CHARTS

1. Decide between $\overline{X}$ and R charts and $\overline{X}$ and S charts. When charts are constructed manually, $\overline{X}$ and R charts are preferable. When charting is automated (and when sample sizes are large) $\overline{X}$ and

S charts are preferred. Also, experience gained from manual charting is useful when beginning a charting operation.

2. Although fewer subgroups can be used, a generally accepted rule is to collect 25 samples before constructing an initial chart with trial control limits.

3. Collect and record data, calculate center lines and control limits, construct charts and plot points

C. DRAW INITIAL CONCLUSIONS

1. Determine control or lack of control

2. Establish what the process is doing relative to what it is supposed to do

3. Take appropriate action

D. CONTINUED USE OF CHARTS

1. Recalculate center line and control limits based on process data with special causes removed

2. Use charts for action on the process on a point-by-point basis in real-time and in terms of pattern buildup

3. Use charts as input to determination regarding specifications (ie., in terms of process capability and decision to revise specifications)

4. Use charts as a basis for possible improvement of the process

INTERPRETING VARIABLES CHARTS

The preceding chapters provide a statistical basis for process control charts, procedures for constructing the principle variables charts, and the basic reasoning for interpreting charted results. Briefly, based on a rational basis of subgrouping, we attempt to isolate purely random variation within subgroups such that non-random or controllable causes, if present, can be detected over and above random variation among subgroups. Broadly speaking, control charts are useful in the following three ways:

1. To identify and solve problems in a process on a one-time basis
2. To determine whether a process is doing what it is supposed to do in terms of aimed-at targets on an on-going basis
3. To establish what a process is doing or what is happening in a process

Ultimately, an up-and-running control chart scheme should be developed in order to determine continually in real-time whether a process is doing what it should be or leading to the elimination of problems when they arise. When used in this way, assessments can be made on a point-by-point basis by comparing each sampled observation with the control limits and reaching a conclusion whether to look for a problem or not. When a single point goes outside a control limit, most often it is reasonable to conclude that an underlying assignable cause was the result. When a single point falls within the control limits, however, basically one concludes that there is insufficient evidence that a change in the process has occurred, even in cases where a change may have occurred.

Obviously, a single result does not provide as much information as an accumulation of points over time. In the latter instance, one has the opportunity for a pattern buildup that can tell more about what has happened in the process. This also would apply when obtaining data from the process on which control limits are calculated in order to solve a specific problem or to determine what a process is doing in terms of estimates of the center lines.

Assuming an accumulation of points is available either in real-time or in the so-called batch mode, the general rules relating to the use of control charts are as follows:

A. Conclude that a process is *in control* or that natural causes of variation are present in the process when:

1. Most of the points fall within the control limits
2. Points are reasonably scattered about the center line with fewer points near the control limits
3. There are no discernable patterns or runs throughout the chart

B. Conclude that a process is *not in control* or that special, or assignable, causes of variation are present in the process when:

1. Points fall outside the control limits
2. Points are not reasonably scattered, such that they tend to be close to the control limits or hug close to the center line
3. Discernable patterns or runs of points appear in the chart

Although the rules apply to both mean and S or R charts, it should be kept in mind that interpretations on an $\overline{X}$ chart should not be made *until* the S or R chart is in control.

With the above general rules in mind, this chapter serves two basic purposes. On the one hand, it provides an understanding of the nature of out of control patterns and ways of interpreting charts. On the other hand, it serves as a reference guide so that patterns in a particular process can more or less be matched as a beginning point for hunting for trouble. Recognize that a particular type of pattern can be the result of numerous causes and that a chart pattern will not indicate the exact nature of the problem in all cases until one develops experience with specific patterns as they relate to a particular process.

BASIC QUESTIONS

Typically, various control chart patterns are presented and then related back to potential causes within a process. Since it is impossible to relate a particular pattern to all specific causes, a better understanding of the meaning of control patterns can be acquired by understanding the basis for the patterns. One then is in a position to assess a situation in terms of the specifics associated with the process control scheme and knowledge of a particular process. It is helpful to consider this in terms of the following five questions:

1. What is the process like at a particular point in time, and what can happen to the process over time?
2. What type of variation is observed at a particular point in time and over time in terms of the method of subgrouping?
3. How is the center line determined, and how are the control limits calculated?
4. What can be concluded about the process mean and standard deviation at a particular point in time and over time on the basis of control chart information?
5. What problems can be detected in the process at various points in time?

By keeping each of these questions in mind when assessing control chart data, one is in a better position to assess information provided by various control chart patterns. We shall discuss each of the items in turn in order to place later material in its proper perspective.

The Nature of A Process

Whenever we address the nature of a process we do so in terms of the *process distribution*. Therefore, when understanding what a process can be like, we can think in terms of the nature of the process distribution at a particular point in time and what happens to the distribution over time. Although we do not observe the theoretical distribution, let us think of an underlying probability distribution that is the basis for generating measurements associated with each quality characteristic. By considering the different forms of this distribution at a point in time and changes over time, we can provide an underlying structure for interpreting control chart patterns.

Exhibit 1 presents examples of the possible forms of process distributions for processes that are homogeneous. We can define a **homogeneous process** as one for which there exists a single system of random variation at a particular point in time. In other words, at a particular point in time, measurements from a homogeneous process are generated from a *single* underlying probability distribution. For example, the measurements of a particular dimension of items produced by a single machine that is operating properly should possess an underlying distribution that is single-peaked and tails-off in either direction of the peak. Depending on the characteristic measured, the distribution could be near symmetric or skewed.

By way of contrast, if the same item is produced on two machines based on separate underlying distributions, say, by differing in terms of the mean or the standard deviation (or shape), the combined output of the process of the two machines would be *heterogeneous*. If the distributions are the same (although there are two machines) we also would refer to the output of the process as homogeneous.

EXHIBIT 1
POSSIBLE FORMS OF UNDERLYING PROCESS PROBABILITY DISTRIBUTIONS
FOR A HOMOGENEOUS PROCESS AT A PARTICULAR POINT IN TIME

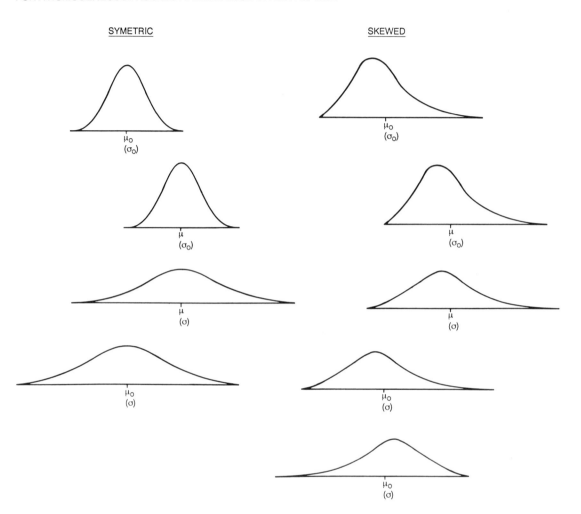

SYMETRIC

μ_0
(σ_0)

μ
(σ_0)

μ
(σ)

μ_0
(σ)

SKEWED

μ_0
(σ_0)

μ
(σ_0)

μ
(σ)

μ_0
(σ)

μ_0
(σ)

Ideally, a process should be homogeneous with respect to every relevant quality characteristic such that the distribution is centered on a targeted or specified mean and the standard deviation either equals or is less than a targeted value. Implied, therefore, is the fact that a process can be homogeneous *but* not on target. In other words, a process with an underlying distribution that is homogeneous at a particular point in time is in a natural state of statistical control about the process mean and standard deviation. When these values equal targeted values, the process is in control about the aimed-at values.

The diagram at the top-most upper left portion of Exhibit 1 depicts a distribution of a homogeneous process operating about the targets μ_0 and σ_0. The remaining three diagrams beneath it also depict homogeneous process distributions; however, these are operating about either a process mean, μ, or standard deviation, σ, or both, which do not equal the target. The diagrams cited all correspond to symmetric distributions, which are reasonable to assume in many quality problems. The right side of the exhibit depicts the same phenomena for skewed distributions in order to provide a more complete picture. We shall, however, spend most of our time concentrating on the symmetric case. Much of what we say, however, applies to both situations.

Process distributions can be non-homogeneous, or heterogeneous, at a particular point in time. In other words, two or more distinct distributions could be operating to generate measurements at any time. This would mean that separate systems of random variation are present. **Exhibit 2** depicts various examples of heterogeneous processes. Consider the clearest case, which is depicted in the upper left portion of the exhibit. In this case, the overall distribution is bimodal, resulting from two separate distributions of similar shape with different means.

As an example, the bimodal distribution could result from two machines producing a similar item, but each machine is adjusted at different levels. The bimodal shape also could result from two machines with identical operating performance characteristics, that are used differently by two operators or are utilizing raw materials from separate vendors that cause variation to exist between the two machines. Actually, such a distribution will result when any second causal system enters a homogeneous process at a particular point in time.

When the underlying distribution is bimodal, it is easily seen that measurements are generated about a mean that is somewhere between the means of the individual distributions comprising the overall distribution. Further, the variability of the measurements is somewhat larger than that which would result from each of the individual contributing distributions.

The example cited deals with a heterogeneous process for which the individual component distributions appear somewhat distinct. Assuming similar variability and shape, as the centers of the distributions become closer, the identities of the individual distributions tend to blur. As indicated, when the means equal one another, the result is a single underlying distribution corresponding to a homogeneous process.

Another possible way in which a process can be heterogeneous based on two underlying systems of variation is in terms of similarly shaped distributions with equal means but unequal variances. This situation is depicted in the two diagrams appearing near the bottom left portion of Exhibit 2. The resulting overall mixed distribution can have the appearance of a homogeneous distribution but will not necessarily have the same functional form, or shape, of either of the two individual contributing distributions. Measurements generated from a corresponding process will have a common mean, but will have a standard deviation somewhere between the standard deviations of the individual component distributions.

Obviously, based on the above observations, it is easy to imagine processes based on a mixture of two distributions such that the means, variances, and shapes of the individual component distributions all are different. This is depicted at the top right portion of the exhibit. The

EXHIBIT 2
EXAMPLES OF UNDERLYING DISTRIBUTIONS FOR A HETEROGENEOUS
PROCESS AT A SINGLE POINT IN TIME

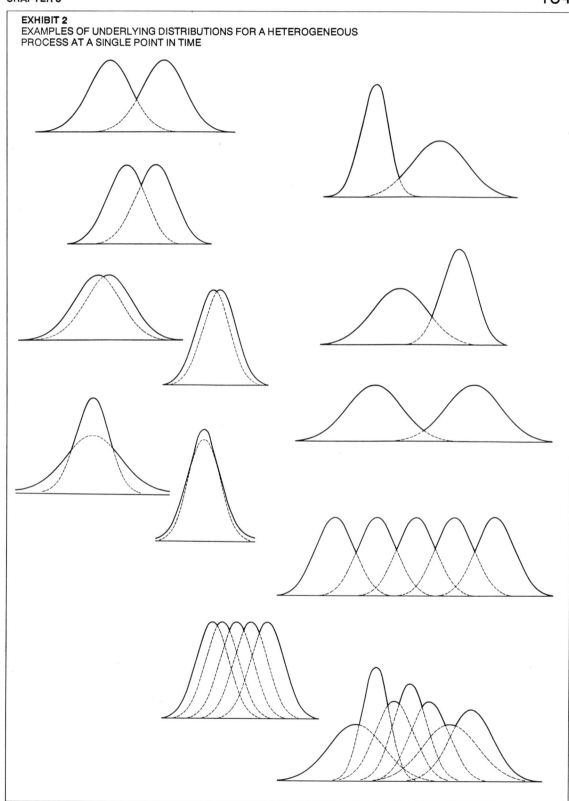

same type of considerations described above apply in these cases and just vary in degree.

At the bottom of Exhibit 2 three diagrams are presented that are selected examples of heterogeneous process distributions composed of more than two individual component distributions. The first of these depicts a heterogeneous distribution made up of five component distributions, each of which maintains some of its identity. This could represent, for example, five machines producing a similar item that is part of a common output. The machines produce measurements that are similarly distributed in terms of shape and variability but different in mean levels. Such a distribution applies to our filling process problem of the previous chapter, where each component distribution corresponds to one of the five nozzles. Conceptually, it is a simple matter to extend the considerations discussed for a mixture of two component distributions to any number including the one with five.

The remaining situations depicted in Exhibit 2 display cases where the component distributions are more diffuse either in terms of proximity of their centers or in terms of differences among centers, variation, and shape. The larger the number and the more varied are the component distributions the more difficult it is to generally describe the properties of the overall distribution actually generating measurements.

When thinking about the nature of a process distribution, keep in mind that it can be homogeneous or heterogeneous at any point in time. When heterogeneous, any number of possibilities exist in terms of the number of contributing component distributions, and the shape, center, and spread of these components. The properties of the overall distribution in terms of mean and standard deviation depend on those of the component distributions. Although charts can be and are constructed on the basis of heterogeneous processes, they really *should* be based on homogeneous process distributions in order to insure that natural variation is measured properly.

The discussion regarding homogeneous and heterogeneous process distributions relates to possible states or the nature of a process at any particular point in time. The concept of a process distribution at a point in time really is a theoretical construct only, since no such thing as an empirical frequency distribution exists at any instant. The process, however, does have a certain status at a point in time that would be the basis for a specific underlying distribution if we knew it.

The next thing necessary to understand the nature of a process is to visualize what can happen to the process distribution when changes in the process occur over time, or from one time point to another. When doing so, it is very important to keep the distinction between a homogeneous and a heterogeneous process in mind. Recall that ideally we should strive to maintain a homogeneous process over time. In such a case, a change in the process will affect some aspect of the process distribution, which can be identified as a single source of special or non-random variation.

When a process is heterogeneous and in the absence of so-called special causes, a source of controllable variation potentially is present that may or may not change over time; one hopes to detect and correct special causes when and if they occur over time. If the basis for the underlying heterogeneity is not eliminated, which in some cases may not be possible, the control problem either can be more difficult or surrounded by greater variation. Consequently, we shall not provide a complete picture of the nature of changes to a process over time. We shall, however, treat the homogeneous case in some detail, which also can be used as a basis for visualizing what can happen in the other cases.

Exhibit 3 provides examples of ways in which a process distribution can change over time, where it is assumed that time passes going down the page. Each half of the exhibit can be considered separately, where each of the dotted rectangles can be used as a reference point with respect to centering and variation.

Appearing at the top left portion of the exhibit are four similar distributions, which also fall within the dotted rectangle. These four distributions characterize a *stable process* over time for which the mean, variation, and shape all remain the same. We shall assume that the amount of variability depicted is due to a single system of chance causes.

By continuing downward on the left portion of the exhibit, the next four distributions have the same shape but are steadily moving outward, indicating a *trend* in the mean of the process. This trend pattern of change continues into the next three distributions, in addition to which there exists an increasing trend in the variability. This is evidenced by the continued widening of the three distributions. A similar increasing trend pattern in variation is depicted in the last four diagrams; in this case, the means are stable, or remain the same. Although it will be addressed at a later point, it is worth noting that if the trend in variability were reversed (depicted as going from the bottom up the page), with constant mean, the pattern would be associated with a process improvement.

The changes just described are in terms of steady and directed changes in the mean or the standard deviation. Other types of changes can occur and are depicted on the right half of Exhibit 3. Again, we start with a stable process depicted by the first three distributions within the dotted box. Instead of a gradual change outward, the next three distributions are an indication of a *sudden sustained shift* in the mean of the process, with no change in variation. Such a change could occur owing to a one-time shift out of a simple adjustment. By observing the next three distributions, which are the same as the stable starting point, we can see what would happen if the process were correctly re-adjusted.

Continuing down the right half of the exhibit, the next thing that is depicted is a sudden sustained shift in variability, shown in the fourth set of three distributions. This shift is evidenced in terms of the three wider distributions all with the same shape and location. The remaining six distributions display a sustained re-adjustment of variation back to the original stable starting point, followed by a sustained sudden shift in mean and variability. Notice, in the latter case not only have the distributions moved to the same location on the left but they all are consistently wider.

Additional types of changes can be described. For example, if one considers the first nine distributions on the right half of Exhibit 3 together, an *oscillatory* pattern in the mean occurs since the mean

EXHIBIT 3
EXAMPLES OF CHANGES IN PROCESS MEAN AND STANDARD
DEVIATION FOR A PROCESS WITH A SINGLE SYMMETRIC
DISTRIBUTION

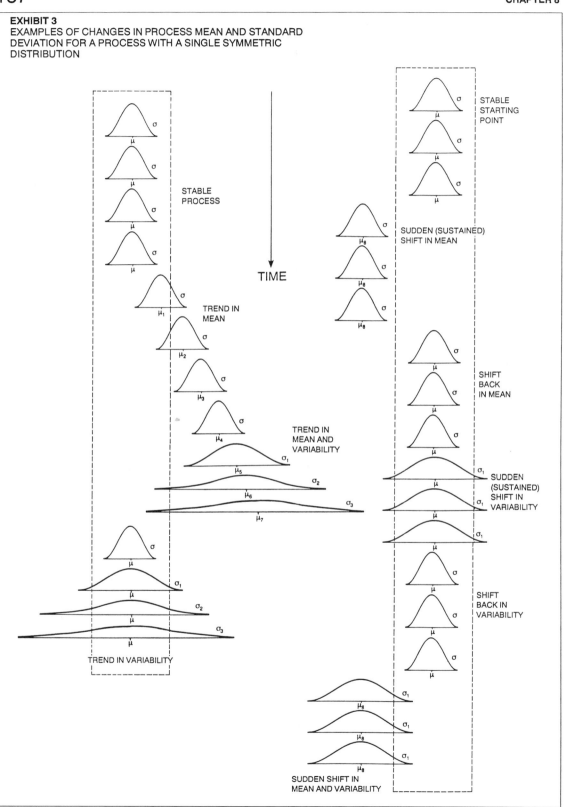

begins at a particular level, moves to another, and returns to the original level. If the mean moves strictly back and forth, the change is referred to as *systematic*. If the movement is not as strict, the change is referred to as *cyclical*. Similar changes can occur with respect to the standard deviation, which is depicted in the last nine distributions, although the last three of these also are accompanied by a sudden shift in the mean.

Patterns of change displayed in the exhibit are in terms of symmetric distributions and are somewhat clear and well defined. Further, they are in terms of either the mean or the standard deviation. The changes described can, of course, occur less regularly or in combinations. Further, changes can occur in the overall shape of the distribution. Similar patterns of change also can occur with respect to skewed process distributions similar to those presented in Exhibit 1.

It should be readily apparent that changes in processes that initially are stable but heterogeneous can occur as well. Depicting these changes in a comprehensive way is more difficult and involved. An easier way to understand this is to re-examine the diagrams presented in Exhibit 2 and imagine all of the possibilities just discussed occurring in combination with each of the individual component distributions comprising the overall heterogeneous distribution. Later, we shall see how an initially heterogeneous process is displayed on a control chart and how changes can be observed. Although such patterns can and do occur, sources of heterogeneity really can be considered as special causes of variation. Consequently, efforts should be directed toward charting homogeneous process characteristics.

Methods of Sampling or Subgrouping

The second relevant question posed in order to understand control chart patterns is in terms of the method of subgrouping. This can affect the values of the center line and the control limits if these are based on process data. Also, the nature of the subgroup can influence what is actually measured and lead to varying patterns.

In order to understand the points to be made, one should constantly be mindful of the discussion of the different types of variation presented at the end of the previous chapter. Whether using process data to construct control limits or to monitor a process in order to detect and eliminate assignable causes, each individual sample or subgroup ideally or ultimately should capture inherent random variation only within each subgroup. On the one hand, if this is accomplished, it is possible to isolate effects that occur among subgroups. On the other hand, individual subgroups may contain controllable or non-random sources of variation within them. Essentially, it is helpful to understand what the subgrouping scheme is designed to measure and detect before one answers questions regarding what is detected.

Control charts mainly are to be used to detect departures in the process due to all non-random causes of variation. Therefore, ideally, when limits or measures of natural variation are established, they should be based on process data corresponding to homogeneous process characteristics from a stable process similar to the one depicted in the upper left portion of Exhibit 3. Whether for constructing limits or for process monitoring, subgroups should, but not always are constructed to, measure random variation within subgroups.

This is easily seen by re-examining Exhibit 3. The key idea to keep in mind is that observations comprising a single subgroup should be taken from a process associated with one of the distributions depicted at a time, rather than across distributions which are different. By doing so, differences among the distributions, or changes to the process, can be detected among subgroups.

Actually, when samples are taken from homogeneous processes which are stable, it doesn't matter how one samples. However, if samples from two or more underlying distributions are used where a change occurs among them, the change will be reflected within the sample. For example, if the frequency or timing of subgrouping is defined such that a sample is selected from a process comprising pairs of the distributions associated with a trend in mean (Exhibit 3), the true change in mean will be understated, and will not be detected as quickly. The measure of subgroup error will reflect the difference among distributions, or the trend effect, and could be greater than the natural amount of variation. Consequently, anything is possible on an S or an R chart.

The preceding comments are based on homogeneous process characteristics, and correspond to the way sampling ultimately should be undertaken. Now consider what happens when samples are taken from heterogeneous processes with distributions at a point in time as depicted in Exhibit 2. If sampled observations are selected uniformly across an entire distribution, the mean will fall in the center and not reflect the individual component distributions when their means differ, and the variation will be larger than that corresponding to the individual distributions. If sampling is done disproportionately across the entire distribution, the individual component distribution corresponding to the larger number of observations will dominate the measures calculated. Changes in proportionality will result in changes in the influence of the underlying process mean and standard deviation on the sample measures observed.

The observations made about heterogeneous processes based on Exhibit 2 apply at a particular point in time and the observations over time if the underlying process distribution remains stable. When changes occur to such processes, the changes will affect the subgroup measures as described earlier but in combination with the effects of heterogeneity and the weightings, or proportionality factors, associated with the individual component distributions.

Methods of Establishing Center Line and Limits

Recall that the center line and limits on a control chart can be based on standards, or targets, or on process data. If the targets are not based on the process data, interpreting the patterns in a control chart truly are in terms of what the process and data are doing relative to the target quantities. The considerations relating to the nature of the process and subgrouping relate only to the sampled observations generated and not to the center line and control limits.

When the center line and control limits are determined from process data, these will be affected by the nature of the process and the method of subgrouping in addition to the patterns observed. Consequently, in such cases one must consider patterns in terms of the way in which the limits were calculated. Ultimately, one should strive toward establishing the center line and limits from stable homogeneous process characteristics.

As an example, observe the first and second groups of four distributions on the left side of Exhibit 3. The first four comprise a homogeneous and stable set of distributions, whereas each of the second four are homogeneous, however, there is an upward trend in the mean; the variation is constant or stable.

First, consider control limits based on the first group of four distributions. Whether subgroups are taken from each individual distribution or from all together, control limits on the mean and the standard deviation chart would correctly reflect what they are supposed to since the process is homogeneous and stable. Based on these limits, if the process changed according to the second set of four distributions and subgroups were taken from each individually, the trend in means would be detected on the mean chart and the standard deviation chart correctly would be in control.

Now consider a process that is composed of the second group of four distributions depicted in Exhibit 3, and imagine the trend pattern *repeating* over time. The effect over time would be an oscillatory pattern with respect to the process mean with stable variation.

Consider two methods of subgrouping: (1) subgroups are taken from each individual distribution, and (2) subgroups are equally represented by the four distributions comprising the so-called trend component. In the first case, the center line on the mean chart will fall somewhere in the middle of the four distributions, which lies above the correct value. Since the variation is measured correctly, the control limits on the mean chart will be of correct width about the center line. The center line and control limits on the standard deviation chart will be correct. When points are plotted on the corresponding charts, the oscillatory pattern in the means will be evident about the higher center line, and the standard deviation chart correctly will demonstrate control.

If, instead, the second subgrouping method is employed, the center line on the mean chart will be about the same; that is, somewhere in the middle of the four distributions. The measure of variation, however, will be larger since it will contain natural variation and the variation among the four distribution means. Consequently, the control limits on the mean chart will be unnecessarily wider. The center line on the standard deviation chart will be higher, and the corresponding limits will be inappropriately wider.

In such a case, the standard deviation chart will appear to be in control and the mean chart will exhibit points closer to the center line and without the correct oscillatory pattern. If the process is supposed to behave like the first four distributions in the exhibit, no indications on the charts will be evident; the mean level and natural variation in the process will be overstated.

In general, the point to be made on the basis of this example is that what is going on in the process and the method of subgrouping affect the values of the center lines, the control limits, and the patterns observed. Obviously, many different configurations can result under the varying types of circumstances described earlier.

Conclusions about Process Characteristics

Once one understands the different forms taken by a process distribution and how it can change over time, the effect and the nature of subgrouping and the control limits, one is in a better position to interpret and use the results of control charting. Remember that we never know fully what is happening to a process distribution. We observe sample estimates, which we use to draw conclusions about central tendency and dispersion of one or more specific process characteristics. The behavior of the sample estimates will reflect what is happening to the process parameters but will be affected by the nature of a particular problem and the way data are collected. Whether the characteristic chosen is useful to solve process problems can be determined by experience.

Conclusions about Process Problems

This and the previous segment concerned with conclusions are introduced in order to emphasize a couple of key points. First, when using control charts to control a process and solve process problems, it is necessary to establish characteristics or variables associated with the process or product that may reveal the existence of problems in the process through changes in the center and dispersion of the corresponding distributions. Second, by concluding that changes occur in the process parameters one may conclude that a problem exists, but not necessarily determine the problem. Control chart information does not automatically identify the problem in the process. One should distinguish between what is happening to the data and what it says about what is happening to the process.

In some cases, the process problem may be obvious, which was the case in the filling problem. This process is simple and self-contained. In other more complicated cases the problem may not be obvious and may require a substantial amount of searching to detect it. For example, in a complicated assembly operation, a control chart at a particular operator location may indicate the existence of a problem. The problem could be due to an adjustment made by the operator or it could be due to problems with raw materials, intermediate parts, or operations long preceding the particular operation.

Over time, it is possible to accumulate knowledge regarding specific types of patterns on $\overline{X}$ or S and R charts and how they relate to particular types of problems in a particular process. Since no two processes are exactly alike, the more knowledge one has about a particular process, the better one is able to detect effectively problems in a process using control chart information. A general guide to types of process problems relating to various control chart patterns can be found in the *Statistical Quality Control Handbook* by the Western Electric Co. [16].

CONTROL CHART PATTERNS

In this section, we shall introduce some examples of common control chart patterns. In all cases, the patterns are somewhat distinct in order to illustrate a specific point. When dealing with actual applications, more than one effect may be operating at the same time and to varying degrees, which may tend to obscure the individual effects.

Our entire discussion is somewhat general since it corresponds to any characteristic of any process. In other words, a given process may be very simple, such as a single machining operation producing a simple part, or the process could be defined in terms of numerous machines, operators, and inspectors producing a complicated assembled part. Although numerous variables may be used to monitor and control a process, we shall consider the charts as if they are related to a single variable treated individually. Ideally, of course, the more simply an individual process is defined for control purposes, the more easily one is able to use control chart information to solve process problems.

When dealing with specific process control problems, it is helpful to keep the five basic questions in mind since the answer to each contributes to controlling effectively a particular process.

It should be noted that the patterns depicted here are no different than ones appearing in numerous documents and texts concerned with process control. Although all of the ideas presented are the same, there is a tendency to be too general and not to define essential concepts precisely. Of key importance is the distinction among the terms stable, control, homogeneous, and change in a process.

We have introduced the term homogeneous specifically in reference to a distribution of a process variable at a particular point, or cross section, in time, which is the result of a single system of chance causes. Typically, this is evidenced as a single-peaked distribution similar to ones depicted in Exhibit 1.

By way of contrast, a process distribution is heterogeneous if more than one system of chance causes are present at a point in time. Typically, this would be evidenced in terms of a multi-peaked, or multi-modal, distribution made up of two or more individual component distributions; the peaks, however, could be obscured if the component distributions are "close" together and tend to overlap or mix greatly. Heterogeneity in variability is less clear-cut.

We can think of a process as being *inherently* heterogeneous in terms of the way it is defined, or it may be heterogeneous as a result of the introduction of an *external* set of chance causes in addition to the basic inherent set. For example, if a process is defined as two machines producing an identical item such that the distributions of a key characteristic are different for each machine, yet the combined output is charted, this can be defined as an inherently heterogeneous process. This is not to say that the process should be defined in this way, but initially it may be for charting purposes.

On the other hand, if we chart a characteristic of a single machine that is in control with no controllable sources of variation present, the defined process initially can be defined as homogeneous. If, however, a portion of raw materials from an alternative manufacturer is used in addition to those existing portions such that the output is affected differently, the process would be heterogeneous due to an assignable cause and would not be inherently heterogeneous. In such a case, the underlying combined process distribution would be bimodal, but the resulting control chart pattern would be different than one corresponding to the case where two differently operating machines represent the process charted.

Previously we defined a process to be in control if the same underlying probability distribution operated on the process over time. That is, we stated that the process is in control if the underlying distribution is stable over time. Implied in this definition is the fact that a process in control must be stable *and* homogeneous. In other words, an inherently heterogeneous process may be stable over time but not in control since it is possible to isolate, or control, the source of heterogeneity.

Aside from the numerous uses of control charts, the primary application is to document the existence of changes to the process, the source of which can be identified and eliminated. A *change* in a process, or its distribution, can occur in one of two fundamental ways:

1. It can occur as a result of an effect that changes the entire overall process distribution either in terms of its mean, variance, shape, or all properties.
2. It can result from the introduction of an added, or external, element of heterogeneity.

Obviously, in order for a control chart to be effective, it must detect changes in a process. Consequently, a control chart pattern will change when a change occurs in a process. In addition to the two types of process changes listed above, control chart *patterns* can change for two additional reasons in cases where the process is heterogeneous:

3. A change in representation occurs in terms of the relative number of items produced by the individual component distributions.
4. A change in representation occurs in the relative number sampled from each of the individual component distributions.

Along the same lines, there is one further distinction that is important to make. The term *mixture* is used frequently both in statistics and quality control applications. On the one hand, the term is used in conjunction with an underlying distribution that is heterogeneous. Therefore, an overall process distribution can be made up of a mixture of any number of individual component distributions. The mixture can be the result of a process that is inherently heterogeneous as defined earlier or it can be heterogeneous as a result of an external source of heterogeneity. In either case, the result of the mixture in the process will be displayed on a control chart. In actuality, many of the control chart patterns illustrated can be attributed to mixtures in process distributions.

There is a tendency, however, within the process control literature to be unclear about a process mixture and a mixture pattern on a control chart. Typically, mixture patterns are presented separately and are related to what we have defined here as an inherently heterogeneous process. Consequently, we shall adhere to this convention and discuss mixture patterns, as such, separately. The key point to keep in mind, however, is that other control chart patterns can be attributed to mixtures in a process resulting from other sources, and that here we have distinguished between processes that are inherently heterogeneous and ones that emanate from other sources.

Various control chart patterns are presented without reference to the type of chart, $\bar{X}$ or S and R chart, the sampling interval, the method of subgrouping, or the manner in which the limits are established. The patterns can arise with respect to charts for central tendency or dispersion and can truly be interpreted in terms of specific problems in terms of actual processes or applications. However, the types of problems can be discussed in terms of the concepts initially presented. It should be kept in mind that when an $\bar{X}$ and an S or R chart are considered together, the S or R chart should be interpreted first in most cases. This results from the fact that limits of an $\bar{X}$ chart are dependent on the corresponding measure of variation and the fact that less meaning can be attached to process centering if the variation in the process is not stable.

Natural Patterns

A natural pattern of variation is one for which uncontrollable chance forces operate on the process; no correctable assignable causes are present. Natural variation in the sample mean, standard deviation, or range is observed in terms of the absence of any recognizable pattern in a sample statistic over a series of plotted points. Natural variation is associated with a stable homogeneous process.

Exhibit 4 provides an example of a natural pattern of variation that also is in control. The example is the same as the simulated filling process problem introduced in Chapter 7.

EXHIBIT 4
EXAMPLE OF A STABLE PROCESS IN CONTROL WITH RESPECT
TO TARGET OR LIMITS BASED ON PROCESS DATA

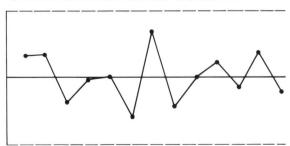

In general, patterns like the one presented in the exhibit are in control in the sense that the points fluctuate about the center line and are within the control limits. If the center line and the control limits are based on process data, we can say that the process naturally is in control.

If, on the other hand, the center line and limits are based on standards or targets, the process would naturally vary at random and would be in control with respect to the specified targets.

The same pattern of variation has been reproduced in **Exhibit 5**; however, the points consistently have been shifted upward so that they do not fluctuate about the center line, and some points fall outside the control limits. This was introduced to make the point that a process can be naturally stable and homogeneous, but not in control with respect to the established center line and control limits.

Exhibit 5
EXAMPLE OF A NATURALLY STABLE PROCESS NOT IN
CONTROL WITH RESPECT TO TARGET

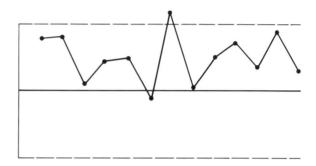

If the pattern observed in Exhibit 5 corresponds to a mean chart, it means that the center of the process distribution is higher than it should be. If the chart represents an initial attempt at charting, a basic question to ask is whether changes in the process can be made to realistically lower the process average to the targeted value. On the contrary, if the process previously was in control about the target, the pattern would constitute a change in the process average that might be corrected with a simple adjustment. Obviously, the change could result from numerous other sources consistently relating to different materials, inspectors, operators, or set-ups.

If, however, the pattern in Exhibit 5 corresponds to an S or R chart, it means that the process dispersion is larger than it is specified to be. Again, if the chart represents an initial charting effort, questions arise regarding the realism of the target or whether the process variation realistically can be reduced to the targeted value. When the pattern observed represents a change from a process previously in control, the reason may possibly be simpler to assess.

Varied Out-of-Control Patterns

Exhibit 6 presents four control chart patterns representing various out-of-control situations. When presented in various texts and quality control manuals, these are referred to, in order, as evidence of complete instability, grouping or bunching, freaks, and a sudden shift. There is no problem with these categorizations; however, it should be recognized that the source of such patterns can vary. Consequently, we shall discuss each briefly with the understanding that other possibilities do exist.

CHAPTER 8

EXHIBIT 6
SOME SAMPLES OF OUT-OF-CONTROL PATTERNS

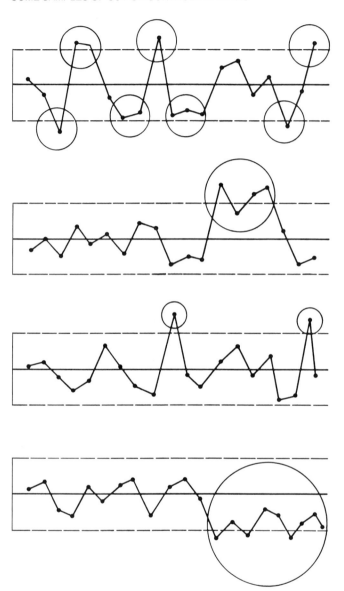

The first case shown in Exhibit 6 illustrates a situation in which there is *no stability* evidenced in terms of erratic fluctuations with points outside both the upper and lower control limits. If the chart is a mean chart, the pattern indicates that the center of the distributions from which samples are taken is changing erratically. If the process is homogeneous at each point in time, the pattern could result from a simple single cause such as constant re-adjustment to random changes in an otherwise stable process. It could also result from the introduction of changes in heterogeneous effects introduced at various points in time. If the process is inherently heterogeneous, the pattern could result from changes in representation of the individual component distributions, changes in representation in the samples, changes in the individual component distributions themselves, or any combination of these. In the case of an S or R chart, the pattern indicates that the spread of the distributions from which samples are taken is changing erratically, and all but the simplest possibilities for means would apply.

The second case depicted in Exhibit 6 commonly is referred to as evidence of *grouping or bunching*. The apparent pattern is one of a clustering of a small group of points within the chart. This would result from the sudden introduction of a set of causes that either is not sustained or for which correction has been made. If one instead of a bunch of points differed from the others, the pattern would be classified as evidence of a freak, which is displayed in the third chart in the exhibit.

Freaks typically are characterized by the occurrence of a single extraneous point that departs from the rest. These freaks could result from an abrupt change in the process that either disappears or is corrected, from an unusual observation in the sample taken from a homogeneous but highly skewed process, or they could be the natural result of chance since occasionally out of control points will result from chance forces alone.

The last pattern depicted in Exhibit 6 indicates evidence of shift in the process center or spread that is *sustained* in one direction. If the shift were in evidence in the center of the chart, it could be viewed as 'grouping or bunching" already introduced. We have encountered a simple example of such a shift in the filling process example of Chapter 7 where the overall fill level was adjusted downward.

Trend Patterns

In general, a *trend* can be defined as a persistent change in a quantity in a single direction, either upward or downward. A trend can occur in the process mean or the process standard deviation either in terms of a complete change in a distribution or through a steadily increasing or decreasing introduction of a heterogeneous influence. A trend also may appear on a control chart as a result of sequentially sampling from individual component distributions of a stable but heterogeneous process with many component distributions.

Exhibit 7 presents three examples of trends. In the first case the trend is upward, which is an indication of either a steadily increasing process mean or standard deviation, or of sampling individual component distributions from low to high values. In the latter instance, although not typically considered as such, the trend could represent a form of mixture pattern.

The remaining two diagrams in the exhibit depict downward trends, reflecting a decreasing mean or standard deviation. In the first of these two cases, the trend is more gradual, and such a pattern is referred to as a gradual change in level, which is nothing more than a slower moving trend. Depending on the amount of inherent random variation present, the gradual trend can be confused with a sustained shift or could result from a sustained shift that is not introduced completely at one time.

EXHIBIT 7
EXAMPLES OF TREND PATTERNS

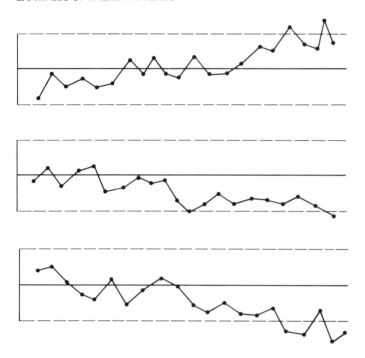

The last diagram in the exhibit is similar to the first, however, the trend is decreasing. In the case of means, a downward trend would indicate that the process mean is decreasing, and would indicate a departure from the targeted process level. In the case of an S or an R chart, the downward trend could indicate continual process improvements in terms of decreased variation.

Oscillatory Patterns

An oscillatory pattern is one in which the points systematically move above and below the center line, or about another level, in a somewhat predictable and nonrandom way. Consequently they result from effects that come and go on a regular basis. Two types of such patterns are depicted in **Exhibit 8**.

In the first case presented in the exhibit, the pattern is *wavelike*, or cyclical, such that points move downward away from the center line, turn and move upward above, turn again and move down toward the center line again. The pattern then repeats itself, but not necessarily with the same amplitude and frequency. Such a pattern could result, for example, from such things as a change in temperature and humidity, operator fatigue and rejuvenation, or shift changes. Which of these would be detected would also be a function of the sampling interval that must vary to pick up the changes in effect over varying time intervals.

EXHIBIT 8
EXAMPLES OF OSCILLATORY PATTERNS

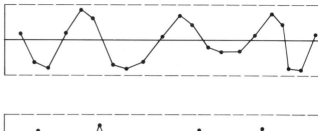

The second pattern appearing in Exhibit 8 also possesses the back and forth, or oscillatory, movement, but it is more abrupt and regular. In this case, a high value is followed by one that is low and vice-versa. If the effect is based on changes in the process, it could represent the same situation described above but with a higher frequency.

A common reason for the *sawtooth* type oscillation is changes in the source of the samples rather than changes in the process. If samples are *systematically* alternated from different sources, a back and forth oscillation will result. We encountered such a pattern in Chapter 7 where filling nozzles numbered 6-10 were mistakenly used in place of nozzles 1-5 on an alternating basis.

Mixture Patterns

We have indicated that we are to consider the term mixture pattern to refer to specific types of patterns appearing on a control chart, although the term mixture may be used in a number of ways. Here, a mixture pattern is one that results from sampling in one of two ways from an inherently heterogeneous process.

Various mixture patterns are depicted in **Exhibit 9**. In the first of these, we see that points oscillate from low to high about the center line with few, if any, points near the center line. Although the pattern does not appear so, it is stable but it does not correspond to one of randomness or control.

The pattern can result by taking samples alternately from individual component distributions comprising a heterogeneous process made up of two heterogeneous components, either with respect to central tendency or dispersion depending on what the chart measures. A similar pattern could result if samples are generated from homogeneous but different distributions at different points in time, such as from a first and second shift. In such a case where it is reasonable to assume that more than one sample per shift is taken, the third pattern appearing in the exhibit would be more reasonable. The points are closer to either control limit but are bunched rather than alternating perfectly.

EXHIBIT 9
MIXTURE PATTERNS

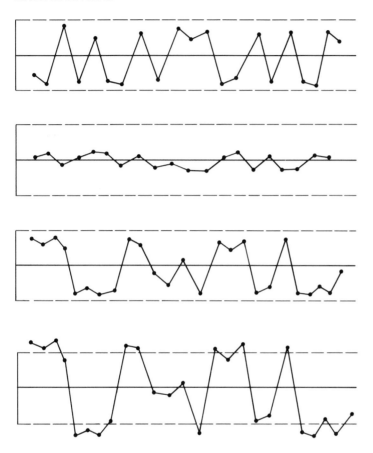

If the mixture patterns just cited appear on a mean chart, it would be an indication that the individual component distributions differ in terms of means. The same would be true on an S or an R chart if the individual component distributions were heterogeneous with respect to dispersion. If separation in terms of means only exists, the plotted variation will reflect the common variation within each of the component distributions; however, the limits on each chart may or may not. If, initially, the control limits are based on a measure of variation calculated from samples representing the mixed process, limits on both charts would be wider than if based on a measure of variation common to the component distributions. As depicted in the exhibit, the center line corresponds to a value midway between the component distributions.

The mixture patterns cited appear very clear. Patterns actually observed can vary to differing degrees depending on the proximity of the component distributions in terms of central tendency or dispersion, representation in terms of the relative number of items from each component, changing representation of each component in the sample, the amount of inherent random variation, and the number of individual component distributions comprising the mixture. The last diagram presented in Exhibit 9 is one example of an unstable mixture where any of the above factors may be present. In general, whenever a mixture pattern is observed, the source of inherent heterogeneity should be determined and eliminated or separate charts should be employed for the separate sources until similarity between the separate distributions can be achieved.

The second chart presented in Exhibit 9 appears much different from the others. The pattern commonly is referred to as a *stratification* pattern, which really is a mixture pattern resulting from a process similar to the one generating the first pattern depicted. The difference in appearance results from the fact that each of the individual component distributions of an inherently heterogeneous process is represented equally in the subgroups, or samples. In the case of a mean chart, the mean of the subgroups would tend to fall in the region of the average of the individual component distributions rather than any of the individual means. In the case of an S or R chart, the subgroup values are consistently large and closer together, assuming equal means and heterogeneity of variation in the individual component distributions alone. In either case, the result is a set of sample points that hug the center line. We already have seen an example of this in the filling example in which each nozzle was represented once in each subgroup where the nozzles are heterogeneous with respect to the average amount filled.

If control limits are constructed on the basis of samples across a number of individual component distributions from an inherently heterogeneous process with respect to mean, the measure of variation used will reflect the amount of variation among the means of the component distributions. Also, plotted points on an S or R chart will reflect this variation. Consequently, the measure of variation can be greater than the inherent chance variation in the process.

ROUGH AND READY RULES

The preceding material presented a number of typical patterns that can be encountered when observing control charts. Some insight also was provided regarding what the patterns reflect about the process distribution or the data used to construct the charts. By examining chart patterns actually observed, it is possible in some circumstances to match the result with those presented and then use the information to hunt for and correct a problem within a particular process.

When using control charts in the batch mode to solve a specific problem or to begin an ongoing control chart effort, complete patterns such as the ones illustrated provide information about the nature of the type of non-randomness present. In the case of an on-going real-time application it is desirable to establish the presence of an unnatural or non-random effect as soon as possible, although a lag will exist between a chart indication and the occurrence of a non-random intervention into the process. Just as we are able to attach a probability statement to a single point outside a control limit, it is possible to establish probabilities associated with particular types of pattern build-up looking backward each time a new point is added to a chart. The resulting rules constitute different types of tests for randomness.

Many different tests for randomness are available, Gibbons [71]. For common control chart applications, some simple practical rules have been adopted that depend on extreme sequences of points too unlikely to occur under the assumption that the process is in control. Although others based on different assumptions can be developed, these rules are based on the assumption of a normal sampling distribution.

The rules are based on sequences of successive points falling on one side of the center line or in zones on a single side in terms of one, two, and three standard errors from the center line. The zones are depicted in **Exhibit 10**. The limits based on two standard errors from the center line sometimes are referred to as warning limits and are used like control limits, but these limits are used to signal possible trouble rather than actually to be a basis for looking for it.

EXHIBIT 10
ZONES FOR ESTABLISHING EXTREME SEQUENCES

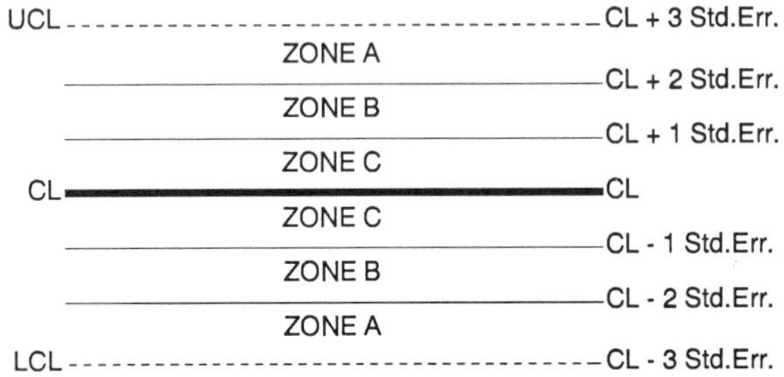

Some of the commonly accepted rules leading to the conclusion of *lack of control* are given as follows:

Rule 1.   One point outside one control limit

Rule 2.   Two out of three consecutive points in one Zone A or beyond on either side of the center line

Rule 3.   Four out of five consecutive points in one Zone B or beyond on either side of the center line

Rule 4a.   Seven consecutive points in one Zone C or beyond on either side of the center line.

Rule 4b.   Eight consecutive points in one Zone C or beyond on either side of the center line.

Additional rules can be found in Grant and Leavenworth [11] and Montgomery [12]. Examples of patterns associated with the above rules are depicted in **Exhibit 11**. In all cases, notice that the points in a particular sequence are circled indicating that a particular condition is met.

EXHIBIT 11
EXAMPLES OF OUT-OF-CONTROL SEQUENCES BASED ON RULES

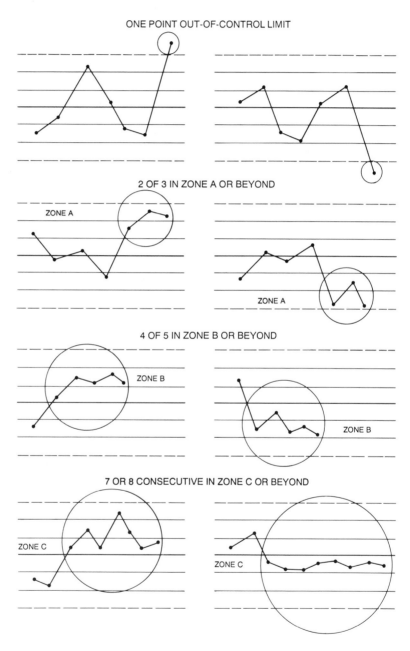

ONE POINT OUT-OF-CONTROL LIMIT

2 OF 3 IN ZONE A OR BEYOND

ZONE A

ZONE A

4 OF 5 IN ZONE B OR BEYOND

ZONE B

ZONE B

7 OR 8 CONSECUTIVE IN ZONE C OR BEYOND

ZONE C

ZONE C

Rules such as the ones presented are guides that may be used as a basis for action. They are not rigid and can be modified to suit the needs of particular environments or processes. Further, since they are based on the assumption of normality, they more strictly apply to mean charts. Since the sampling distribution of the range and the sample standard deviation are not normal, the related probabilities do not apply directly for these statistics. Although the appropriate distributions are available to calculate the exact probabilities, the rules are useful in these instances since the exact probabilities typically are of the same small order of magnitude.

Probabilities Associated With the Rules (Optional)

Based on the way control limits are calculated as three standard errors from the center line, assuming normality, we have seen that the probability of a point falling outside a single control limit equals 0.00135. Similarly, we could use the Table of Areas of the Normal Curve to find the areas or probabilities associated with each of the zones. These are presented in **Exhibit 12**.

EXHIBIT 12
PROBABILITIES ASSOCIATED WITH CONTROL
CHART ZONES BASED ON NORMAL CURVE

	.00135
ZONE A	.02134
ZONE B	.13591
ZONE C	.34134
ZONE C	.34134
ZONE B	.13591
ZONE A	.02134
	.00135

Using the simple rules for combining probabilities presented in Chapter 4 or the binomial distribution that was briefly introduced, the probabilities associated with the out-of-control sequences can be developed. For example, consider Rule 1 associated with two of three points in Zone A or beyond. The probability of a single point in A or beyond on one side of the chart is the sum of the probabilities in Zone A and the out of control region. That is,

$$P(\text{Zone A or beyond on one side}) = .02134 + .00135$$

$$= .02269$$

Assuming no change in the value of the center line, the probability of one sequence of two points in A or beyond and one that is not, say {In In, Not In} is given as

$$P(I,I,NI) = (.02269)\,(.02269)\,(1 - .02269)$$

$$= (.02269)^2(.97731)$$

$$= .000503$$

Since in a given set of three points, obtaining two in A or beyond and one not can occur in three different ways, the probability of two out of three points in Zone A or beyond is

$$P(2 \text{ of } 3 \text{ in A or beyond on one side}) = 3(.000503)$$

$$= .001509$$

The probability of obtaining two out of three successive points in Zone A or beyond on either side of the center line, therefore, is twice the above probability, or

$$P(2 \text{ of } 3 \text{ in A or beyond on either side}) = 2(.001509)$$
$$= 0.003$$

Effectively, this probability is similar to the probability of obtaining a single point outside of either of the two control limits, or roughly 3 in 1000. In other words, it represents a conservatively small level of significance, or probability of committing a Type I error for the specific decision rule to conclude that there is a shift in the center line when any two of three consecutive points fall in Zone A or beyond on either side of the center line. This corresponds to a two tailed test; if a shift only in one direction is of interest, then half the given probability, or 0.0016, would apply. A summary of the probability calculations, based on the values given in Exhibit 12, for the rule discussed in detail above and the remaining rules is as follows:

$$P(2 \text{ of } 3 \text{ in A or beyond}) = 2[3(.02134 + .00135)(1 - .02134 - .00135)]$$

$$= 2[.001596]$$

$$= 0.0030$$

$$P(4 \text{ of } 5 \text{ in B or beyond}) = 2[5(.13591 + .02134 + .00135)^4$$
$$\times (1 - .13591 - .02134 - .00135)]$$

$$= 2[.002662]$$

$$= 0.0053$$

$$P(7 \text{ of } 7 \text{ in C or beyond}) = 2[(.34134 + .13591 + .02134 + .00135)^7]$$

$$= 2[(.5)^7]$$

$$= 2[.007813]$$

$$= 0.0156$$

$$P(8 \text{ of } 8 \text{ in C or beyond}) = 2[(.34134 + .13591 + .02134 + .00135)^8]$$

INTERPRETING
VARIABLES CHARTS

$$= 2[(.5)^8]$$

$$= 2[.003906]$$

$$= 0.0078$$

The results with the rules are presented in **Exhibit 13**.

EXHIBIT 13
PROBABILITY OF TYPE I ERROR FOR SIMPLE OUT-
OF-CONTROL PATTERN RULES

		One Particular Side	Either Side
Rule 1:	1 Point Out-of-Control Limit	.0014	.0027
Rule 2:	2 of 3 Points in Zone A or Beyond	.0016	.0032
Rule 3:	4 of 5 Points in Zone B or Beyond	.0027	.0053
Rule 4a:	7 Consecutive Points in Zone C or Beyond	.0078	.0156
Rule 4b:	8 Consecutive Points in Zone C or Beyond	.0039	.0078

Regardless of any particular rule used, the reasoning behind its use is that the probability of observing the specific sequence embodied within the rule is so small that chance alone could not have produced such a sequence under the assumption that the value at the center line remains the same. Consequently, a change must have occurred when such a sequence is observed.

By examining the results in Exhibit 13, we can see that the probabilities associated with all of the rules except Rule 4a are, although different, all very small and of the same order of magnitude. In the case of Rule 4a (i.e., 7 consecutive points in C or beyond) the two-sided probability of approximately 0.02 is somewhat larger than the others; however, it is considered to be reasonably small in terms of general statistical applications. Whether to use such a rule corresponding to similar or other probability levels is based on an individual's preference.

A few commonly used and simple rules have been presented. Others are available. Further, any number of rules can be developed based on similar reasoning. For example, one could develop a rule for so-called stratification patterns in terms of a consecutive run of points that oscillate about or hug the center line within both regions designated as Zone C, instead of either one or beyond. Assuming normality, probabilities of selected consecutive sequences of points in either Zone C are calculated as follows:

$$8 \text{ Consecutive Points } = (.34134 + .34134)^8$$
$$= .0472$$

$$10 \text{ Consecutive Points } = (.34134 + .34134)^{10}$$
$$= .0220$$

$$12 \text{ Consecutive Points} = (.34134 + .34134)^{12}$$
$$= .0102$$

$$15 \text{ Consecutive Points} = (.34134 + .34134)^{12}$$
$$= .0033$$

Any one of these sequences would lead to a decision rule based on commonly accepted levels of significance. Fifteen consecutive points hugging the center line within Zone C, however, would be necessary to achieve a level of significance of roughly 3 in 1000, equivalent to that associated with the fundamental rule of one point outside a control limit.

Parenthetically, it is worth commenting on the nature of the calculated probabilities associated with the rules presented. When these are presented as specific decision rules, the calculated probabilities represent the probability of committing a Type I error relating to the hypothesis that there is no change in the value at the center line. The specific type sequence stated in the rule acts much like any critical value associated with any test of an hypothesis.

Instead of establishing the rule before the fact, it is possible to observe any sequence of points on a control chart and calculate the probability of observing that sequence or one more extreme, under the assumption that the center line remains the same. The resulting probability would constitute a p-value as defined in Chapter 5, which could be compared to a pre-established level of significance in order to decide whether a change has occurred or not.

ADDITIONAL CONSIDERATIONS

This chapter focuses on various types of out-of-control chart patterns, with emphasis on understanding the statistical basis for these patterns. By understanding the fundamentals, one then should be able to apply the ideas to particular process environments. This, however, can only be acquired through continued experience with charting specific processes with which one gains continued familiarity.

It should be recognized clearly that process control is not a one-shot deal in which a chart is developed and a process automatically becomes controlled. Whether one uses targets or process data at the outset, it is necessary to make initial decisions on process definition, process characteristics to measure, and the method and timing of subgroups, all of which then results in charts in batch mode as a starting point. This potentially can lead to elimination of problems in the process in addition to sources of inherent heterogeneity.

By continued re-definition and re-assessment of the control limits one strives to chart essential process characteristics, which can be assumed to be homogeneous and properly centered with an allowable amount of chance error. The method and timing of subgrouping should be re-considered with the ultimate goal of capturing chance error within each subgroup. By charting homogeneous processes using subgroups that are properly constructed and timed, changes in chart patterns on a real-time basis can then be meaningfully associated with correctable problems relating to changes in the process.

Whether implicit or explicit, assessment of control chart patterns throughout should be done with the five basic questions in mind. By doing so, one can more systematically work toward eliminating process problems, a task that is interrelated with the nature of the process and the means of collecting and processing data.

The control chart patterns presented can appear on an $\overline{X}$ chart or on an S or R chart. Obviously, the effect on the distribution of a process characteristic is different depending on the chart possessing the particular pattern. As indicated, both types of charts should be used when working with variables type data.

When interpreting such charts, the chart for variability should demonstrate control *before* interpreting the mean chart. On the one hand, this results from the fact that the variation in sample means is affected by the amount of inherent variability associated with the measured characteristic. Also, the limits on a mean chart are calculated on the basis of a measure of this inherent variation. Consequently, a mean may be incorrectly assessed as in or out of control when based on limits that do not properly reflect the true variation in a process.

Control charts as presented in this and the previous chapter are very simple devices that are used to monitor process characteristics and ultimately control processes. How easily they can be applied depends, in large part, on the nature and complexity of the problem or overall process to be controlled. Obviously, our filling problem is much simpler to deal with than, say, the complete process needed to manufacture a modern-day computerized module that monitors and controls a car engine, made up of three to four-hundred small finely tooled parts and firmware.

In the latter case, not only must many charts be used but they must be used creatively. Many problems exist where control charts may be used effectively in conjunction with a variety of other statistical methods. For example, print quality of carton-type containers is dependent on pH-level and moisture content of the board, which is delivered and processed in large rolls, such that level and content may not be uniform throughout the rolls. To varying extents, these rolls are not subject to the control of the carton manufacturer. However, in the case of automated printing operations, the process can be adjusted to accommodate for variations in level and content based on results of other types of analyses.

By undertaking analyses that capitalize on the dependency between quality and content, process adjustment rules may be developed that not only replace operator judgment but provide a more objective basis for making process adjustments, in addition to those made on the basis of a control chart.

Another simple case relates to gradual tool wear evidenced in terms of a trend pattern on a control chart. By setting limits about an established trend estimate instead of a constant center line, a basis for deciding when to replace or sharpen a tool can be established in addition to accounting for the tool wear in assessing whether other problems in a process exist.

The Filling Process Problem

Owing to its simplicity, the filling process problem introduced in Chapter 7 provides an excellent basis for summarizing a number of ideas presented in this chapter in order to unify some of the concepts discussed.

If handled properly, there are a number of distinct aspects to the overall problem. As the problem was originally handled, the primary goal was stated as fill all containers in such a way that the label claim limit of 425 grams is not violated. By setting the targets as stated initially, the goal was met since all individual containers were filled in excess of the label claim limit.

By examining control charts of the process, however, it becomes clear that the process was not in control. On the one hand, without stability it is not possible to count on its ability to consistently meet the label claim. Furthermore, at least at this point, it appears that the problem is mostly one of adjustment. By controlling the process not only is it possible to determine when to adjust and when to leave the process alone, but it is possible to establish targets that accommodate the label claim and reduce overfill, or its cost. Further analysis and on-going charting could lead to improvements in the process, reducing the natural variation while concomitantly resulting in further reductions in overfill.

Patterns Illustrated by the Filling Problem

Now that we have introduced various types of control chart patterns, it is instructive to summarize the patterns that were evident in the filling example, which is a simple but real application. By examining Exhibit 2 in Chapter 7, the first noticeable departure from stability is at the beginning of the mean chart in terms of the first three points. The first two are very high above the center line and the third is somewhat lower. Whether within the control limits or not, this is a common start-up pattern in which the initial setting overshoots the target, followed by an adjustment in the opposite direction.

Further examination of Exhibits 2 and 3 in Chapter 7 provides a sudden shift pattern in the mean chart and a sawtooth, or systematic pattern, mostly in the standard deviation and range charts. With points out of the natural control limits eliminated, we observe a mixture pattern in terms of stratification about the center line in the means of Exhibit 6. Finally, in Exhibit 8 we have an example of what the charts should look like in a state of control with a targeted mean of 442 grams assuming natural variation is equal to a standard deviation of 2.3 grams. No information directly is given about meeting specifications on fill levels of individual containers.

The Basic Questions and the Filling Problem

Now that we have discussed various types of control chart patterns and ones related to the filling process problem, it also is instructive to re-introduce the basic questions presented earlier in this chapter and relate them to the filling problem. The five questions provide a structure for assessing control chart information; however, it is virtually impossible to apply them generally without a specific application or context. The filling process problem does, however, provide an opportunity to illustrate their usefulness in a simple applied setting.

The actual filling operation consists of numerous filling machines operating for two daily shifts five days a week. For charting purposes, the process could be defined in terms of all the filling machines, each machine individually, or each nozzle of each machine separately. Ours was defined in terms of an individual machine with 10 nozzles.

Recall that when speaking of the nature of the process we must think in terms of the distribution(s) of a relevant process characteristic. In our case, we are interested in the distribution of the weights of the contents of filled glass containers. At any particular point in time, a separate distribution of weights could be associated with each nozzle. Consequently, if the output of the filling machine is considered without considering the nozzles separately, the distribution of the output potentially is inherently heterogeneous. Depending on the nature of the system used, each individual component distribution could change over time either in the same way or differently, with respect to mean, variation, or both. Each of the individual component distributions could be homogeneous or heterogeneous at any point in time depending on the nature of the underlying filling machine or the composition of the contents of the filling substance.

Since subgroups are composed of the output of the nozzles from a single filling machine, the variation present within a subgroup at a particular point in time is composed of natural variation and potentially variation among the nozzles, in addition to possible sources of variation within the filling machine and the composition of the filling substance. The latter should, but may not, be constant across the nozzles. If additional machines were considered as part of the same process, variations among machines and nozzles among machines would represent an added source of potential variation.

Over time, the method of subgrouping would allow for changes in the machine or the composition of the filling substance to be detected. Further, changes in the relative number sampled from each nozzle would be observable over time.

The basic problem as stated in Example 1 of Chapter 7 was to insure against underfill relative to the label claim of 425 grams. A targeted mean of 442 grams and a targeted standard deviation of 4.5 grams were conservatively established to insure against violating the label claim. This was based on the reasoning that three standard deviations below the mean results in an individual container content weight of 428.5 grams, which allows a 3.5 gram margin of safety over the label claim. A nominal upper specification limit for overfill at three standard deviations above the mean, equal to 455.5 grams, also results from the stated targets.

The center lines, therefore, for the two variables charts were initially based on the set targets. Control limits were calculated as three standard errors based on the targeted value of the standard deviation equal to 4.5 grams. Implied, therefore, in these values is the fact that process control in terms of the corresponding charts is in terms of the question, "how is the process behaving with respect to the specified targets?" Parenthetically, the problem really should be posed as what should be done, if anything, to control the process; when in control how does the process behave naturally; and what targets if subject to control should be set in order to meet the label claim limit and minimize overfill and cost or waste? We shall address these points in more detail in the next chapter concerned with specifications and process capability.

By examining the original charts appearing in Exhibits 2 and 3 of Chapter 7, which are based on the targets, we can conclude that the limits are too wide for a point to fall outside when the process is not in control. Real-time application of the charts on a point-by-point basis would be effectual only with an accumulation of points. By timing the subgroups a half-hour apart, half a working shift could expire before trouble would be sought.

By examining the mean chart, we can conclude on the basis of patterns within the limits that the mean contents from the machine (or across nozzles) was adjusted above the target, re-adjusted to below, re-adjusted again, and then let to remain at a value below the target. The standard deviation appears to remain stable at a value below the target; however, the chart does not indicate control. The systematic oscillation appearing on the S or R chart does not reflect an alternating process standard deviation but a mistaken alternation between sets of nozzles forming the subgroups.

Although other problems in the process may exist, the main problem detected on the basis of the initial charts is one of adjustment. On the one hand, it appears that the process will hold to a machine adjustment. On the other hand, the process has been over-and-under adjusted without any measurable basis nor with respect to the specified targets. More realistic targets should be set, and the control chart should be used as a basis for adjustment in the future. The basis for subgrouping should be reconsidered in order to possibly get a truer measure of natural variation. This would be useful in order to establish limits, detect other problems as they arise, assess more clearly the balance between the label claim constraint and problems of overfill, and form the basis for possible process improvements. After introducing the basic ideas relating to process capability, we shall address these issues again.

PROCESS CAPABILITY

The previous two chapters focused on process control, or on methods to detect the existence of controllable sources of variation with the goal of eliminating these sources from a process. Little or no emphasis was placed on meeting product *specifications*. Ultimately, it is the quality of the output or product of a process which is of concern.

Process capability is the term used to refer to the reproducibility or uniformity of a process with respect to product quality. Ideally, the capability of a process would be characterized in terms of the distribution of each relevant quality characteristic at every cross section, or point, in time. This would provide a complete picture regarding the nature of variation that is present in a process.

Unfortunately, it is not possible to know the underlying distribution at every point in time. Consequently, it is necessary to estimate either the distribution or various characteristics of it by using data accumulated over time. Process capability really only has meaning when related to *controlled processes*, or ones that are homogeneous and stable over time.

When working with process capability, there are two basic questions that can be answered:

1. What is the process doing, or what is the process distribution like, under a state of control?
2. What is it possible for the process to do, or what can the process distribution be after eliminating further sources of variation, or making improvements to the process?

Although both involve measurement, the second question can involve complete capability studies with many forms of analysis and experimentation in order to obtain an answer. We shall focus mostly on the first of the two and provide some insight at the end regarding the need for capability studies.

The symbols used in this chapter are defined as follows:

μ	= the process mean
μ_0	= a targeted process mean
$\bar{\bar{X}}$	= the mean of a set of sample means
$\hat{\mu}$	= an estimate of the process mean
$\bar{X}$	= the mean of a sample
σ	= the process standard deviation
$\bar{S}$	= the mean of a set of standard deviations
$\hat{\sigma}$	= an estimate of the process standard deviation
$\bar{R}$	= the mean of a set of sample ranges
S	= the standard deviation of a sample
R	= the range of a sample
USL	= the upper specification limit
LSL	= the lower specification limit
UNL	= the upper natural limit of a process
LNL	= the lower natural limit of a process
C_p	= the process capability ratio
C_{pU}	= one-sided, upper process capability ratio
C_{pL}	= one-sided, lower process capability ratio
C_{pk}	= the "C-P-K" index

ESTABLISHING CAPABILITY

There are a number of ways of measuring or establishing capability. Essentially these involve estimating the process distribution empirically with a histogram, fitting a theoretical curve or distribution to process data, establishing intervals within which all or most of the product will fall, or using an index. Directly or indirectly, these methods relate the process to its ability to meet specifications and to the fraction defective or non-conforming product.

Using a Histogram

One way to obtain a picture of the process capability is to estimate the process distribution with a frequency distribution or histogram. As an example, let us use the data associated with the simulated filling process weights presented in Chapter 7. The data are reproduced in **Exhibit 1** for convenience.

Note. In the case of the original filling problem introduced in Chapter 7, the targeted mean and standard deviation equal to 442 and 4.5 grams, respectively, suggest specification limits of individual container content weights of 428.5 and 455.5 grams which correspond to the targetted mean plus-and-minus three standard deviations. The original intent was to allow a margin of error between the label claim of 425 grams and the lower specification limit. The upper specification limit arbitrarily became the limit for overfill. For purposes of illustration in the following examples, we shall assume *two-sided* specifications of 428.5 and 455.5 grams which consider both underfill and overfill.

EXHIBIT 1
SIMULATED FILLING PROCESS WEIGHTS BASED ON NORMAL PROCESS WITH $\mu = 442$ AND $\sigma = 2.3$

Sample Number	Nozzle 1	Nozzle 2	Nozzle 3	Nozzle 4	Nozzle 5	$\overline{X}$	S	R
1	444.6	442.6	441.8	441.8	445.1	443.2	1.57	3.3
2	439.9	447.2	442.6	442.9	444.1	443.3	2.65	7.3
3	440.1	441.1	438.9	443.7	438.9	441.0	2.07	4.8
4	442.9	442.6	440.5	442.4	442.2	442.1	.94	2.4
5	446.0	438.1	445.1	442.1	439.8	442.2	3.37	7.9
6	444.2	440.0	439.1	441.0	437.2	440.3	2.59	7.0
7	445.4	443.7	445.8	441.9	444.4	444.2	1.55	3.9
8	442.6	441.9	439.5	441.2	439.3	440.9	1.46	3.3
9	441.6	439.4	441.7	446.2	442.2	442.2	2.47	6.8
10	443.3	443.5	441.6	438.8	446.6	442.8	2.86	7.8
11	441.0	438.7	444.9	444.8	439.1	441.7	3.00	6.2
12	440.7	442.5	447.2	442.7	443.0	443.2	2.40	6.5
13	441.7	440.4	442.1	440.7	441.7	441.3	.73	1.7
					Totals	5748.4	27.66	68.9

Since we know that the results are in control, a frequency distribution provides a basis for observing the nature of the distribution. Using the procedures presented in Chapter 1, a convenient class interval size of 1.25 grams is chosen. Since the lowest observed weight equals 437.2, we set the lower limit of the first class equal to 437. The resulting frequency distribution is given as follows:

Weight (g)	Number of Containers	Relative Frequency
437.00 but less than 438.25	2	.031
438.25 but less than 439.50	9	.138
439.50 but less than 440.75	8	.123
440.75 but less than 442.00	13	.200
442.00 but less than 443.25	14	.215
443.25 but less than 444.50	7	.108
444.50 but less than 445.75	6	.092
445.75 but less than 447.00	4	.062
447.00 but less than 448.25	2	.031
Total	65	1.000

The resulting histogram appears as

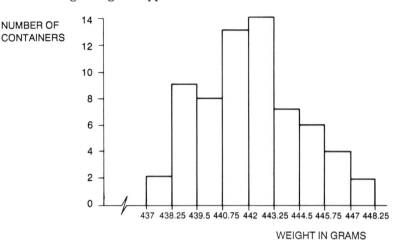

NUMBER OF
CONTAINERS

WEIGHT IN GRAMS

Assuming a state of statistical control, which is valid in this particular case, the histogram provides somewhat of a shadowed image of the underlying process distribution. This results from the fact that the observed frequency distribution is subject to chance variation, just as are sample means, standard deviations, and ranges.

If the histogram alone is used to establish capability, basically one can get an immediate visual impression regarding the distribution shape and how the distribution relates to product specifications. In this case, the distribution appears a little skewed to the right such that the distribution of weights falls within the limits of 437 and 448.25 grams. These values are within the specified limits of 428.5 and 455.5 grams. Assuming the histogram reflects the process distribution, the conclusion is that the process can be assumed to be capable of meeting the specifications.

For the moment, we are considering the histogram approach as one method of establishing the capability of the process in terms of what it is doing in a state of control. Before discussing the approach, let us introduce other available approaches.

Curve Fitting

Another way of establishing capability is by fitting a theoretical probability distribution to process data and determining how the distribution relates to product specifications. This method has the advantage of providing greater generality to the result and a basis for calculating the probability or percent of product outside the specification limits. If used without other information about the process, the approach has the disadvantage of masking the true behavior of the process by using the resulting distribution without continued observation of the behavior of the process data.

The conventional approach is to assume normality, fit a normal distribution to the process data, and display the resulting normal curve. It is possible to fit other types of distributions that are skewed to varying degrees. Here, we shall illustrate the more conventional normal approach. The fitting procedure for most distributional forms is easily conducted with a variety of available software packages.

In order to understand what is actually done when fitting a probability distribution to a set of data, we shall briefly outline the normal case. Since the normal distribution is characterized by the two parameters, μ and σ, which are the mean and standard deviation, estimates of these values must be obtained. These values can be substituted into the formula for the normal curve presented in Chapter 4 using selected values of X, or weights in this case. A smoothed curve then can be traced through the selected coordinate values that result.

As an example, again consider the controlled filling process data presented in Exhibit 1. The first thing to be done is to estimate μ and σ, estimates of which we shall designate as $\hat{\mu}$ and $\hat{\sigma}$, respectively. Our best estimate, $\hat{\mu}$, is $\bar{\bar{X}}$ which is given as

$$\hat{\mu} = \bar{\bar{X}}$$

$$= \frac{\Sigma \bar{X}}{k}$$

$$= \frac{5748.4}{13}$$

$$= 442.2$$

where $\Sigma \bar{X}$ is obtained from Exhibit 1.

Recall, the process standard deviation can be estimated in one of two ways, either using $\bar{S}$ or $\bar{R}$, depending on the measure used for control chart purposes. Since we have both sets of information, for illustrative purposes we obtain

$$\bar{S} = \frac{\Sigma S}{k}$$

$$= \frac{27.66}{13}$$

$$= 2.128$$

$$\bar{R} = \frac{\Sigma R}{k}$$

$$= \frac{68.9}{13}$$

$$= 5.3$$

and

$$\hat{\sigma} = \frac{\bar{S}}{c_4}$$

$$= \frac{2.128}{9400}$$

$$= 2.26$$

or

$$\hat{\sigma} = \frac{\overline{R}}{d_2}$$

$$= \frac{5.3}{2.326}$$

$$= 2.28$$

Obviously, only one of these values would be calculated as an estimate in a particular application, although here we see that the results are very close. Arbitrarily let us use $\hat{\sigma}$ equal to 2.26 based on $\overline{S}$.

Although in today's computing environment, the approach is cumbersome, effectively the fitting process reduces to substituting $\hat{\mu}$ and $\hat{\sigma}$ into the following formula for the normal

$$f(x) = \frac{1}{\sqrt{2\pi}\ \hat{\sigma}}\ e^{-\frac{1}{2}\left(\frac{x-\hat{\mu}}{\hat{\sigma}}\right)^2}$$

and calculating f(x) for selected values of X. The values of X selected appear in the following table along with the calculated values of f(x).

Selected Values of X	$z = \dfrac{X - \hat{\mu}}{\hat{\sigma}}$	f(X)	$\Phi = f(z)$
435.42	−3	.0020	.00443
437.68	−2	.0239	.05399
439.94	−1	.1071	.24197
442.2	0	.1765	.39894
444.46	1	.1071	.24197
446.72	2	.0239	.05399
448.98	3	.0020	.00443

Also included in the table are the corresponding multiples of the standard deviation, z, which correspond to the standardized normal distribution

$$\Phi = f(z)$$

$$= \frac{1}{\sqrt{2\pi}}\ e^{-\frac{1}{2}z^2}$$

and the values of Φ, or f(z). Φ is referred to as the ordinate of the standardized normal. Tables of Φ are available, and Φ also can automatically be obtained with some scientific calculators merely by entering the appropriate multiple. This becomes convenient since dividing Φ by σ yields the corresponding value of f(x). That is,

$$f(x) = \frac{\Phi}{\sigma}$$

which can be verified in the above table.

Values in the table are plotted below together with a smoothed curve through the plotted points.

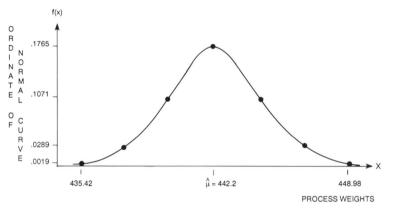

The distribution is centered on the estimated mean of 442.2, and the 3-sigma limits that are provided obviously fall within the specification limits of 428.5 and 455.5 grams. Frequently, a graph such as the one presented above is used to demonstrate the capability of a process. Although it appears very objective, a question always arises regarding how well the curve represents the data and the underlying process distribution.

In order to get some idea of the degree to which the data are represented by the fitted curve, probabilities from the fitted normal are compared with the corresponding relative frequencies in **Exhibit 2**. Two frequency distributions based on the same data are presented in the exhibit. The first is the one developed in the previous portion of this chapter, while the second is based on a distribution constructed using interval widths equal to the estimated standard deviation. Since we have introduced multiples of the standard deviation in order to establish coordinates for the fitted normal, it is convenient to use this as a basis for introducing an alternative histogram in order to gain insight into the nature of the results of the fitting process.

In general, when a probability distribution is fitted to a set of data, we do not know the form of the underlying distribution. Consequently, one of the problems addressed is to reach a conclusion regarding the distribution form based on the sample data. This will be addressed to some extent at the very end of the chapter. In this case, we know the underlying distribution to be normal since it was generated in this way.

Because we are dealing with continuous measurements, it is necessary to group the data in order to construct a frequency distribution. Since no one particular method of grouping is necessarily the only correct one, the resulting distribution constructed will be affected by the method of grouping in addition to sampling variation and the form of the underlying distribution.

EXHIBIT 2
COMPARISON OF RELATIVE FREQUENCIES AND FITTED
NORMAL PROBABILITIES

Process Weights (Interval Size Equal to 1.25)	Relative Frequency	Probability Based on Normal
437.00 but less than 438.25	.031	.030
438.25 but less than 439.50	.138	.076
439.50 but less than 440.75	.123	.144
440.75 but less than 442.00	.200	.236
442.00 but less than 443.25	.215	.182
443.25 but less than 444.50	.108	.167
444.50 but less than 445.75	.092	.096
445.75 but less than 447.00	.062	.041
447.00 but less than 448.25	.031	.013
Total	1.000	.985

Process Weights (Interval Size Equal to $\hat{\sigma}$ = 2.26)	Relative Frequency	Probability Based on Normal
435.42 but less than 437.68	.015	.021
437.68 but less than 439.94	.185	.136
439.94 but less than 442.20	.323	.341
442.20 but less than 444.46	.292	.341
444.46 but less than 446.72	.154	.136
446.72 but less than 448.98	.031	.021
Total	1.000	.996

By examining Exhibit 2, we see overall that differences obviously exist between the observed relative frequencies and the theoretical probabilities. In the second example, the theoretical values appear to be closer; the larger the class interval the greater the smoothing effect, at the expense of greater detail.

The effect of fitting becomes more apparent when presented graphically, which is done in **Exhibit 3**. The normal curve is the same in both cases; however, the histograms appear different. The one based on the larger interval appears more symmetric, whereas the other appears a little bumpy and skewed. It should be noted that the height of the bars of the histogram plotted in the exhibit are *not* equal to the relative frequencies presented in the earlier exhibit. The relative frequencies have been rescaled to be comparable to the ordinates of the normal curve such that the area of each bar equals the corresponding relative frequency.

The curve fitting approach has been introduced in this portion in order to provide awareness of the method. By comparing the fitted curve with the histogram representing the underlying data, one gets a better idea of the nature of a fitted curve. Without being constantly mindful of the data and the nature of the process, the fitted curve too simply deals with the problem. The real benefit lies in the use of a fitted distribution to estimate defective rates, but this should only be done with a clear recognition and understanding of the process distribution that it represents.

EXHIBIT 3
FITTED NORMAL DISTRIBUTION SUPERIMPOSED ON ALTERNATIVE HISTOGRAMS
OF CONTROLLED FILLING PROCESS DATA

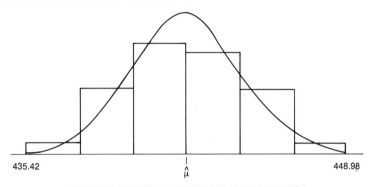

435.42 $\overset{\wedge}{\mu}$ 448.98

INTERVAL SIZE EQUAL TO ONE STANDARD DEVIATION

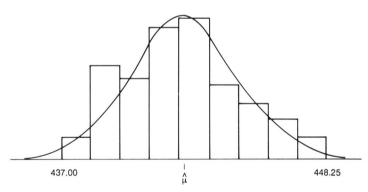

437.00 $\overset{\wedge}{\mu}$ 448.25

INTERVAL SIZE EQUAL TO 1.25 GRAMS

Using an Interval

Another way to establish the capability of a process is to estimate the natural limits of the process, which are the largest and smallest dimensions produced by the process, respectively. The conventional approach used in process control is to assume normality and calculate the 3-sigma limits based on the estimated process mean and standard deviation. That is, estimate the natural limits using the expression

$$\hat{\mu} \pm 3\hat{\sigma}$$

Actually, under the assumption of normality, this approach provides as much information as the fitted normal curve when used independently of the underlying data or histogram. Recall that approximately 99.7

percent of the observations fall within three standard deviations of the mean of a normal distribution. Of course, higher multiples than 3, which correspond to higher probabilities can be used. Three standard deviations, however, is the accepted procedure.

In order to illustrate the procedure, assuming an interval alone were of interest, one would first obtain estimates $\hat{\mu}$ and $\hat{\sigma}$ and substitute in the above formula as follows:

$$\hat{\mu} \pm 3\hat{\sigma}$$

$$442.2 \pm 3(2.26)$$

$$\pm 6.78$$

$$435.42 - 448.98$$

Again, if compared to the specification limits of 428.5 and 455.5 grams, we see that the natural limits fall within. Therefore, the conclusion based on this information would be that the natural process limits fall within specifications and that no unacceptable items are produced by the process while in a state of control.

Capability Indexes

The three approaches presented above, namely, the histogram, curve fitting, and using an interval, all primarily aim at establishing the natural tolerances of a process. Minimally, we have alluded to the fact that these tolerances should fall within specification limits, or design tolerances. Although more needs to be said, we shall introduce certain measures, or indexes, which relate the natural tolerances to the specification limits. Like any other index, they represent *summary* guides that do not completely solve the problem and should be used with caution.

The first of these indexes, which sometimes is referred to as the process capability ratio is designated as C_p and is given by the formula

$$C_p = \frac{USL - LSL}{6\hat{\sigma}}$$

where USL and LSL represent the upper and lower specification limits, respectively, and $\hat{\sigma}$ is the estimated process standard deviation.

The quantity, $6\hat{\sigma}$, appearing in the denominator represents an estimate of the natural spread of the process assuming normality. In other words, if we designate UNL and LNL as the upper and lower *natural limits* of the process, these are estimated as

$$UNL = \hat{\mu} + 3\hat{\sigma}$$

$$LNL = \hat{\mu} - 3\hat{\sigma}$$

The natural spread of the process, therefore, is estimated as the difference between the two limits as

$$\text{UNL} - \text{LNL} = \hat{\mu} + 3\hat{\sigma} - (\hat{\mu} - 3\hat{\sigma})$$

$$= \hat{\mu} + 3\hat{\sigma} - \hat{\mu} + 3\hat{\sigma}$$

$$= 3\hat{\sigma} + 3\hat{\sigma}$$

$$= 6\hat{\sigma}$$

The C_p index will equal 1 when the natural spread exactly equals the spread between the specification limits. Assuming, therefore, that a normal process is properly centered at a targeted mean, or exactly midway between the specification limits, approximately 3 in 1000 items will be non-conforming, or out of the specification limits, when the C_p index equals *one*. In general, values of C_p equal to or greater than one suggest that a process has the *potential* or is capable of meeting specifications. The further the process mean is from the targeted mean, the larger the ratio must be for the process to meet specifications and not produce non-conforming product.

The C_p index as defined above considers two-sided specifications only and does not directly incorporate process centering into the index. Additional indexes, referred to as C_{pU}, C_{pL}, and C_{pk}, which do account for this and which relate more to current performance are defined as follows:

$$C_{pU} = \frac{\text{USL} - \hat{\mu}}{3\hat{\sigma}}$$

$$C_{pL} = \frac{\hat{\mu} - \text{LSL}}{3\hat{\sigma}}$$

$$C_{pk} = \min\left(C_{pL}, C_{pU}\right)$$

C_{pU} and C_{pL} relate the difference between the current estimate of the process mean, $\hat{\mu}$, and either the upper or lower specification limit to half the estimated natural process spread, $3\hat{\sigma}$. When a two-sided specification is in force, both C_{pU} and C_{pL} are calculated and the smaller of the two is the C_{pk} index.

In the case of all three indexes, a value of *one* or greater suggests that the process is *currently capable* of meeting specifications. When less than one, there are two possibilities. First, if the process mean is equal to the targeted value, the result suggests that the process variability is too large, which may or may not be easily correctable. This also may be the case if the process mean does not equal the targeted mean.

The second possibility, when the process mean does not equal the target, is that a shift in the process mean, possibly through a simple adjustment, will result in an index equal to one or more, depending on whether there exists a two or one-sided specification limit. The more each of these indexes exceeds one, the more the process mean may change while producing items that potentially remain within specifications.

Recognize that with respect to the C_p index, a value of one or more suggests that a process is *potentially* capable since the process variation is acceptable as long as the centering is controlled at the proper level.

On the other hand, a C_p of less than one *automatically* suggests that process variation is unacceptable. In the case of C_{pU}, C_{pL}, and C_{pk}, a value less than one suggests that an acceptable value *could* result if the process is properly centered provided that the variability is at an acceptable level.

In order to illustrate how the various indexes are calculated, consider the filling data already presented in Exhibit 1. The process mean and standard deviation have been estimated to be

$$\hat{\mu} = 442.2$$

$$\hat{\sigma} = 2.26$$

Specifications were given as

$$USL = 455.5$$

$$LSL = 428.5$$

The various indexes are calculated as follows:

$$C_p = \frac{USL - LSL}{6\hat{\sigma}}$$

$$= \frac{455.5 - 428.5}{6(2.26)}$$

$$= \frac{27}{13.56}$$

$$= 1.991$$

$$C_{pU} = \frac{USL - \hat{\mu}}{3\hat{\sigma}}$$

$$= \frac{455.5 - 442.2}{3(2.26)}$$

$$= \frac{13.3}{6.78}$$

$$= 1.962$$

$$C_{pL} = \frac{\hat{\mu} - LSL}{3\hat{\sigma}}$$

$$= \frac{442.2 - 428.5}{3(2.26)}$$

$$= \frac{13.7}{6.78}$$

$$= 2.021$$

$$C_{pk} = \min(C_{pL}, C_{pU})$$

$$= \min(2.021, \ 1.962)$$

$$= 1.962$$

Since we know beforehand that the process data evidenced control about the targeted mean, on the surface a C_p index equal to 1.99 more than suggests that, in this particular case, the process is capable because the natural spread of the process is close to half the spread between the specification limits. Based on a C_{pk} equal 1.962, we can automatically conclude that the process is capable of meeting specifications. The C_p and C_{pk} indexes are close in value since the process is normal and properly centered.

If, instead of the two-sided specifications, suppose only underfill were a problem and we used the label claim limit of 425 grams as a one-sided lower specification limit, or

$$LSL = 425$$

Of the various indexes, only C_{pL} is appropriate and is calculated as

$$C_{pL} = \frac{\hat{\mu} - LSL}{3\hat{\sigma}}$$

$$= \frac{442.2 - 425}{3(2.26)}$$

$$= \frac{17.2}{6.78}$$

$$= 2.537$$

If an index were used in the original problem posed in Chapter 7 where underfill alone were considered, this would be the one. The conclusion based on this value would have been that, assuming control, the process is capable of meeting the label claim limit all the time. The corresponding probability of non-conformance would equal 1-in-100-trillion. Rather than using the simulated data, if we would have used the results of the actual charted data, the C_{pL} index would equal 1.951, which corresponds to a probability of nonconformance of roughly 2-in-1-billion!

ADDITIONAL IDEAS

Four commonly accepted procedures for establishing the capability of a process for meeting specifications have been presented. Emphasis was placed on what the process is doing in a state of control rather than what it can do if additional changes are made. Before discussing the latter, let us discuss some issues related to what we have done.

The histogram approach is the most general since it does not assume normality, as do the last three methods. Further, the histogram supplies more information about the process data than do the other approaches. Since a fitted curve of any form is based on a minimal set of characteristics or parameters, use of a smoothed curve alone tends to obscure features of the data that may be important. The advantage of a fitted curve is that it provides a convenient way of establishing generalized probability statements regarding the fraction conforming and non-conforming. If probability estimates are not of interest, the use of an interval to establish the natural tolerance limits accomplishes the goal of determining whether specifications can be met based on the same set of process data.

Of the capability indexes presented, C_{pU}, C_{pL}, and C_{pk} provide more information than the C_p index since they are related to existing process centering in addition to the specification limits. These indexes do not really provide any more information than the normal curve fit or an interval estimate based on the assumption of normality, since a correspondence exists between fraction non-conforming and values of the C_{pk} or corresponding one-sided indexes. Multiplying these by 3 yields the value for the standardized normal variate, z, in the one-sided case, or when the estimated process mean is centered between the upper and lower specification limits in the two-sided case.

Exhibit 4 presents the number of non-conforming items per million produced corresponding to selected values of C_{pL} or C_{pU} and C_{pk}, which are indicated in terms of one-sided or two-sided specifications. A normal centered process that consistently produces a capability index of 1 will produce either 1350 or 2700 defectives per million produced, depending on whether specifications are one or two-sided. The larger the value of the index, the further the specification limits are from the existing process mean, and the smaller the number of parts per million defective.

EXHIBIT 4
NUMBER NON-CONFORMING ITEMS PER MILLION (ppm)
FOR SELECTED VALUES OF CAPABILITY
INDEXES FOR A NORMAL CENTERED PROCESS

Capability Index	No. Of Standard Deviations From Mean	One-Sided Specification	Two-Sided Specification
.50	1.50	66810	133620
.75	2.25	12220	24440
.90	2.70	3470	6940
1.00	3.00	1350	2700
1.10	3.30	483	967
1.25	3.75	88	177
1.33	3.99	33	66
1.40	4.20	13	27
1.45	4.35	7	14
1.50	4.50	3.4	6.8
1.60	4.80	.8	1.6
1.67	5.01	.27	.55
1.80	5.40	.03	.07
2.00	6.00	.0009	.002

Guidelines have been established relating to the capability ratios in order to ensure against non-conforming product. The most commonly accepted value is 1.33. When safety and extra strength are important, higher values are suggested. Also, when dealing with a new as opposed to an existing process, higher values are recommended. In such cases, capability ratios between 1.45 and 1.67 can be considered.

If the process is not normal, the probability of non-conformance can be larger than the ones on which the values in Exhibit 4 are based. Consequently, the parts per million figures all will be larger based on the nature of the departure between the actual process distribution and the normal. If evidence suggests a non-normal process distribution, obviously one should aim toward higher capability ratios corresponding to an accepted level of non-conformance, especially if the functional form of the process distribution has not been established.

It should be noted that in any presentation such as this, there is a tendency to focus on "the process" or "the variable" and discuss guidelines in terms of these singular concepts. In an actual application there may exist more than one or many variables associated with the acceptability of a product. In such cases, capability should be considered with respect to all variables with the goal of improving the process with respect to each. The capability of the process in terms of product acceptability, however, ultimately is determined in terms of the characteristic with the *lowest* capability ratio, or the poorest performance.

Reconsidering the Fundamentals

It has been stated repeatedly that methods such as the ones introduced can be used as guides for action. They do not, however, provide conclusive information regarding the capability of a process in meeting product specifications. On the one hand, the results associated with each of the methods are subject to *sampling variation*. Obviously, the larger the amount of data used, the less sampling variation that exists. On the other hand, we are dealing with a *dynamic* decision environment in which changes constantly are occurring within a process over time. Consequently, a correct decision regarding process capability at one point in time may not apply to future time points.

In order to reinforce these ideas, it is instructive to return to the concept of an underlying process distribution. Recall that this is a theoretical construct that drives our results, but about which we do not have complete information. In other words, we never really know the exact form of the process distribution at any point in time. The concept, however, does provide a basis for dealing with resulting process data.

Exhibit 5 presents diagrams of various theoretical distributions. Two sets are given, such that one corresponds to the symmetric case and the other when skewed. Although many process control procedures assume symmetry, the general concepts apply to all cases. Superimposed on both sets of diagrams are specification limits. The endpoints of each distribution are assumed to be the natural tolerances produced by a process.

At a given point in time, we assume that there exists an underlying process distribution. If we assume that the process is homogeneous, the distribution could appear as any of the ones shown in Exhibit 5. If the distribution remains the same over time, the process is in a state of natural control. It may, however, not be in control with respect to targeted values of the mean and standard deviation, nor may the process be capable of meeting specifications all of the time.

EXHIBIT 5
HOMOGENEOUS PROCESS DISTRIBUTIONS IN RELATION
TO PRODUCT SPECIFICATIONS

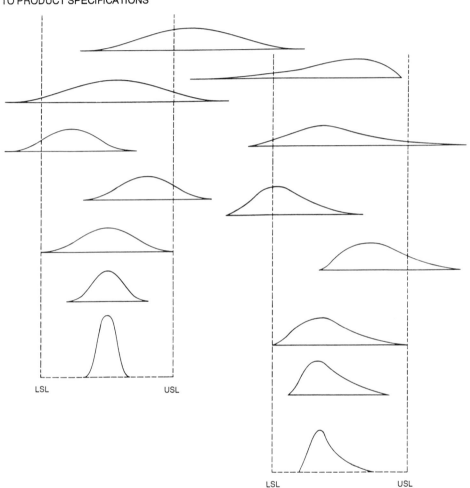

Assuming control, if the underlying process distribution were like the ones shown at the very top of the exhibit, we see that it is not centered between the specification limits. If it were, the variation is too great to produce all items within specifications, which is shown in the second diagram going down the page.

With regard to the third and fourth cases depicted, the natural spread exactly equals the spread between the tolerances; however, neither are centered between the tolerances. When centered properly, as in the fifth case, the process is completely capable and all items will meet specifications. Small changes, however, in the process mean or standard deviation will result in non-conforming items.

By continually improving the process through continued reduction in variation, process distributions such as the last two shown could result. In such cases, the process mean or standard deviation could go out of control, yet the process may still produce all conforming product.

When one is interested in determining the capability of a process, interest lies in capturing the essential features of the process distribution as depicted. Since process data are used in order to establish capability, the information is incomplete. As we have said, the data are subject to sampling variation, and the process continually experiences changes over time. Consequently, whatever method is used, it must be applied on a *continual* basis.

The most information is obtained by using histograms, or their equivalents, along with control chart information. Once control is established on the basis of the charts, observing histograms over time provides a visual picture of the nature of a process by reflecting the nature of the underlying distribution. The other methods, while useful, summarize the information in the histogram to varying degrees.

The Filling Process Problem

Before closing, it is instructive to reintroduce the filling problem. Now that we have introduced control charts, their interpretation, and the establishment of capability, we are in a position to tie the material together in order to demonstrate the way in which the concepts may be used effectively.

The methods for establishing capability were illustrated in terms of the simulated filling process data of Chapter 7, which were based on the original targeted mean of 442 grams and a standard deviation equal to 2.3 grams. The standard deviation was based on an estimate from the process data after removing points outside the natural control limits calculated from the process data.

Although the standard deviation of 2.3 grams appears to be a more realistic estimate of the natural process variability, more really needs to be done in order to determine the true measure of error. Also, the measure of error, equal to 4.5 grams, used initially to establish the specification limits and as a basis for control chart limits, was arbitrarily determined and was overstated to insure against violating the label claim limit. Consequently, the label claim was never violated, even with points out of control. By setting an upper specification limit based on the same error considerations, the problem of overfill was not properly addressed. By re-examining the data and clearly addressing the relevant issues, it is possible to deal with the label claim and the problem of overfill more effectively.

Exhibit 6 presents dotplots of the original filling process data provided in Chapter 7. A *dotplot* provides basically the same information about the patterns in a set of data as a histogram without frequencies indicated on a vertical scale. In the case of the filling data, the integer part of the weights are presented on the horizontal scale and the decimal part is plotted as a dot above the integer associated with it. The decimal value is not recorded. For example, values from 436.0 to 436.9 would be positioned above 436 while values from 437.0 to 437.9 position above

437; hence the value of 436.7 would be represented as one of the dots above the value 436. The more dots above a particular integer part, the higher the frequency. In the cases depicted, the dotplots are equivalent to histograms with an interval width equal to one gram. In a sense, therefore, the dotplot is more detailed and clearer than a histogram since there is less of a grouping effect. Further, the dotplot is very convenient to use when there are small amounts of data.

EXHIBIT 6
DOTPLOTS OF ORIGINAL FILLING PROCESS DATA AGGREGATED
AND PLOTTED BY INITIAL NOZZLE DESIGNATION NO. 1-5

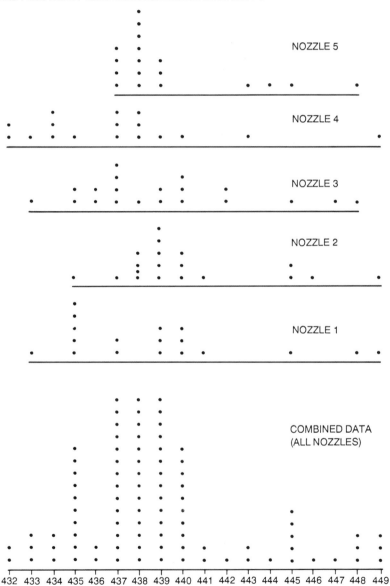

At the very bottom of Exhibit 6 is a dotplot of the original filling process data for all nozzles combined. Values cover a range roughly between 432 and 449, with the bulk clustering within the vicinity of 435 to 440. In addition to the relatively large amount of variation, there is evidence of multi-modality, indicating the possibility of heterogeneity in the process.

Appearing above the graph cited are the dotplots for each of the originally designated nozzles numbered 1-5. Notice that these vary in appearance in a number of ways. Aside from sampling variation, the plots appear differently with respect to the centering of the bulk of the dots or observations and with respect to variation. As an example, consider Nozzle 4, whose data are more spread out than the others. Also, some evidence of bimodality is present here, as well as in other cases.

Some of the extreme variation can be accounted for in terms of the values resulting from over-and-under adjustment that was evident in terms of some points outside the natural control limits on the mean chart. When the individual values corresponding to these samples are eliminated, the resulting dotplots appear as those in **Exhibit 7**. In all cases depicted, the variation appears diminished. The combined distribution still appears multi-modal and the individual nozzle distributions still vary with respect to central tendency and dispersion, indicating heterogeneity among nozzles with respect to central tendency and dispersion.

If you will recall, Nozzles 6-10 were not isolated and were mistakenly labeled as 1-5. The five individual distributions are further broken up into dotplots of the ten individual nozzles in **Exhibit 8**. Although the number of data points corresponding to each nozzle is sparse, it is evident that differences in central tendency and variability exist among the nozzles, although the variation within each nozzle appears less in each case.

In order to appreciate the kinds of things happening, consider the original distribution of observations from Nozzle 4 in Exhibit 7. When separated as Nozzle 4 and Nozzle 9 in Exhibit 8, the two clusters appearing in the original plot now appear as two distinct distributions, whereby observations from Nozzle 9 are far to the left of those of Nozzle 4. The difference between the central tendency corresponding to the two nozzles created the appearance of greater variation in the nozzle originally designated as Number 4.

Many of the observations made on the basis of the dotplots already have been suggested when we discussed the interpretation of the control charts. Recognize that the control charts do not point to the specific nozzles which are different from each other. Moreover, by introducing the individual plots, we are able to emphasize, in a simple way, the need for further analysis in order to establish possible additional needs for control and a second aspect of process capability. That is, determining how capable the process can be made to be beyond the initial capability established.

With respect to our example it appears that, in addition to the overall filling machine setting, the individual nozzles should be adjusted in order to provide a more uniform output. In the particular case, separate control charts were placed on each individual nozzle until it

**PROCESS
CAPABILITY**

was understood how to stabilize the output of each. Once this was accomplished, new targets were established and a single control chart was instituted using a sample size of 10 with one observation per nozzle. The targeted mean and standard deviation were changed to 431 grams and 1.5 grams, respectively. By overcoming the tendency to over-and under adjust without a measurable basis and eliminating much of the nozzle-to-nozzle variation, the basic measure of chance error was reduced. Some insight regarding the implications of this will be illustrated.

EXHIBIT 7
DOTPLOTS OF FILLING PROCESS DATA WITH POINTS
OUT-OF-NATURAL-CONTROL LIMITS REMOVED

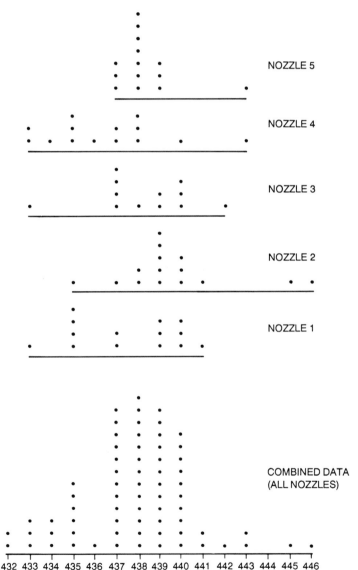

EXHIBIT 8
DOTPLOTS OF FILLING PROCESS DATA WITH POINTS
OUT-OF-NATURAL-CONTROL LIMITS REMOVED FOR NOZZLES
NO. 1-10 ISOLATED

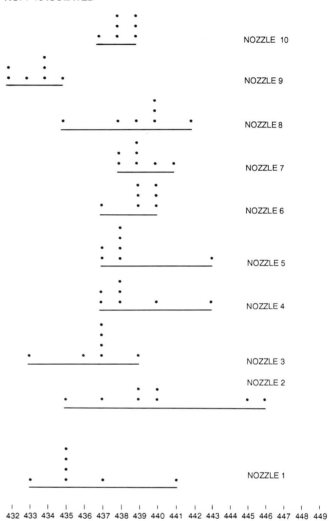

An example of the output of a sequence of twenty-five samples is provided in **Exhibit 9**. In addition to the sample means and standard deviations, the mean and standard deviation associated with individual nozzles also are provided. Notice that the values of the means are quite close. Also, differences exist among the standard deviations; however, this can be attributed to sampling variation.

Results associated with the process data are presented in **Exhibit 10.** By examining the control charts at the top of the exhibit, it appears clear that the process is in control with respect to centering and dispersion about the established targets. The control charts are placed in the same exhibit as the capability information that follows in order to emphasize that control should be demonstrated before establishing capability with any of the methods introduced.

A histogram constructed on the basis of all the data is presented. The histogram is single peaked with no evidence of heterogeneity present. Although masked by grouping, the lowest and highest weights actually observed equal 427.2 and 435.8 grams, respectively. A normal curve fitted to the data is superimposed on the histogram with no unreasonable departures apparent. The histogram has been rescaled so that areas of the bars equal the relative frequency and, therefore, a vertical scale has not been given.

EXHIBIT 9
CONTROLLED FILLING PROCESS DATA FOR NOZZLES 1-10 BASED ON
TARGETS $\mu_0 = 431$ AND $\sigma_0 = 1.5$ GRAMS

Sample Number	Nozzle										$\overline{X}$	S
	1	2	3	4	5	6	7	8	9	10		
1	428.6	432.7	432.0	433.5	430.4	430.5	428.3	431.1	430.8	430.6	430.8	1.63
2	428.5	432.2	433.4	430.0	432.3	429.1	433.2	431.1	430.9	430.9	431.2	1.65
3	429.0	431.9	433.0	431.2	427.7	433.2	433.0	430.5	428.6	430.7	430.9	1.97
4	432.6	431.8	430.9	431.1	430.0	431.8	429.9	431.1	433.0	430.0	431.2	1.09
5	429.0	431.3	431.6	431.8	429.5	430.6	431.1	429.5	430.5	428.0	430.3	1.24
6	429.8	431.4	430.8	432.0	432.1	432.2	432.8	431.1	431.5	428.0	431.3	1.17
7	430.0	433.0	431.8	430.4	429.4	430.7	429.7	432.2	429.1	432.8	430.9	1.44
8	430.8	429.9	429.4	429.6	432.1	430.6	431.2	432.8	428.8	433.9	430.9	1.62
9	430.5	432.8	432.4	431.2	435.8	430.6	433.2	429.3	432.9	430.9	432.0	1.84
10	432.1	430.0	430.4	431.2	431.7	429.8	431.9	432.7	432.5	429.7	431.2	1.13
11	430.9	429.5	429.0	430.0	431.3	431.9	428.9	430.1	430.2	430.6	430.2	0.96
12	431.4	431.5	430.5	431.0	430.0	434.1	431.8	432.3	431.9	429.8	431.4	1.25
13	430.3	430.9	431.5	431.7	430.9	429.2	432.2	430.6	433.1	429.4	431.0	1.22
14	427.9	429.6	429.0	434.0	432.6	429.0	430.1	427.6	429.5	431.7	430.1	2.07
15	431.0	429.4	432.8	429.9	430.2	433.9	430.6	430.3	431.4	430.1	431.0	1.41
16	430.5	429.6	433.5	429.2	427.6	430.8	433.3	430.8	432.8	432.2	431.0	1.93
17	432.1	429.9	427.2	432.9	431.5	432.9	432.0	433.5	429.9	430.0	431.2	1.91
18	432.7	430.4	431.6	431.8	431.3	429.2	431.2	432.9	431.5	428.9	431.2	1.31
19	430.9	430.0	431.8	430.9	429.2	432.9	430.6	430.7	430.4	433.7	431.1	1.34
20	431.3	431.2	432.3	428.2	432.6	430.8	431.1	430.4	429.8	430.0	430.8	1.26
21	431.4	431.0	432.5	432.3	430.9	429.8	431.8	431.3	430.7	431.0	431.3	0.80
22	433.0	430.6	430.7	430.5	432.0	430.3	430.4	431.1	431.6	432.8	431.3	1.01
23	431.2	430.6	428.9	429.7	439.8	432.6	429.6	431.8	427.6	430.8	430.4	1.43
24	431.0	429.0	429.4	429.6	427.4	431.5	431.4	431.1	428.4	431.8	430.1	1.49
25	431.5	434.4	430.7	428.6	430.5	430.2	432.7	430.2	432.4	432.0	431.3	1.62
Mean	430.7	431.0	431.1	430.9	430.8	431.1	431.3	431.1	430.8	430.8	431.0	1.44
Std.Dev.	1.36	1.33	1.59	1.44	1.83	1.51	1.39	1.28	1.58	1.49		

Of the available capability indexes, the C_{pL} was calculated since the real concern was associated with not falling below the label claim limit of 425 grams. Based on this value, the C_{pL} index equals 1.33. This exceeds one and complies with the generally accepted guidelines for capability. Implied is that the label claim limit is roughly four standard

leviations from the process mean and, assuming normality, ap-
roximately three items of every 100 thousand produced will fall below
he limit.

It is of interest to note that the value of C_{pL} given above is less than
he one calculated on the basis of the original process data presented
ie., $C_{pL} = 1.95$). Although the variation was larger, the higher value of
he index results from the fact that the process was operating closer to
a higher targeted process mean. The fact that the process was not in
control implies that one cannot count on the result, and that controlling
he process at a lower targeted mean with less variation provides a more
uniform product with less overfill.

Exhibit 10
RESULTS OF PROCESS CAPABILITY ANALYSIS, RECTIFICATION
AND REDEFINITION—PROCESS IN CONTROL

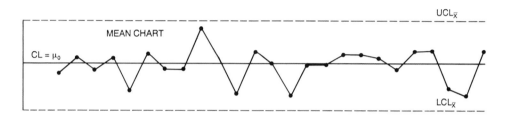

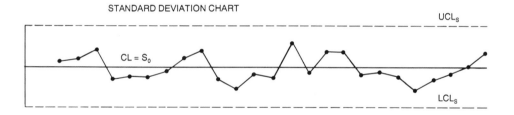

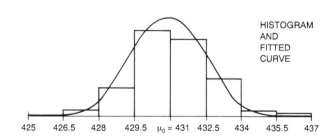

GIVEN:

$\mu_0 = 431$ GMS
$\sigma_0 = 1.5$ GMS

$\overline{\overline{X}} = 431.0$ GMS
$\hat{\sigma} = 1.48$

INDEX:

$$C_{pL} = \frac{\hat{\mu} - LSL}{3\hat{\sigma}}$$

$= 1.33$

ESTIMATED NATURAL TOLERANCE:

$\hat{\mu} - 3\hat{\sigma} = 426.6$
$\hat{\mu} + 3\hat{\sigma} = 435.5$

In order to appreciate what has happened as a result of controlling the process at the new targets, consider **Exhibit 11**. There five dotplots are presented. The one at the bottom is a reproduction of the dotplot corresponding to the initial process data that was completely out of control and based on the initial targets. This has been included as a frame of reference for the other plots.

Appearing immediately above is a dotplot of the process data corresponding to a random sample of three of the ten nozzles. Since basically the *same pattern* would be provided if all ten were considered, in the interest of space, a sample was used. The dotplots of the individual nozzles sampled appears at the top.

The first important point to note is that the current distribution appears as single peaked, indicating that the process is *homogeneous*. Second, the center of the distribution is *shifted* much farther to the left than the original, and the distribution is much *tighter*, indicating considerably less variation. Third, although closer to the lower specification limit, or the label claim, *none* of the points is below the label claim limit.

When examining the individual nozzle distributions, they appear different. The differences, however, are due to sampling variation. More notable, though, is the fact that they appear *aligned* above the center of the combined distribution, and the variation is basically the same. The patterns indicating heterogeneity in Exhibits 6, 7, and 8 no longer are present. The same is true of the remaining seven distributions not displayed.

Exhibit 12 summarizes what has been accomplished by comparing the underlying distribution associated with the initially established and arbitrary specifications, the resulting distribution after over-and-under adjustment was eliminated, and the final distribution resulting from a capability analysis and process improvement. The message embedded within the exhibit is *fundamental*.

In the first case, the average contents were targeted at a very high value with a conservatively large amount of error. At such levels, the process could be very much out of control without violating the label claim. By eliminating the adjustment problem, it was found that the initially targeted mean really was not met; however, the variation also is somewhat lower than anticipated. Assuming the degree of heterogeneity among nozzles remains stable, the process could be operated at this level and meet the label claim. However, by assessing the differences among the nozzles, improvements were instituted that allow for a still lower process mean based on a lower assessed value of inherent or chance variation.

Based on the initial targets, the average overfill per container is roughly eleven grams higher than what it has been demonstrated to be and still meet the label claim. Effectively, average overfill has been reduced by roughly seven grams from what the process was producing. Implied, however, in the final distribution presented is the fact that the process must be more closely monitored and controlled in order to meet the label claim. Increased variation will result in a wider distribution, requiring a higher mean in order to meet the label claim. Continued analysis based on the nature of the contents, nozzle design, nozzle

EXHIBIT 11
DOTPLOTS COMPARING CONTROLLED AND IMPROVED PROCESS,
INDIVIDUAL NOZZLES AND ORIGINAL PROCESS

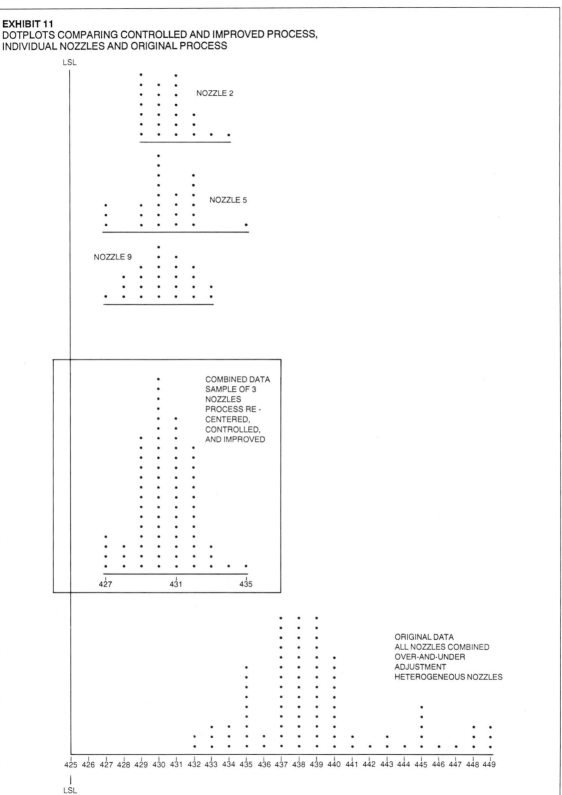

adjustments, and machine settings could lead, however, to still lower amounts of variation.

EXHIBIT 12
COMPARISON OF PROCESS DISTRIBUTIONS BASED ON ARBITRARY
SPECIFICATIONS, INITIAL CONTROL, AND PROCESS CAPABILITY
ANALYSIS AND IMPROVEMENT

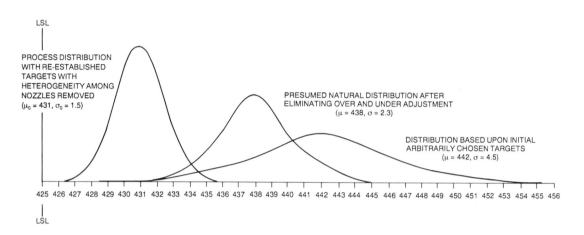

Reduced variation in this case leads to one of two possible situations. On the one hand, if the targeted mean is maintained, there is a larger margin of error between the distribution of weights and the label claim. The mean, then, can move out-of-control within limits without worrying as much about the label claim. On the other hand, the targeted mean level could be reduced, which aims at less overfill, but which must be more tightly controlled because a lower shift in the distribution would lead to violations of the claim.

By viewing the filling problem in the way that we have, it really is not necessary to place a specified upper limit on the weight of the contents. If, however, as a result of the process improvements the container size is reduced in the future, some consideration may be given to a maximum weight.

Assuming variation is under control, two-sided *control limits* on the mean chart are still useful. When out-of-control is indicated at the lower end, the potential problem is one of violating the label claim. On the contrary, out-of-control at the higher end is related to problems of overfill. In the case of variation, two-sided control limits are useful also. Out-of-control at the high end is potentially a problem both in terms of overfill and violation of the label claim. At the low end, if sustained, possible improvements in terms of reduced variation may be signaled.

In summary, process capability analysis is an important component of any quality control effort when used in conjunction with control charts. Mainly, capability analysis is useful in the following ways:

1. Provides a basis for establishing how well a process meets specifications

2. Provides a basis for predicting how well a process will meet specifications

3. Provides input into subgrouping decisions for purposes of control

4. Provides basis for re-assessing, establishing, and modifying specifications

5. Suggests measures for controlling a process

6. Provides a basis for further statistical studies that can lead to the reduction of variability and process improvement

7. Can be used as a tool to assess vendor performance

Like any method of statistical analysis, process capability procedures and measures should not be used blindly. A good illustration of this is presented in the filling problem for which a substantially large capability index could have been calculated on the basis of the original data collected. If taken at face value, the conclusion would have been that no problem existed. This conclusion results from the fact that the problem originally was not defined properly. Knowledge of the process coupled with a clear definition of goals is essential when assessing any measures of process performance.

Some Final Comments

The filling process problem used throughout this module was introduced because of its simplicity in the sense that it represents a well-defined and self-contained process. The output of this process constitutes a stream of production based on the existence of multiple filling heads. Streams of output can result in many ways based on output from multiple machines, operators, shifts, and so on.

The concept of a stream was useful to our presentation because we were able to use the results to demonstrate how alignment of the means and variation of the components of the stream leads to better defined targets and reduced variation. Further, we were able to illustrate this with the simplest tools of data analysis that are based on material introduced at the outset in Chapter 1. Many more statistical tools are available which can aid in determining factors affecting variation and how to reduce it. The most important thing to obtain from the use of the dotplots is a stronger grasp of the concept of variation and how it can be examined and managed; the ideas presented go far *beyond* the particular problem illustrated.

Although every process control problem is not the same, certain general principles and concepts are common to them all. All rest with an understanding of the nature and the management of variation in key process characteristics and involve the following four components in order to be complete:

1. Measurement
2. Control
3. Capability
4. Improvement

Simply stated, it is not possible to understand objectively or to manage a process unless measurable evidence in the proper form is collected and analyzed. Based on the use of measured data, it is possible to manage the variation in a process, ultimately to provide quality product at lower cost. If process control is used *effectively*, it is possible to reduce or eliminate many different types of inspection activities, since all of the required information about the product is embedded within the control and capability data and emphasis is shifted to the process producing acceptable items. As the capability of a process becomes better with respect to essential product characteristics and major special causes are eliminated, less and less concern may be needed with respect to small out-of-control indications in terms of their effect on specifications. When tolerances are tightened, more responsive control procedures cited later may be sought in order to detect small changes.

Key to a total quality control effort is *continued improvement*. Assuming properly defined targeted levels are maintained, quality improvement is more precisely defined in terms of the reduction of variation. Although it is not possible to eliminate variation entirely, it is possible to continually chip away at the sources and make it smaller and smaller. By viewing the problem in this way, quality is redefined dynamically over time rather than in terms of a static set of requirements.

ADDITIONAL CONTROL CHARTS

Module Summary

The preceding modules deliver the message relating to the main theme of the book. Fundamental statistical principles associated with the primary process control tool have been presented. Since variables charts are the most useful for process control, the basic variables charts were emphasized. The underlying reasoning applies to all charts.

In addition to the variables charts presented in Module C, additional charts with different features that also can meet different needs are available. Moreover, charts dealing with different data types are useful in different circumstances. This module introduces additional control charts that are useful in different circumstances.

Chapter 10 introduces np- and p-charts, which are applied to qualitative process or product information that are dealt with in terms of counts rather than measured quantities. The np- and p-charts are associated with characteristics that are either present or absent, which include defective or non-conforming product. A brief discussion of Pareto diagrams, a useful problem-solving tool, is given at the end of the chapter. C and u-charts, which are considered in Chapter 11, are associated with the number of abnormalities, which includes defects, or non-conformities, and errors. Some attention is given in both chapters to related material already presented in Module C.

Chapter 12 mainly is concerned with two additional types of variables control charts. The first of these are the moving average and moving range charts. In addition to being more responsive to change than the mean and standard deviation or range charts already discussed, they can be used in situations where it either is expedient or necessary to use a single observation at any given point in time.

Median charts as alternatives to mean charts also are considered. Although a median chart does not add to the ability to provide more information, median charts are very convenient to use manually since no computation is required with respect to individual subgroup observations. Additionally, more sophisticated types of variables charts are cited at the end of the chapter.

ATTRIBUTE DATA, np- AND p-CHARTS

The previous module was concerned with process control problems in terms of measurements, or variables data. Measured quantities, when available, provide the most information; however, in some cases these either are not available or the requirements of a particular problem suggest other alternatives. In such cases, qualitative data of one form or another may be used.

Quality characteristics always exist that cannot be measured or are too costly or time-consuming to measure. In such cases, information from gauges or visual inspections result in conclusions that items do not conform to standards or specifications on a qualitative basis. The ultimate concern associated with a final product is whether it meets the needs of its intended purpose or not. When confronted with such situations, attribute control charts can be used as aids in process control.

In other cases where measurements can be taken on a truly variable basis, attribute charts could be used; however, they are not as sensitive as variables charts and do not provide the same level of information about a process as do variables charts. They can, however, be employed as useful supplements as quality summaries when used in conjunction with variables charts.

In particular, original specifications can be designated on a measured scale, yet acceptance can be on a pass-fail, go-no-go, or present-absent basis. The most common example considers an item either as acceptable or unacceptable in terms of specifications, and we consider it to be defective, or non-conforming. In other cases, an item may possess imperfections in terms of one or more defects, or non-conformities. An item may contain a number of defects and not be defective.

The terms *defective* and *defect* have been commonly used up to the present. However, due to inconsistencies regarding the terminology,

243

standards are in the process of changing and the terms *non-conforming* and *non-conformity*, respectively, are used in their place. The real problem is based on legal issues associated with the everyday use of the term defective, which implies that a product is not suited for its intended use. It is possible,however,that a product may be suited to a particular use but is defective with respect to a particular specification. Recognizing the relation among the various terms, we shall adopt the use of the terms non-conforming and non-conformity to mean unacceptable and imperfection, respectively.

When working with non-conforming items or product and non-conformities quantitatively, we deal with *counts* or the number of non-conforming items or non-conformities. Therefore, control charts associated with non-conforming items either can be constructed in terms of the number or fraction or percent non-conforming. In the case of non-conformities we deal with the number per unit of space or time. Although the underlying distribution theory departs from that of variables, basic control chart procedures in these two cases reduces to the procedures we have discussed with some modifications.

In this chapter we introduce control charts for items that are defective, rejected, or non-conforming. Defects, or non-conformities, are considered in Chapter 11. Since many of the basic principles underlying control charting have been presented in Chapters 6-9 concerned with measurable variables, these principles will be reviewed but not repeated in detail with respect to attribute charts. The relation between the two types of charts and departures between the two will be highlighted. To a great extent, this chapter does stand alone; however, familiarity with earlier material is useful in order to understand all of the ideas presented.

Control charts for non-conforming items can be in the form of the number *or* the fraction non-conforming in a sample in terms of an np or a p-chart, respectively. For purposes of process control, the four components measurement, control, capability, and improvement apply to non-conforming items, and will be considered throughout this chapter. Actually, since we are to deal with qualitative or categorical data, it would be more precise to substitute the term "quantify" for the term "measurement"; a similar connotation, however, is implied. Moreover, due to the nature of the qualitative information used, a direct relationship can be established between the corresponding control chart information and process capability and product acceptance. This relationship and the problem of product acceptance is addressed at the end of the chapter along with other related issues.

The Nature of Non-Conforming Data

Before introducing the control charts, it is useful to discuss the nature of the various situations from which attribute data are generated. By doing so, it is easier to place the subsequent material in perspective, especially in terms of what can and cannot be accomplished with attribute charts. Of importance is the distinction between the nature of an individual item produced by a particular process and the nature of the sample data used for charting purposes.

It was clear when using variables control charts that each chart corresponded to a single relevant quality characteristic. If more than one characteristic was necessary to monitor and control a process, separate charts would be maintained on each characteristic. This, however, is not necessarily the case with respect to attribute charts.

An individual universe element or process item can be designated as non-conforming in a number of different ways:

1. Clearly categorical

   (a) Based on a single qualitative characteristic that is present or absent

   (b) Based on more than one qualitative characteristic that is present or absent

   (c) Based on a single type of non-conformity that occurs more than once and exceeds some pre-established numerical threshold

   (d) Based on more than one type of non-conformity, the number of which either in combination visually or physically is unacceptable or exceeds some preestablished numerical threshold regarding the total number of non-conformities

2. Measurable characteristic that is translated into a present-or-absent condition

   (a) Based on a single measurable characteristic

   (b) Based on more than one measurable characteristic

3. Realizable combinations of the above possibilities

Before considering the importance of introducing the above list, let us consider some examples of each case delineated. In the first case, regarding a clearly categorical characteristic, a filled container could have a cracked bottle cap, or a printed document or carton could have a visually unacceptable color tone. A device simply may not operate. These are examples for which a single quality characteristic is either present or not. An example in which more than one qualitative characteristic each of which is present or absent is a filled container with a cracked cap whose label has an unacceptable color tone.

Recognize that in the above examples, a crack in a bottle cap and an unacceptable color tone can be considered as imperfections, or non-conformities. Although a container that contains each imperfection can be considered to possess two imperfections, each is considered to be either present or absent, or as a single non-conformity of each type.

By way of contrast, a dishwasher cabinet or housing could possess one or more paint blisters; either case could lead to the dishwasher's acceptance or unacceptance depending on pre-designated criteria. The dishwasher could be returned for refinishing with one blister, two blisters, or ten blisters. In such cases, the pre-designated criterion determines the present or absent state leading to lack of acceptability.

We can take the above example a step further by considering blisters, nicks, and dents, all of which simply are considered as imperfections without regard to type. A dishwasher, therefore, can be considered unacceptable, or non-conforming, if it possesses any number of any type or combination of imperfections.

The second basic way in which an item can be considered to be non-conforming is if an item does not fall within measured design specifications, or tolerances. Although more information is provided by working with the measurements directly, treating an item as acceptable or unacceptable can be advantageous, especially when interested in developing a running commentary on product acceptability based on multiple quality characteristics. For controlling processes in terms of identifying assignable causes of variation, however, the designation really is beneficial in cases where a single qualitative characteristic is used and is the only alternative.

Four main points relate to the above discussion:

1. Unlike variables data, an item can be characterized as non-conforming based on different designations and types of information

2. When items are designated as non-conforming in terms of more than one non-conformity or in terms of a measured characteristic, alternative charting procedures are available beside those discussed in this chapter

3. When more than one type of quality characteristic and combinations of the alternative possibilities are considered, the corresponding control charts are less suited for process control purposes but still may be useful for ongoing monitoring and product acceptance

4. Control charts for non-conforming product are based on samples of items and are plotted in terms of the number of fraction non-conforming in the sample. The individual ways in which items are designated as non-conforming may or may not be identified.

The remaining portion of this chapter presents, in turn, the construction of control charts associated with non-conforming items, interpretation of corresponding control chart patterns, process capability and improvement, and the relation of such charts to the problem of lot acceptance. When necessary, the distinction among the different types of data and situations underlying the designation of non-conformance will be highlighted. Otherwise, the designation will not be assumed to be relevant to the illustration of material presented.

The symbols used in this chapter are defined as follows:

r = number of units rejected, or the number non-conforming in the sample

p = fraction rejected or non-conforming in a sample

P = process fraction non-conforming

P_0 = standard or targeted fraction non-conforming

n = sample size

k   = number of samples

$\bar{p}$ = the mean fraction non-conforming based on a series of samples

μ_r = mean of the sampling distribution of the number non-conforming

μ_p = mean of the sampling distribution of the sample fraction non-conforming

σ_r = the standard error of the number non-conforming

σ_p = the standard error of the sampling distribution of the sample fraction non-conforming

$\bar{n}$ = average sample size

CL   = center line

UCL   = upper control limit

LCL   = lower control limit

CONTROL CHARTS FOR THE NUMBER AND FRACTION NON-CONFORMING

In this segment we introduce the control charts that apply in cases where one of two possible states or outcomes is observable with respect to a single item or unit, and a sample of such items or units is selected at various time points associated with a particular process. More typically, we are interested in the output of a process in terms of that which is non-conforming. A non-conforming item or unit could be composed of any number of non-conformities or could be based on numerical tolerances.

Background (Optional)

The basic tool for establishing the center line and control limits on a control chart is the sampling distribution of the sample statistic observed. In the case of non-conforming items, the basic quantity of interest is the *number* of non-conforming items in a sample. As illustrated in Chapter 4, the underlying distribution of the number of non-conforming items, r, in a random sample of size n is given by the binomial distribution

$$f(r) = \begin{bmatrix} n \\ r \end{bmatrix} P^r (1 - P)^{n-r}; \; r = 0, 1, 2, \ldots, n$$

where P equals the process fraction non-conforming. The expression $\begin{bmatrix} n \\ r \end{bmatrix}$ represents the number of *combinations* of n things combined r at a time ways, and is given as

$$\begin{bmatrix} n \\ r \end{bmatrix} = \frac{n!}{r!(n-r)!}$$

The exclamation mark after each of the symbols indicates the *factorial* operation, meaning that the quantity preceding it is multiplied by all successive integers less than it. Values of the binomial distribution for selected values of n and P are tabulated in Table A-2 in the Appendix.

The binomial distribution is a sampling distribution in the sense defined in Chapter 5 and, therefore, automatically applies if we assume independent sampling with replacement. The mean and standard error of the binomial are given as

$$\mu_r = nP$$

$$\sigma_r = \sqrt{nP(1 - P)}$$

By dividing the number non-conforming by the sample size, we can obtain expressions for the fraction non-conforming, p, in the sample as

$$\mu_p = P$$

$$\sigma_p = \sqrt{\frac{P(1 - P)}{n}}$$

If we were to interpret the means in the two cases, the mean of r equals the process fraction non-conforming multiplied by the sample size, and the mean of the sample fraction, p, equals the process fraction, P. As the sample size is increased for any P, both distributions become more symmetric and can be demonstrated to approach the normal curve in the limit. This provides a reasonable basis to use the normal as an approximation for large samples. For details regarding the binomial distribution and the normal approximation to the binomial, the reader is referred to *Statistics for Decision Making* by Ronald Gulezian, Chapters 9 and 11.

Depending on one's personal preference, a control chart can be constructed for the number *or* the fraction non-conforming in a sample. Both provide the same basic information about a process. Charts based on the number are referred to as np-charts and those based on the fraction are referred to as p-charts.

Constructing np-Charts

An np-chart is one form of attribute chart that is used to chart processes involving qualitative characteristics of items that are designated as non-conforming or not. On an np-chart, the *number* of non-conforming items, r, in a sample of size n is plotted.

Just as in the case of variables charts we can distinguish between *two cases*, charts based on targets, or standards, and ones based on process data. Control limits conventionally are constructed as three standard errors from the center line. When process data are used, the mean and standard error of the underlying binomial distribution must be estimated from process data. Formulas for the two cases are given below.

Based on standard:

Center line, $CL_r = nP_0$

Upper control limit, $UCL_r = CL_r + 3\sigma_r$

$$= nP_0 + 3\sqrt{nP_0(1 - P_0)}$$

Lower control limit, $LCL_r = CL_r - 3\sigma_r$

$$= nP_0 - 3\sqrt{nP(1 - P_0)}$$

The symbol P_0 stands for the targeted process fraction non-conforming and σ_r represents the standard error of the number non-conforming in a sample.

Based on process data:

Center line, $CL_r = n\bar{p}$

Upper control limit, $UCL_r = n\bar{p} + 3\sqrt{n\bar{p}(1 - \bar{p})}$

Lower control limit, $LCL_r = n\bar{p} - 3\sqrt{n\bar{p}(1 - \bar{p})}$

In this case, estimates of the process fraction and the standard error are required. All are based on an estimate of the process faction non-conforming, which is given as $\bar{p}$, or

$$\bar{p} = \frac{\Sigma r}{\Sigma n}$$

The quantity $\bar{p}$ represents the *average* of the sample fractions and is found by dividing the total non-conforming in *all* samples, Σr, by the total number of observations in all samples, Σn. This quantity is equivalent to finding the weighted average of the individual sample fractions non-conforming. In either case, since the standard error is dependent on the fraction non-conforming, only *one* control chart is required, which provides all of the essential information.

Example 1 As part of a final check on completed product produced by a carton manufacturing process, a control chart monitoring the fraction of product that is non-conforming is maintained at the end of the entire process. A carton is determined to be non-conforming if, on a judgmental basis, it has too many imperfections such as wrinkles, spots, holes, or poor color tone. Aside from other imperfections, squareness alone is a key determinant of conformance since if not met, cartons will jam the customer's machinery. The control chart is maintained in order to monitor the fraction non-conforming and to aid overall in documenting the existence of problems along the line that otherwise may not be identified by other means.

Appearing below are data collected every two hours covering five working shifts. Equal samples of 150 cartons drawn at random were used. The data are to be used to illustrate construction of np-charts.

Carton Non-Conformance Data

Sample Number	Sample Size, n	Number Rejected, r	Fraction Rejected, p
1	150	4	.027
2	150	2	.013
3	150	4	.027
4	150	4	.027
5	150	8	.053
6	150	4	.027
7	150	4	.027
8	150	3	.020
9	150	5	.033
10	150	3	.020
11	150	5	.033
12	150	2	.013
13	150	3	.020
14	150	5	.033
15	150	2	.013
16	150	8	.053
17	150	4	.027
18	150	3	.020
19	150	8	.053
20	150	6	.040
Total	3000	87	

If a *standard* fraction non-conforming of 0.03 is adopted, the center line and control limits are found as follows:

$$CL_r = nP_0$$

$$= 150(.03)$$

$$= 4.5$$

$$UCL_r = nnP_0 + 3\sqrt{nP_0(1 - P_0)}$$

$$= 4.5 + 3\sqrt{150(.03)(.97)}$$

$$= 4.5 + 3(2.0893)$$

$$= 10.77 \text{ or } 11 \text{ rounded}$$

$$LCL_r = nP_0 - 3\sqrt{nP_0(1 - P_0)}$$

$$= 4.5 - 3(2.0893)$$

$$= -1.77 \text{ "or zero"}$$

Since the procedure employed really is an approximation, non-integer results can be obtained for the control limits. These can be rounded to the nearest integer. Further, if *a negative* lower control limit is calculated, this automatically can be changed to zero in most cases, since negative numbers are not appropriate in this case. A non-integer center line is acceptable since it corresponds to an average and does not have to be an integer.

If instead, the center line and control limits were based on the *process data*, we first must estimate the process fraction non-conforming. This is done as

$$\bar{p} = \frac{87}{3000}$$

$$= 0.029$$

Based on this result, we have

$$CL_r = n\bar{p}$$

$$= 150(.029)$$

$$= 4.35$$

$$UCL_r = n\bar{p} + 3\sqrt{np(1-\bar{p})}$$

$$= 4.35 + 3\sqrt{150(.029)(.971)}$$

$$= 4.35 + 3\sqrt{(2.0552)}$$

$$= 10.52 \text{ or } 11 \text{ rounded}$$

$$LCL_r = n\bar{p} - 3\sqrt{n\bar{p}(1-\bar{p})}$$

$$= 4.35 - 3(2.0552)$$

$$= -1.8156 \text{ "or zero"}$$

Since the control limits in this example are the same using both approaches and the center lines are very close, a plotted chart is illustrated in **Exhibit 1** in the case of the target only. In general, considerations regarding the two cases are similar to those discussed with respect to variables data. The limits are similar here since the estimated fraction and the standard are so close. Since control limits are dependent on these values, the limits would differ if more of a difference existed between the estimated and targeted fraction non-conforming.

By examining the exhibit, it is clear that the chart indicates control. No points fall outside the control limits, and there are no apparent patterns within.

EXHIBIT 1
EXAMPLE OF AN np-CHART

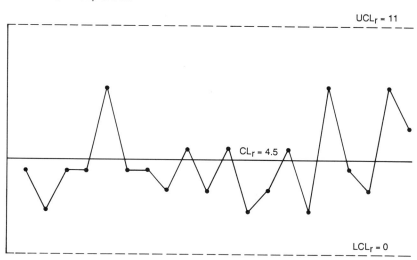

Constructing p-Charts

In the case of an np-chart, it is not necessary to make any calculations before plotting points on the chart. Although it is necessary to calculate fractions, some prefer to observe fraction rejected or non-conforming because they feel it is more informative. Further, fractions are independent of scaling considerations. The following formulas are used as a basis for p-charts:

Based on standard:

$$\text{Center line, } CL_p = P_0$$

$$\text{Upper control limit, } UCL_p = P_0 + 3\sqrt{\frac{P_0(1 - P_0)}{n}}$$

$$\text{Lower control limit, } LCL_p = P_0 - 3\sqrt{\frac{P_0(1 - P_0)}{n}}$$

Based on process data:

$$\text{Center line, } CL_p = \bar{p}$$

$$\text{Upper control limit, } UCL_p = \bar{p} + 3\sqrt{\frac{\bar{p}(1 - \bar{p})}{n}}$$

$$\text{Lower control limit, } LCL_p = \bar{p} - 3\sqrt{\frac{\bar{p}(1 - \bar{p})}{n}}$$

The terms P_0, n, and $\bar{p}$ are defined similarly.

Example 2   Use the information provided in Example 1 to establish control limits and construct a p-chart.

Based on the *standard* fraction non-conforming, 0.03, the center line and control limits are found as follows:

$$CL_p = P_0$$

$$= .03$$

$$UCLp = P_0 + 3\sqrt{\frac{P_0(1 - P_0)}{n}}$$

$$= .03 + 3\sqrt{\frac{.03(.97)}{150}}$$

$$= .03 + 3(.0139)$$

$$= .07$$

$$LCL_p = P_0 - 3\sqrt{\frac{P_0(1 - P_0)}{n}}$$

$$= .03 - 3\sqrt{\frac{.03(.97)}{150}}$$

$$= .03 - 3(.0139)$$

$$= -.1 \text{ "or zero"}$$

Again, since the lower limit cannot be negative, it automatically is changed to zero.

If *process data* were to be used instead, the results would be obtained as follows:

$$CL_p = \bar{p}$$

$$= \frac{\Sigma r}{\Sigma n}$$

$$= \frac{87}{3000}$$

$$= .029$$

$$UCL_p = \bar{p} + 3\sqrt{\frac{\bar{p}(1 - \bar{p})}{n}}$$

$$= .029 + 3\sqrt{\frac{.029(.971)}{150}}$$

$$= .029 + 3(.0137)$$

$$= .07$$

$$LCL_p = \bar{p} - \sqrt{\frac{\bar{p}(1 - \bar{p})}{n}}$$

$$= .029 - 3(.0137)$$

$$= -.01 \text{ "or zero"}$$

Using the values of the fraction rejected, p, in Example 1 and the first set of results calculated, the corresponding control chart is presented in **Exhibit 2**. Since the two sets of results are almost identical, only one chart is given. Although the scale is different, the pattern on the p-chart is the same as the one on the np-chart shown in Exhibit 1.

EXHIBIT 2
EXAMPLE OF A p-CHART

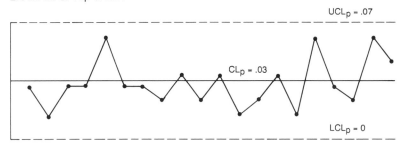

Unequal Subgroup Sizes

When working with variables control charts we implicitly assumed that the subgroup size was fixed at some small number. In such cases it is reasonable to assume fixed subgroup sizes, although it is possible that either information is lost or a change in subgroup size could occur. In the case of p- or np-charts, *larger* sample sizes are necessary and control over the size of the subgroups is *not* as great. Further, given the summary nature of p and np-charts, they tend to be of greater value in assessing quality of output for larger segments of production. Since production runs will vary in size, assuming constant subgroup or sample sizes is not as realistic.

Since the standard errors on which limits are based vary with subgroup size, it is necessary to account for this. Three alternative approaches are available in order to deal with unequal subgroups:

1. Recompute control limits for different subgroup sizes.
2. Construct control limits on the basis of an average sample size, $\bar{n}$, and recalculate a control limit when a point falls close to the ones based on the average sample size.
3. Construct two sets of control limits based on the largest and smallest sample sizes observed.

Since we have introduced considerations relating to batch as opposed to real-time applications in Chapter 7, it is not necessary to discuss them here. It should be recognized, however, that alternatives (2) and (3) require an accumulation of process data before limits can be constructed and would apply to observing a chart retrospectively. Aside from its more confusing appearance, and the inconvenience when applied manually, a chart based on recalculated limits can be used in real-time based on a targeted value of the fraction rejected. When using a computer, the varying limits are tabulated automatically.

When any one of the three alternatives are applied, the formulas already given for the two cases, based on targets and process data, apply.

We shall provide an illustration of the first alternative in which limits are recalculated for different sample sizes using the p-chart. A p-chart is more manageable because it is independent of *scale* considerations and imparts more information when the subgroup size is variable. Also, the center line remains *constant*, which would not be the case on an np-chart.

Example 3 Assume we are confronted with a situation similar to that of Example 1 and have collected data regarding the number of rejected items in 20 subgroups, but of varying sizes. The subgroup size, n, the number rejected, r, and the fraction rejected, p, are given in **Exhibit 3** along with additional information.

EXHIBIT 3
EXAMPLE OF RECALCULATION OF CONTROL LIMITS FOR p-CHART WITH VARYING SUBGROUP SIZE

Sample Number	n	r	p	$3\sqrt{\bar{p}(1-\bar{p})}$	$3\dfrac{\sqrt{\bar{p}(1-\bar{p})}}{\sqrt{n}}$	LCL_p	UCL_p
1	180	9	.050	.6898	.0514	.005	.107
2	204	11	.054	.6898	.0483	.008	.104
3	199	8	.040	.6898	.0489	.007	.104
4	213	11	.052	.6898	.0473	.009	.103
5	212	13	.061	.6898	.0474	.009	.103
6	211	7	.033	.6898	.0475	.009	.103
7	216	11	.051	.6898	.0469	.009	.103
8	182	10	.055	.6898	.0511	.005	.107
9	212	15	.071	.6898	.0474	.009	.103
10	222	13	.056	.6898	.0463	.010	.102
11	172	12	.070	.6898	.0526	.003	.109
12	212	14	.067	.6898	.0474	.009	.103
13	205	9	.044	.6898	.0482	.008	.104
14	203	8	.039	.6898	.0484	.008	.104
15	199	10	.050	.6898	.0489	.007	.105
16	199	15	.075	.6898	.0489	.007	.105
17	185	7	.038	.6898	.0507	.005	.107
18	201	14	.070	.6898	.0487	.007	.105
19	196	15	.077	.6898	.0493	.007	.105
20	202	15	.074	.6898	.0485	.007	.105
Total	4025	227	($\bar{p}$ = .056)			(.007)	(.105)

Based on the use of the process data the center line is calculated as the weighted average of fraction rejected, which reduces to

$$CL_p = \frac{\Sigma r}{\Sigma n}$$

$$= \frac{227}{4025}$$

$$= 0.056$$

The control limits are found using the formulas given above, which are

$$UCL_p = \bar{p} + 3\sqrt{\frac{\bar{p}(1 - \bar{p})}{n}}$$

$$LCL = \bar{p} - 3\sqrt{\frac{\bar{p}(1 - \bar{p})}{n}}$$

Since only the subgroup size, n, varies in each case it is convenient to isolate n in the standard error and calculate $3\sqrt{\bar{p}(1 - \bar{p})}$ independently, since this remains constant. That is,

$$3\sqrt{\bar{p}(1 - \bar{p})} = 3\sqrt{.056(.944)}$$

$$= .6898$$

Three times the standard error can be re-written as

$$3\sqrt{\frac{\bar{p}(1 - \bar{p})}{n}} = 3\frac{\sqrt{\bar{p}(1 - \bar{p})}}{\sqrt{n}}$$

which means that the above constant term need only be divided by $\sqrt{n}$ for each subgroup that changes in size. This, then, is added and subtracted to the value of the center line. These considerations only are of importance as a convenience if limits are calculated manually.

Recalculated values of three times the standard error and the control limits appear in Exhibit 3 along with the original data. The control chart with variable limits is presented in **Exhibit 4**.

EXHIBIT 4
p-CHART WITH RECALCULATED CONTROL
LIMITS BASED ON VARIABLE SUBGROUP SIZE

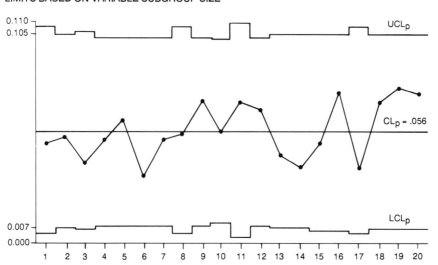

It should become clear that the use of variable limits really is useful when considering whether a *single point* is in or out of control. Patterns within the limits also can be observed; however, one must be *careful* when there are large departures among the subgroup sizes since the reliability of each result will vary also. In the example used, the departures among the subgroup sizes do not add much more information when compared with the use of the average sample size, equal to 201.25, resulting in control limits equal to 0.007 and 0.105. These are indicated at the bottom of Exhibit 3.

Interpreting np and p-Charts

We have introduced both np- and p-charts since one or the other may be used as a matter of personal preference since both impart the same basic information. Because fractions are independent of scale and the fraction non-conforming, as opposed to the number, more clearly indicates the degree of product acceptability on a relative basis, p-charts overall are more useful. Consequently, we shall focus on the interpretation of p-charts in this section with the understanding that similar conclusions may be drawn from np-charts.

Before discussing the interpretation of p-charts, it is important to distinguish between two situations, at a minimum, in order to have a clearer understanding of p-charts. Based on the list presented earlier, we may broadly distinguish between the following situations:

1. Non-conformance is based on a *single* quality characteristic either on a present or absent basis or on the basis of a single measured characteristic in terms of specific tolerances.

2. Non-conformance is based on *more than one* or numerous quality characteristics quantified on a present-or-absent basis, number of imperfections or non-conformities, or more than one set of tolerances related to the same item.

Although p-charts can be used for the purpose of process control, for detecting assignable causes in both cases, they are more useful in the first case than in the second. As more and more quality characteristics contribute to non-conformance the fraction non-conforming becomes a summary measure, say, as in the case of a complicated assembly, and p-charts become more useful as supplements to other charts in terms of providing a running commentary on quality and levels of acceptable product. Consequently, for most of our discussion of p-chart interpretation we shall assume that it is concerned with process control based on a single quality characteristic. Multiple characteristics and acceptance problems will be considered briefly where appropriate near the end of the chapter.

In order to interpret results on a p-chart, it is important to understand the nature of the inputs from the process to the data observed, just as was the case with respect to variables data. In addition to answers to certain basic questions that can be addressed similar to those of Chapter 8, an added distinction should be made in order to fully understand a p-chart.

For the moment, let us assume that we are dealing with a single product characteristic. Based on this characteristic, the item is non-conforming (N) or conforming (C), which is determined either on the basis of a measurable standard or on the basis of a purely qualitative categorization. **Exhibit 5** depicts examples of some alternative process states under the two circumstances.

EXHIBIT 5
EXAMPLES OF SIMPLE ALTERNATIVE PROCESS STATES UNDERLYING
USE OF p-CHARTS BASED ON A SINGLE QUALITY CHARACTERISTIC

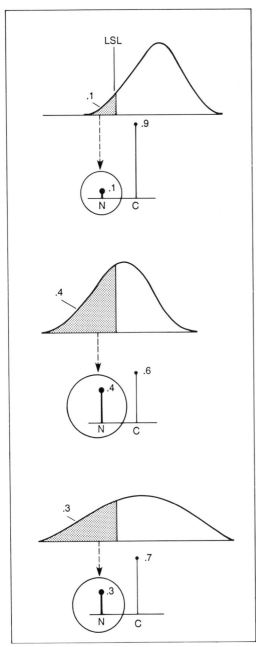

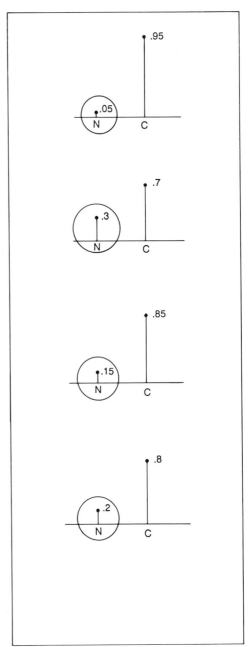

FRACTION NON-CONFORMING REFLECTING
MEASUREABLE SPECIFICATION

FRACTION NON-CONFORMING BASED
PURELY ON QUALITATIVE STATE(S)

Consider the left-hand portion of the exhibit that depicts possibilities relating to an underlying measurable quality characteristic. The distributions used assume that the characteristic is homogeneous with a single lower specification limit (LSL). By examining the top-most diagram, we see that 10 percent (shaded tail-area region) of the items are non-conforming. In terms of fraction non-conforming, the distribution forming the basis for the p-chart is binary and is shown immediately below. In other words, the input into a p-chart is a sample of N's and C's reflecting the two possible states. If in control, the fraction of N's will vary at random about a center line of 0.10.

Now consider the second curve depicted on the left. The distribution is of the same shape; however, the mean has shifted downward resulting in 40 percent non-conforming. The p-chart only will reflect a change in the fraction non-conforming, not that the mean has shifted. The information charted again gets its information, so to speak, from the binary distribution below the curve. Similarly, the fraction non-conforming in the third set of diagrams, equal to 0.3, results from increased variation in the process, which is evidenced by an increased fraction non-conforming on the p-chart but not by an increased process variance. Already, it becomes apparent that while some information about process change is provided, the information is less than provided by variables charts. Recognize that opposite type shifts in the mean and variance could offset one another. That is, the variance could increase and the mean could shift upward to a point where no change would occur in the fraction non-conforming. The p-chart, in such a case, would indicate no change, whereas both changes would be evident on a mean and a standard deviation chart.

In the above example, only one homogeneous characteristic was considered. Based on the changes cited, if one were ultimately interested in correcting a process problem, a complicating factor has been introduced since the p-chart does not isolate as much information about the nature of the change in the process characteristic. When the underlying basis for non-conformance is a measurement, the problem can get complicated in two fundamental ways. First, all of the considerations relating to process heterogeneity, types of changes over time, method of subgrouping, and changes in proportionality factors with respect to subgroups and heterogeneous influences all will have an effect on the basic measurement and the single measure used in a p-chart.

The second way in which problems associated with a process can get clouded is when more than one quality characteristic is considered. If all are measurable, all of the considerations cited above would apply to each and would operate together in terms of their effect on the fraction non-conforming.

In order to emphasize the distinction we introduced at the outset, the right side of Exhibit 5 displays underlying distributions for purely qualitative quality characteristics. In each case, the fraction non-conforming is the only thing underlying the chart. If one quality characteristic is monitored on a chart, in the absence of subgrouping influences or process mixtures, the p-chart will reflect changes appropriately with respect to the fraction non-conforming.

For example, if the basis for rejection is a cracked cap and the corresponding control chart monitors the fraction of caps that are cracked, the control chart can be used to pick up increases and decreases in the overall fraction. When out-of-control patterns are observed reasons for the increased cracks can be sought such as heating conditions, vender quality, installation machine problems, and so on. Mixture patterns resulting from different suppliers, installation machines, and methods of subgrouping, to name a few, can be observed and corrected

The main point to be made is that in cases where a single qualitative characteristic is monitored with a p-chart, it can be used much like other charts we have introduced. When the underlying characteristic is purely qualitative, stability in the fraction non-conforming or corresponding changes are observed. When the underlying basis for non-conformance is measurable, information from a chart relates directly to the fraction non-conforming and not to the underlying distribution of measurements

As was the case with respect to variables, it is helpful to address five basic questions when interpreting p-charts:

1. What is the nature of the process at a point in time and over time?
2. How are samples or subgroups selected or generated?
3. How are the center line and control limits established: in terms of a standard or target or based on process data?
4. What conclusions can be made about the process fraction non-conforming with respect to a particular quality characteristic?
5. What problems or assignable causes in the process can be uncovered and eliminated?

At any point in time, a process can be homogeneous or heterogeneous with respect to a particular fraction non-conforming. A process is *homogeneous* if a single set of chance causes operate to generate a single fraction non-conforming. For example, a single machine producing items may produce a certain fraction that are unacceptable on a regular basis without any discernable problems associated with the machine. If, on the other hand, total process output is based on the output of two machines each with different fractions, the fraction non-conforming for the entire process would be the weighted average of the two fractions and the overall process would be *heterogeneous*. As homogeneity is defined here, the concept applies to a single characteristic with respect to two machines, not two characteristics from one machine which corresponds to another problem. Instead of an entire distribution discussed in earlier chapters, we can express the concept of homogeneity in terms of a *single* fraction.

In the above example, changes over time in the process fraction non-conforming can occur due to changes resulting from either machine or both. If both machines are producing the same fraction and experience no problems, a new batch of raw materials used on one of the machines that results in a differing fraction would constitute an *externally* generated heterogeneous element. At a point in time, if the machines produce differing fractions using the same raw materials, we refer to the overall process as *inherently* heterogeneous.

For purposes of process control, the same principle of subgrouping introduced before holds. That is, any assignable causes of variation affecting a process should occur among subgroups and only chance should operate within. This means that subgroups containing items from an inherently heterogeneous process will contain an assignable cause of variation that will persist over time unless the source of heterogeneity is eliminated. In general, changes in homogeneous processes based on properly drawn subgroups will result in non-random patterns on a p-chart. Parenthetically, if a p-chart is used as a basis for lot acceptance, samples or subgroups should be as representative of a lot as possible. This would mean that all heterogeneous influences should be represented within a subgroup. More will be said about this later.

When considering the way in which the center line and control limits are established, the question is whether process data are used or whether a target or standard is used. If process data are used, basically one is describing what the process is doing in terms of fraction non-conforming, as opposed to how the process performs relative to the targeted fraction.

It should be noted that when a target is used, it is *implied* that one is willing to accept a certain fraction non-conforming. The real goal should be to maintain stability and then systematically seek ways to maintain lower and lower targets, meaning that constant improvements to the process are being made.

Before attempting to identify and eliminate assignable causes of variation in a process, one should ask what is happening to the process fraction non-conforming. In other words, has the fraction gone up or down, or is a pattern due to changes in representation either in terms of heterogeneous influences or in terms of the proportional representation of these influences in the subgroups. Once this is established, one is in a position to begin to look for problems or reasons for improvement based on the nature of a particular process and one's experience with various p-chart patterns as they relate to a particular process.

Generally speaking, patterns on a p-chart can be categorized as they were in Chapter 8 for variables as follows:

1. Natural pattern
2. Complete instability
3. Grouping or bunching, freaks, and sudden shifts
4. Sustained shift
5. Trends
6. Oscillatory patterns
7. Mixture (stable, unstable, and stratification)

Graphical displays of the various patterns are similar to those displayed in Exhibits 4-9 in Chapter 8.

Rules associated with out-of-control patterns based on zones presented in Chapter 8 also may be applied to p-charts. In such cases, more attention must be paid to the sample size since the zone rules are based on the assumption of normality, which may not hold if the sample size is not sufficiently large for a given process fraction, especially when the fraction is very small. Exact probabilities and rules can be established, however, by using the binomial distribution directly. Also, large variations in subgroup sizes would require recalculation of the zone limits, which although doable, is not as easily interpretable.

Process Capability

When we presented the concept of process capability for variables in Chapter 9, we introduced four alternative commonly accepted procedures for assessing capability: (1) histogram, (2) curve fit, (3) interval estimate, and (4) indices. All were aimed at determining the extent to which items meet specifications, or fall within tolerances.

Another *alternative* when using a measured characteristic is to employ the value of $\bar{p}$ used as a center line on a chart corresponding to a controlled process. The value of $\bar{p}$ can be viewed as an estimate of the shaded tail areas in the distributions used as illustrations on the left side of Exhibit 5.

For quality characteristics that truly are qualitatively recorded on a present-absent basis, $\bar{p}$ becomes an *automatic* measure of process capability, and indicates the fraction non-conforming produced by the process. The result only is meaningful for *controlled* processes.

Although not consistent with current thinking, if a standard fraction, P_0, is used, indications of control about the standard would indicate that the process is producing at an "acceptable quality level." The term is one that is used more in connection with acceptance and inspection than process control.

Rather than maintain an acceptable quality level, in keeping with modern practices, one should continue to use a chart to maintain control and also seek ways to *improve* a process on a continual basis by reducing the value of the center line and, therefore, increasing the capability of the process by producing fewer non-conforming items.

ADDITIONAL IDEAS

Many of the considerations relating to p-charts are similar to those associated with variables charts, especially in terms of the format of a chart and patterns appearing on a chart. Fundamentally, p-charts differ in terms of the underlying sampling properties, the fact that one instead of two charts is necessary, the nature of the underlying data, and that a single p-chart can be based on more than one quality characteristic. Due to the nature of the underlying input data and the potential summary nature of a p-chart some additional considerations relating to subgrouping and the applicability of the charts should be discussed before closing the chapter.

Sampling Considerations and Subgrouping

A major difference between p-charts and variables charts that is important to consider can be discussed in terms of the subgroup size required to obtain an effective chart. In the case of variables charts, subgroups of 3 or 5 have been demonstrated to be very useful in identifying changes and process problems. Much *larger samples*, however, are necessary to establish a process fraction non-conforming and detect changes in a process, even in cases where a single quality characteristic is considered.

For example, suppose a sample of 5 were used and the fraction non-conforming is stable and is as high as 0.05. In a sample of 5, the possible outcomes in terms of the number non-conforming are the integers 0-5. The possible fraction non-conforming are these values divided by 5, or the following:

$$0 \quad .2 \quad .4 \quad .6 \quad .8 \quad 1$$

None of these values equals 0.05. The closest results possible in a single sample are either zero and 0.2, which occur with probabilities of .7738 and .2036, respectively.

In order to obtain one non-conforming item, on the average, in a sample drawn from a process with 5 percent non-conforming, it is necessary to use a sample of 20. For a smaller process fraction non-conforming, the subgroup size must be larger. If, for example, the 3-sigma criterion or 3 in 1000 criterion were used, it is necessary to use a sample of 333 in order to expect one non-conforming item in the sample.

There are two important implications associated with the need for large sample sizes when using p-charts. First, unless data are collected and processed in automated form, the *timeliness* of control chart information suggested earlier possibly is lost in collecting and processing large samples. Second, for purposes of process control, it is difficult for operators or workers directly related to the process to interrupt their work in order to *collect* the data. This is not the case when samples of 3 or 5 are required.

Another consideration relating to subgroup size is that for small values of the process fraction non-conforming or in the presence of continued process improvement resulting in lower process fractions, larger sample sizes may be required in order to achieve a reasonable approximation to the underlying binomial by the normal curve. As mentioned, exact limits based on the binomial may be used that involve direct application of the binomial distribution.

The basic premise underlying design and construction of subgroups for purposes of process control is that only common causes of variation are captured within a subgroup and special causes, when present, should occur only among subgroups. This premise, then, applies to the use of p-charts for purposes of control.

Since larger samples are required which also provide an opportunity for variation in sample size from time point to time point, care must also be exercised when observing p-charts since so-called freaks can occur as a result of sharp increases in the sample size. Further, continually changing sample sizes can result in erratic patterns that are attributable to corresponding fluctuations in the standard error rather than special causes of variation.

Multiple Quality Characteristics

When more than one quality characteristic is used as a basis for non-conformance and an item can be rejected on the basis of any one of the characteristics, the process control problem is not as clear. This especially would be the case if quantitative and purely qualitative characteristics were considered.

Consider a simple problem related to our earlier filling process problem in order to reinforce some of the ideas we have presented. First, consider the original filling problem for which it was necessary to maintain a fill level above the label claim of 425 grams. With the process completely out of control as it was when we first looked at it, if a p-chart alone were used to monitor the fraction underfilled, the chart would have demonstrated *no* underfill consistently and would not have led to any corrective action.

Now assume that we have controlled and improved the filling process, and variables charts are maintained in order to control the filling process. Assume further that the contents of a container are of an acceptable quality level determined by other means, and that a p-chart is implemented on the final delivered item. Criteria for non-conformance are underfill below the label claim, a cracked cap, improperly positioned optical scan, chipped container, blurred label, or visibly unacceptable packaging. Any one, or all, of these conditions can render the final product as unacceptable.

When a p-chart applied to this situation indicates an out-of- control condition, it could result from any one or all of the characteristics contributing to the condition. Further, the p-chart does not tell us which cause or causes for lack of control. Before looking to the process for problems one would have to use additional information and/or other problem-solving tools to determine which of the conditions contributes to the product's unacceptability; different parts of the overall process presumably would be involved with each of the unacceptable conditions. In such a case, or one that is more complex, the p-chart does provide a running commentary on *overall product acceptability* or process capability and aids in documenting the need for corrective action, when appropriate. When p-charts are used in terms of numerous quality characteristics, the p-chart becomes more useful as an overall summary of product quality and whether a particular level of product quality is maintained. As such, these are of great benefit in dealing with overall product acceptance regarding the quality of final product or assemblies and for monitoring the level of acceptability of vendor's raw materials and products.

When used in this way, they can be applied to entire production lots. In such cases, if samples are drawn from the lots, more care must be exercised to insure that the samples are drawn randomly. This results from the fact that *representation* across the entire lot, which no doubt contains heterogeneous influences, is necessary. For lot acceptance purposes, the premise underlying the sampling approach is completely different from that used for purposes of control. Control charts still may be applied for on-going monitoring and capability assessment and as a basis for more refined charting and process studies.

Further, the power or operating characteristic, introduced at the end of Chapter 5, should be considered with charts applied to acceptance problems in order to protect against incorrectly accepting unacceptable lots based on one's own or a customer's requirements. Although the concepts are applicable to variables charts, they are not as important to consider if they are used exclusively for the purpose of process control. Details regarding lot acceptance problems and control charts that also are considered for acceptance purposes are presented in Grant and Leavenworth [11].

Using p-Charts

Due to the fact that the final information used in a p-chart is simpler and less informative than that for variables charts but can be based on alternative underlying types of information, p-charts can be useful in a number of different ways. At this point, it is convenient to summarize the alternative uses in the form of a list that follows:

1. As a process control tool, most benefit can be obtained when a single clearly qualitative quality characteristic is considered alone.
2. In cases where measurable characteristics are available for control purposes, more information for purposes of control is available in terms of variables; however, a p-chart can be used to accompany variables charts as a running record of capability and improvement in terms of product acceptability.
3. Before using other types of control charts, p-charts based on varying criteria may be used to begin to pinpoint areas or characteristics requiring control.
4. With respect to final product or complicated assemblies with many relevant quality characteristics, p-charts are useful to document product quality levels and possibly provide indications of the existence of assignable causes that can be tracked with more detailed methods.
5. p-charts may be useful in conjunction with predesignated acceptance criteria for lot screening and monitoring of inputs from vendors.

The above list includes cases where multiple quality characteristics are considered; when considered together, these could include characteristics that are unacceptable on a purely present-absent basis, measured characteristics, and any number of imperfections. When a single quality characteristic is quantifiable in terms of the number of imperfections, neither p-charts nor variables are appropriate. In this case, c- or u-charts are applicable and are considered in the next chapter.

Pareto Diagrams (Optional)

When solving quality problems numerous tools besides control charts, both statistical and otherwise, can be employed. Many of these can be called on an as-needed basis and may involve advanced training in statistics and data analysis.

As part of the more recent quality control movement, some generally applicable problem-solving tools have been incorporated into the quality control literature and have been demonstrated to be useful. Although it can be used in many ways, due to its qualitative nature, it is appropriate to introduce one of these tools here.

A very simple and obvious, but often neglected, way of identifying problems or prioritizing problems is through *Pareto analysis*. This is based on the work of the Italian economist Vilfredo Pareto who introduced the idea that a few vital factors tend to account for a large part of the outcomes. For example, a small faction of customers account for a large portion of a company's sales, or a small number of problems contribute in large part to the number of non-conforming product.

Later the concept became popular in terms of the so-called 80/20 rule, which suggests that approximately 80 percent of the outcomes, problems, or costs originate from 20 percent of the possible sources. This could be reformulated as a 90/10 rule or a 70/30 rule! Overall, the point is that, in reality, many situations exist for which a large portion of outcomes can be attributed to a small portion of sources. Consequently, this provides a simple basis on which to prioritize projects when solving problems.

As an example, consider a manufacturer of quality cartons that is interested in launching a total quality control system within its plant. Recognizing that it is impossible to implement a complete program at one time owing to a lack of backup personnel and the realities of the situation, it is decided to select a highly visible project with the greatest benefit as a start.

Main activities involved in carton manufacture in the plant are divided along departmental lines. Consequently, it is felt that the most sensible thing to do is to identify a project within a single department, or major activity. Scrap is the single most identifiable contributor to cost, and the overall goal is to eliminate as much scrap as possible.

Scrap occurs in two basic ways. First, board can be discarded within each of the departments as each main activity is undertaken or it can occur when production is completed. Since accurate scrap rates are not available during production, data at the end of production aggregated over all jobs by month are used.

Exhibit 6 presents data regarding the number of discarded cartons for a sample of three individual months, broken down by department. The totals for each department along with the corresponding percent of the overall total of discarded cartons also is reported.

Appearing below the basic data is a chart that summarizes the data. On the left vertical scale appear the number of discarded cartons whereas the right-most scale provides the percent of the total. Bars similar to those of a histogram are used to represent the frequency associated with each of the departments designated on the horizontal axis. The main feature of the chart is that the bars correspond to frequencies in *descending* order, from left to right. The resulting graph is referred to as a Pareto diagram.

By quickly glancing at the diagram it becomes obvious that printing appears to be the major contributor to discarded cartons, and therefore seems to be the reasonable candidate within which to begin to undertake a control chart project.

The example used to illustrate Pareto analysis is very simple and dealt with the identification of a beginning point to start a quality control endeavor. Obviously, the device can be used in many situations in order to prioritize options in terms of the way in which problems are identified and where to concentrate one's effort. The technique also can be used in conjunction with other methods in the solution of problems such as in hunting down assignable causes once a control chart indicates that such causes are present in a process.

EXHIBIT 6
EXAMPLE OF PARETO ANALYSIS AND DIAGRAM
Rejected Cartons

Department	NUMBER DISCARDED				
	April	May	June	Total	Percent
Printing	56000	42000	48000	146000	50.3
Stripping	14400	12300	16500	43200	14.9
Cutting and Creasing	6500	12700	9600	28800	10.0
Fold and Glue	19200	22300	16100	57600	19.9
Shipping	4200	5400	4800	14400	5.0
				290000	100.0

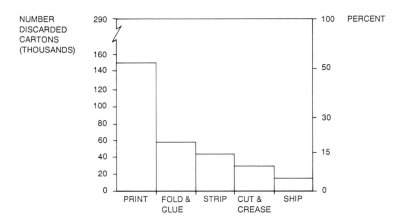

It should be recognized that the Pareto diagram is somewhat of a rough-and-ready tool that is one-dimensional in nature in the sense that only a single set of origins or sources is considered. It does provide a concise way of presenting essential information but is not universally applicable without being accompanied by other forms of more in-depth data analysis in some instances.

C-CHARTS AND U-CHARTS FOR NON-CONFORMITIES

The preceding chapter was concerned with items or product that are non-comforming, which can be established in terms of one or more non-conformities determined on a qualitative or a measurable basis. Considering items in this way focuses directly on product acceptability.

Instead of being interested in an entire item as non-conforming, it may be of interest to focus on the number of imperfections or non-conformities in an item, which may or may not lead to rejection. Minor cosmetic imperfections in a dishwasher, though undesirable and indicative of potential process problems, may not warrant discarding the entire unit.

In other cases, the composition of the output may be different, such as in the case of large rolls of cloth or of rolls of paper. Here, the produced unit ultimately may be used in pieces in terms of the final product. In either case, imperfections such as snags, blemishes, or discolorations occur in a continuum, so to speak, such that the non-occurrence of imperfections cannot be counted. A snag, once defined, can be identified and enumerated. "Non-snags" are virtually limitless and are not enumerable. The same reasoning can also be applied to the number of customer complaints or to the number of invoice errors.

In this chapter we are interested in control charts for the *number* of imperfections or non-conformities in the output of a process, such that non-conformities may or may not lead to rejection of the output. In addition, different types of non-conformities may be considered together and do not have to be of one type. Too little solder and discolorations can be considered together. The basic charts used in cases like this are referred to as a c-chart and a u-chart. The *c-chart* applies in cases where

there is a single observational unit of the same type or size, whereas th
u-chart applies when more than one unit is observed and is identifie
for charting purposes.

The symbols used in this chapter are defined as follows:

c	=	the number of non-conformities in a sample
u	=	the number of non-conformities per unit in a sample
μ	=	the process average number of non-conformities
μ_0	=	standard or targeted average number of non-conform ities
$\bar{c}$	=	average number of non-conformities in a series o samples
$\bar{u}$	=	average number of non-conformities per unit in a serie of samples
n	=	number of observational units
k	=	number of samples
CL	=	center line
UCL	=	upper control limit
LCL	=	lower control limit

Background (Optional)

In order to establish a basis for constructing control charts for th
number of non-conformities, we need to know something about it
sampling properties, namely, the sampling distribution of the numbe
of occurrences of isolated events in a continuum. The underlying mode
is referred to as the Poisson distribution, which was illustrated briefl
in Chapter 4.

The Poisson probability distribution is a specific distribution tha
takes the form

$$f(c) = \frac{\mu^c e^{-\mu}}{c!},$$

$$c = 0, 1, 2, ...$$

where "c" is the variable representing the number of non-conformities i
a sample. The quantity "e" is a natural constant that is the base of th
natural logarithms, and μ is the average number of non-conformitie
corresponding to the sampling unit. The term "c!" is read "c-factorial" an
represents the term c multiplied by all successive integers less than it.

The Poisson also is a sampling distribution in the sense defined i
Chapter 5 and applies to the number of non-conformities in a fixe
observational unit such that non-conformities are generated indepen
dently. Like any probability distribution, the Poisson has a mean and
standard deviation, which for the Poisson are given as

$$\mu_c = \mu$$

$$\sigma_c = \sqrt{\mu}$$

n other words, the single identifying parameter, μ, is the mean or expected number of non-conformities, and also the variance. The standard error, therefore, equals the square root of μ. As μ becomes larger, the Poisson tends toward symmetry and eventually approaches a normal curve.

Note. A difference exists in terms of the nature of the inspection unit used here as opposed to the other cases we have seen. As presented above, the distribution applies to *one* inspection unit. The number of non-conformities, c, corresponds to the number observable for one unit. Hence, in the case of dishwashers, it is the number of imperfections or non-conformities in one or more dishwashers. In the case of a large roll of cloth, the inspection unit may be any number of square feet. In both cases the area of opportunity, so to speak, must remain *constant.* In the case of dishwashers, μ is the average number of non-conformities per given number of dishwashers, whereas in the roll of cloth μ is the average number of non-conformities per given number of square feet. In either case, in order for the distribution to apply as given, once the inspection unit is chosen, it is used throughout.

CONSTRUCTING CONTROL CHARTS FOR THE NUMBER OF NON-CONFORMITIES

C-Charts

Control charts for the number of non-conformities, c, based on a single inspection unit of constant size can be constructed in terms of a *standard* average, μ_0, or in terms of *process data*, in which case the average must be estimated. Formulas for the two cases are given below.

Based on standard:

$$\text{Center line, } CL_c = \mu_0$$

$$\text{Upper control limit, } UCL_c = \mu_0 + 3\sigma_c$$

$$= \mu_0 + 3\sqrt{\mu_0}$$

$$\text{Lower control limit, } LCL_c = \mu_0 - 3\sigma_c$$

$$= \mu_0 - 3\sqrt{\mu_0}$$

The symbol μ_0 represents the targeted or standard average number of non-conformities per unit.

Based on process data:

$$\text{Central line, } CL_c = \bar{c}$$

$$\text{Upper control limit, } UCL_c = \bar{c} + 3\sqrt{\bar{c}}$$

Lower control limit, $LCL = \bar{c} - 3\sqrt{\bar{c}}$

Here, $\bar{c}$ represents the estimated average number of non-conformitie observed and is found using the formula

$$\bar{c} = \frac{\Sigma c}{k}$$

In words, the average number of non-conformities observed equals the sum of the non-conformities from all samples divided by the number o samples, k. In either of the two cases, since the standard error is relate to the targeted or estimated average, only one control chart is require to provide all of the essential information.

Example 1 Twenty samples of 10 square feet of cloth are taken a one-half hour intervals, and the following number of im perfections per sample have been recorded:

Sample Number	No. of Imperfections	Sample Number	No. of Imperfections
1	1	11	8
2	4	12	2
3	6	13	3
4	9	14	2
5	4	15	5
6	2	16	5
7	7	17	3
8	8	18	8
9	7	19	8
10	2	20	2

Assume, for purposes of illustration, that a targeted mean equal to 3 imperfections has been established.

Based on the given standard, we find

$$CL_c = \mu_0$$

$$= 3$$

$$UCL_c = \mu_0 + 3\sigma c$$

$$= 3 + 3\sqrt{3}$$

$$= 3 + 5.2$$

$$= 8.2 \text{ or } 8 \text{ rounded}$$

$$LCL_c = \mu_0 - 3\sigma_c$$

$$= 3 - 3\sqrt{3}$$

$$= 3-5.2$$

$$= -2.2 \text{ "or zero"}$$

Since the number of non-conformities cannot be *negative*, it is conventional to round negative lower limits up to zero. The corresponding control chart with center line, control limits and plotted points appears in **Exhibit 1**.

EXHIBIT 1
EXAMPLE OF A C-CHART BASED ON TARGET $\mu_0 = 3$ Cloth Example

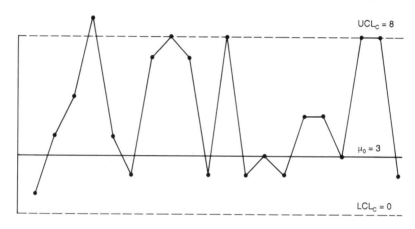

Example 2   When a target is not available or not to be used, we need to use the process data to estimate the center line and control limits. With respect to the data presented in Example 1, the process mean is estimated as

$$\bar{c} = \frac{\Sigma c}{k}$$

$$= \frac{1 + 4 + \cdots + 8 + 2}{20}$$

$$= \frac{96}{20}$$

$$= 4.8$$

Therefore,

$$CL = \bar{c}$$

$$= 4.8$$

$$UCL_c = \bar{c} + 3\sqrt{\bar{c}}$$

$$= 4.8 + 3\sqrt{4.8}$$

$$= 4.8 + 3(2.19)$$

**C-CHARTS AND U-CHARTS
FOR NON-CONFORMITIES**

$$= 11.37 \text{ or } 11 \text{ rounded}$$

$$\text{LCL}_c = \overline{c} - 3\sqrt{\overline{c}}$$

$$= 4.8 - 3\sqrt{4.8}$$

$$= 4.8 - 3(2.19)$$

$$= -1.77 \text{ "or zero"}$$

Since the number of non-conformities cannot be negative the lower limit is set at zero. The resulting control chart is displayed in **Exhibit 2**.

EXHIBIT 2
AN EXAMPLE OF A C-CHART BASED ON PROCESS DATA
Cloth Example

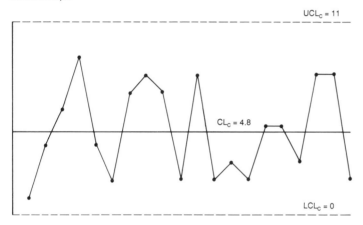

Note. Although the data points plotted on the charts appearing in Exhibit 1 and 2 are the same, the impressions obtained from the two charts could be different. The samples are generated from a naturally controlled process about a mean of 5. This is borne out in Exhibit 2 where the center line estimated from the process, 4.8, is close to the inputted value. Further, since the standard error is a function of the mean, the control limits are further from the center line.

In the case of the established standard equal to 3, not only is the center line lower but the limits are narrower in Exhibit 1. Therefore, the chart based on the lower target exhibits points out of control and, on the average, are shifted upward.

Another point worth noting is that in our example, a constant sample of 10 square feet was employed. Although any *constant* sample or area of opportunity could be employed, such as one square foot, *lower* actual averages may lead to very few, if any imperfections. By increasing the area of opportunity and using appropriate control limits based on a higher average as a center line, an opportunity to observe varying non-zero numbers of non-conformities is provided. As long as the number of product units or the area of opportunity is constant, no special problems result.

U-Charts

When the area of opportunity or the size of the observational or inspection unit varies, a c-chart could be constructed with variable limits; however, the center line must be recalculated also. The result is a chart that is somewhat confusing. An alternative can be used by constructing the chart on the basis of a common unit of observation, which would provide a constant center line. Such a chart is referred to as a u-chart.

A u-chart is similar to a c-chart; however, all results are divided through by the number of basic observational units. That is, the number of non-conformities per basic observational unit, u, is given as the number of non-conformities, c, divided by n, or

$$u = \frac{c}{n}$$

The center line, therefore, is found as the average value of u, or

$$CL_u = \bar{u}$$

$$= \frac{\Sigma c}{\Sigma n}$$

where Σc is the total number of non-conformities in all samples and Σn equals the total number of basic observational units in all samples. Control limits for the u-chart are found as

$$UCL_u = \bar{u} + 3\frac{\sqrt{\bar{u}}}{\sqrt{n}}$$

$$UCL_u = \bar{u} - 3\frac{\sqrt{\bar{u}}}{\sqrt{n}}$$

The reasoning behind the calculation of the estimated standard error, $\sqrt{\bar{u}}/\sqrt{n}$, is similar to that of mean charts for which the estimated process standard deviation is divided by the square root of the sample size. The distribution of u is not Poisson as in the case of c; the results are approximate but have been shown to be useful.

When the inspection unit is variable in size, the limits can be re-calculated just as in the case of a p-chart with variable sample sizes. The corresponding u-chart has a similar appearance. The other alternatives cited earlier with respect to p-charts with variable sample size apply also; that is, using an average sample size or two sets of limits that correspond to the largest and smallest units of observation. A similar limitation applies in the sense that they really are applicable on a point-by-point basis. Further details regarding u-charts can be found in Grant and Leavenworth, [11].

Example 3   Appearing in **Exhibit 3** are the number of imperfections associated with 20 samples of cloth, based on varying amounts of cloth, or sample sizes.

CHAPTER 11

EXHIBIT 3
EXAMPLE OF U-CHART CONTROL LIMIT CALCULATIONS
BASED ON VARIABLE SUBGROUPS

Sample Number	Number of Square feet, n	Number of Imperfections, c	Number of Imperfections Per Square Foot, u	$\frac{3\sqrt{\bar{u}}}{\sqrt{n}}$	LCL_u	UCL_u
1	10	1	.1	.620	0	1.05
2	20	8	.4	.438	.01	.87
3	15	9	.6	.506	0	.93
4	10	9	.9	.620	0	1.05
5	30	12	.4	.358	.07	.78
6	20	4	.2	.438	.01	.87
7	10	7	.7	.620	0	1.05
8	10	8	.8	.620	0	1.05
9	10	7	.7	.620	0	1.05
10	20	4	.2	.438	.01	.87
11	30	24	.8	.358	.07	.78
12	10	2	.2	.620	0	1.05
13	40	12	.3	.310	.12	.74
14	40	8	.2	.310	.12	.74
15	10	5	.5	.620	0	1.05
16	30	15	.5	.358	.07	.78
17	20	6	.3	.438	.01	.87
18	20	16	.8	.438	.01	.87
19	10	8	.8	.620	0	1.05
20	40	8	.2	.310	.12	.74
Total	405	173	-		(0)	(.87)
Mean	$\bar{n} = 20.25$		$\bar{u} = .427$			

Accompanying the basic data are values of u, recalculated values of 3 times the standard error and the variable control limits. In cases where the lower control limits are calculated as negative values, these have been changed to zero. The center line is found as

$$CL_u = \bar{u}$$

$$= \frac{\Sigma c}{\Sigma n}$$

$$= \frac{173}{405}$$

$$= .427$$

The corresponding control chart based on the variable limits is presented in **Exhibit 4**.

If the average sample size were used to establish constant limits, the results would be zero and 0.87, respectively. The upper control limit of 0.87 is plotted as a dotted line in the exhibit. Notice the point corresponding to Sample 4 is outside this limit; in such a case, if the average sample size were used, the recalculated limit shows that the point is in control. The reverse is true for Sample 11.

EXHIBIT 4
EXAMPLE OF A U-CHART WITH VARYING SUBGROUP SIZE
Cloth Example

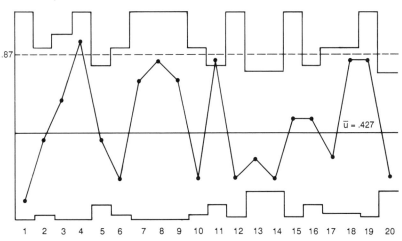

ADDITIONAL IDEAS

Many of the ideas that apply to control charts generally apply to control charts for non-conformities. Much of what we have said about underlying principles and interpretation apply to c and u-charts with some key differences:

1. The underlying sampling properties of the number of non-conformities are different

2. One chart similar to the fraction non-conforming is necessary

3. As the mean number of process non-conformities increases, the variation and scatter on the chart becomes greater

4. The number of non-conformities may correspond to one quality characteristic or many, which may correspond to qualitative or measurable quality characteristics

5. When a measurable quality characteristic is used as a basis for identifying a non-conformity in an item, only one non-conformity would be counted and would be used in conjunction with other types of non-conformities relating to other characteristics. Otherwise, the item would be considered as non-conforming and a p-chart would be used.

Although similar considerations apply to both c- and u-charts, with some minor departures, we shall refer only to c-charts in the discussion that follows.

Interpreting C-Charts

When a single quality characteristic is used as a basis for a c-chart, interpretation is similar to that of a p-chart introduced in Chapter 10 for a clearly qualitative characteristic, except that one is observing the number of non-conformities in an item or sample of product rather than the number of non-conforming items in a sample.

At any point in time a process may be homogeneous or heterogeneous, which is expressed in terms of the mean number of non-conformities per unit of measurement or observation. In both cases heterogeneous influences can be introduced at various points in time. The heterogeneous elements can be considered as special causes of variation, in addition to others, which would be evident on a chart in terms of different patterns depending on whether they appear within a sample or among samples. Visual patterns similar to the ones depicted in Chapter 8 can be observed on a c-chart and have a similar basis in terms of the mean number of non-conformities.

Similar ideas apply to sampling considerations, also. On the one hand, as the average number of non-conformities is reduced, say through process improvements, larger sample sizes or larger samples of product are required to detect non-conformities. In such cases, larger samples of product also are necessary to assume near normality. Exact limits also are calculable based on the Poisson distribution directly.

When multiple quality characteristics of varying kinds are used as a basis for enumerating non-conformities, a c-chart for purposes of process control is not as clear as in the case of a single quality characteristic. When multiple characteristics are considered, a c-chart becomes useful as a summary relating to *overall product quality* and whether particular quality levels are maintained or improved in terms of the average number of non-conformities. The value of $\bar{c}$ automatically becomes a measure of *capability* or the level of acceptability in terms of product non-conformities or imperfections.

Using C-Charts

Due to the specialized nature of c-charts, they have not been used as extensively as the other types of charts presented. They do have benefits in certain applications, some of which are listed as follows:

1. To function as a process control tool when a single clearly qualitative and relevant quality characteristic is considered, which is important to consider in terms of numbers of non-conformities rather than just acceptability

2. To establish current quality levels, possible control problems, and the degree of overall quality improvement in cases where non-conformities of all kinds are to be eliminated

3. To obtain information regarding inspection practices and irregularities with respect to final product

4. To lead to overall quality improvement and greater product acceptability in cases where some non-conformities are permissible

5. To use in sampling acceptance based on defects per unit (see Grant and Levenworth [11] for more details)

OTHER VARIABLES CHARTS

The variables charts considered in Module C are the most commonly used. There are, however, other variables charts available. Some are more sophisticated in the sense that they are more responsive to process changes, while others are suited to special circumstances. Also, alternatives exist that are simpler to implement manually.

In this chapter, we shall focus on two such types of charts, the *moving average* and *moving range* chart, and the *median* chart. Moving average charts are useful in cases where a single observation at each time point is observed and as a more responsive alternative to the simple mean chart. Median charts that are based on the sample median simply are more easily implemented alternatives to mean charts. An overview of other types of variable charts is given at the end of the chapter.

The symbols used in this chapter are defined as follows:

MA = moving average

$\bar{\bar{X}}$ = mean of a series of observations, or sample averages

MR = moving range

$\overline{MR}$ = the mean of a series of moving ranges

$\tilde{X}$ = sample median

$\bar{\bar{X}}$ = the arithmetic mean of a series of sample medians

n = sample or subgroup size

k = number of individual samples observed or the number of subgroups

UCL = upper control limit

LCL = lower control limit

CL = center line

279

A_2 = tabulated factor used to calculate limits on a mean chart.

D_3 = tabulated factor used to calculate a lower limit on a range chart

D_4 = tabulated factor used to calculate an upper limit on a range chart

A_6 = tabulated factor used to calculate control limits on a median chart based on the sample range

MOVING AVERAGE AND MOVING RANGE CHARTS

Moving average charts are useful in cases where a single observation as opposed to a sample is observed at each time point. Cases such as this might result from situations where it is too costly or time consuming to process more observations or when automatic recording devices are programmed to collect one observation at a time. Also, if the time interval between produced items is long, the timeliness of selecting an entire subgroup is lost.

Numerous other situations exist where it is inappropriate to consider more than one observation. The most notable case in point is related to chemical processes in which large volumes of continuously flowing fluid must meet requirements regarding composition or concentration. In such instances, a volume of product is extracted at a point in time and analyzed. The volume of product extracted constitutes a sample of one. Other obvious examples are moisture, temperature and pressure for which one value exists at a point in time.

It is possible (and often the case) that the individual observations are charted over time. As a process control tool, so-called *individuals charts* are not very responsive. An alternative based on individual observations is a moving average chart or one related to it.

A **moving average** simply is an arithmetic mean calculated on the basis of a series of observations by successively dropping an observation at the beginning and adding one at the end and re-computing the mean. **Exhibit 1** provides a series of 30 observations representing the water content in a salt solution, expressed as a percent. Along with other information, moving averages based on three observations at a time are given.

In order to understand how the moving averages are calculated, consider the first three values. The mean of these three is the first moving average, and is calculated as follows:

$$MA_1 = \frac{13.6 + 15.5 + 13.8}{3}$$

$$= 42.9$$

$$= 14.3$$

The second moving average is found by dropping the first value in the series and including the fourth as

$$MA_2 = \frac{15.5 + 13.8 + 14.1}{3}$$

$$= \frac{43.4}{3}$$

$$= 14.5$$

EXHIBIT 1
EXAMPLE OF A MOVING AVERAGE AND MOVING RANGE

Observation Number	Water Content	Moving Total	Moving Average (MA)	Moving Range (MR)
1	13.6			
2	15.5			
3	13.8	42.9	14.3	1.9
4	14.1	43.4	14.5	1.7
5	13.9	41.8	13.9	.3
6	14.0	42.0	14.0	.2
7	15.1	43.0	14.3	1.2
8	14.7	43.8	14.6	1.1
9	17.3	47.1	15.7	2.6
10	14.6	46.6	15.5	2.7
11	17.8	49.7	16.6	3.2
12	15.2	47.6	15.9	3.2
13	14.3	47.3	15.8	3.5
14	14.9	44.4	14.8	.9
15	14.4	43.6	14.5	.6
16	12.6	41.9	14.0	2.3
17	16.8	43.8	14.6	4.2
18	14.4	43.8	14.6	4.2
19	15.3	46.5	15.5	2.4
20	15.3	45.0	15.0	.9
21	15.6	46.2	15.4	.3
22	15.4	46.3	15.4	.3
23	15.0	46.0	15.3	.6
24	13.9	44.3	14.8	1.5
25	14.8	43.7	14.6	1.1
26	15.2	43.9	14.6	1.3
27	14.8	44.8	14.9	.4
28	16.2	46.2	15.4	1.4
29	16.0	47.0	15.7	1.4
30	15.7	47.5	15.8	.5
Total	450.6			45.8

The remaining values in the moving average column are found similarly by dropping one and adding another observation and successively calculating an average. Notice that there are two less moving averages than there are observations. This results from the fact that these values are used to calculate a single average. The number lost always will be one less than the number comprising the average.

If the calculation is performed manually, it is convenient to calcu
late moving totals, which are displayed in the second column of the
exhibit, and then obtain the moving average by dividing by the ap
propriate number, which in this case equals three. Also, when moving
averages are used in most statistical applications,the average is center
ed on the middle of the observations composing the particular average
In this example,the averages are aligned with the last value appearing
in the average, which is more conventional in process control applica‑
tions. By plotting the results correspondingly, a sense of greater timeli
ness is associated with the results.

Since measures of variability must be based on *at least two* obser‑
vations, a moving measure of variation is needed, whether an in‑
dividuals chart or a moving average chart is used. The most convenient
measure to use manually is the *moving range*. Values of the moving
range associated with the observations in Exhibit 1 also are presented
The first two are calculated below in order to illustrate the concept.

$$MR_1 = 15.5 - 13.6$$

$$= 1.9$$

$$MR_2 = 15.5 - 13.8$$

$$= 1.6$$

Using the ideas presented, we are in a position to construct a moving
average and an accompanying moving range chart.

Control Limits for the Moving Average and Moving Range

The conventional approach for calculating control limits for a mov‑
ing average chart follows along the same lines as for mean charts
presented in Chapter 7. Assuming limits are based on process data
limits are found as

$$UCL_{MA} = \bar{\bar{X}} + A_2 \overline{MR}$$

$$UCL_{MA} = \bar{\bar{X}} - A_2 \overline{MR}$$

where $\bar{\bar{X}}$ is the mean of the individual observations and $\overline{MR}$ is the mean
of the moving ranges. This is treated similar to $\bar{R}$, and consequently
factor A_2 used earlier applies. Recognize that $\overline{MR}$ is calculated similar
to $\bar{R}$ but the divisor is not equal to the original number of observations
since a loss occurs when calculating the moving range.

In the example we used in Exhibit 1, the value of $\overline{MR}$ is calculated as

$$\overline{MR}_3 = \frac{\Sigma MR}{k - 2}$$

where the "3" indicates that the moving range is calculated on the basis
of successive groups of three observations, $\overline{MR}$ is the mean of the moving
range, MR, and k is the number of individual sampled observations. In

general for any number of values for calculating the range, n, $\overline{MR}$ can be found as

$$\overline{MR}_n = \frac{\Sigma MR}{k - (n - 1)}$$

$$= \frac{\Sigma MR}{k - n + 1}$$

Control limits for the moving range chart are based on formulas used earlier with $\overline{MR}$ used in place of $\overline{R}$. That is,

$$UCL_{MR} = D_4\overline{MR}$$

$$LCL_{MR} = D_3\overline{MR}$$

where D_3 and D_4 are tabulated factors provided in the Appendix. In cases where targeted values are given, control limits may be calculated using the formulas given in Chapter 7 for variables.

Example 1   Use the information provided in Exhibit 1 to develop a moving average and moving range chart based on samples of three.

$$CL_{MA} = \overline{\overline{X}}$$

$$= \frac{450.6}{30}$$

$$= 15.02$$

$$\overline{MR} = \frac{\Sigma R}{k - n + 1}$$

$$= \frac{45.8}{30 - 3 + 1}$$

$$= \frac{45.8}{28}$$

$$= 1.64$$

$$UCL_{MA} = \overline{\overline{X}} + A_2 \overline{MR}$$

$$= 15.02 + 1.023(1.64)$$

$$= 16.7$$

$$LCL_{MA} = \overline{\overline{X}} - A_2 \overline{MR}$$

$$= 15.02 - 1.023 (1.64)$$

$$= 13.3$$

$$UCL_{MR} = D_4 \overline{MR}$$

$$= 2.574(1.64)$$

$$= 4.2$$

$$\text{LCL}_{\text{MR}} = D_3\overline{\text{MR}}$$

$$= 0(1.64)$$

$$= 0$$

A_2, D_3, and D_4 are found based on a sample size of 3 from **Table C-2** in the Appendix.

Exhibit 2 presents completed control charts based on the above calculations and the results provided in Exhibit 1. Accompanying the control charts is a plot of the original series without limits imposed on the graph. The original series is presented in order to get a better idea of the way in which a moving average operates.

The original series actually is random but possesses some high spikes near the middle of the series. The original observations ranged between 12.6 and 17.8. If limits were placed on the series, the 3-sigma limits would be 12.1 and 17.9.

By observing the moving average chart, it can be seen that the corresponding series is smoother than the original. The larger the number of observations used in the averaging process, the smoother the moving average chart will be in appearance. If trends, cycles, or sustained shifts in level were present in the original series, these would be picked up more quickly on a moving average chart than on a regular mean chart.

The range chart appears to be somewhat unruly, in part due to the specific nature of the original series. Without knowledge of the way the original series was generated, the two points exactly equal to the upper control limit may suggest looking for an assignable cause. Seemingly apparent also is a trend pattern and a sudden shift.

It should be recognized that individual points on a moving average and moving range chart are not independent, as they are on a simple mean and range chart. Consecutive points as well as triples in this case contain either two or one similar observations so that successive points tend to be tied to one another, so to speak. Consequently, there will be a tendency for wave-like movements in the moving series. Also, as a result of this, the rules presented in Chapter 8 regarding sequences of points indicating lack of control do not directly apply. Similar rules can be developed that are based on longer sequences.

MEDIAN CHARTS

We mentioned at the outset that median charts can be used as an alternative to a mean chart in order to simplify the manual effort involved in constructing a control chart. Median charts are useful in

EXHIBIT 2
EXAMPLE OF MOVING AVERAGE AND MOVING RANGE CHART WITH ORIGINAL SERIES

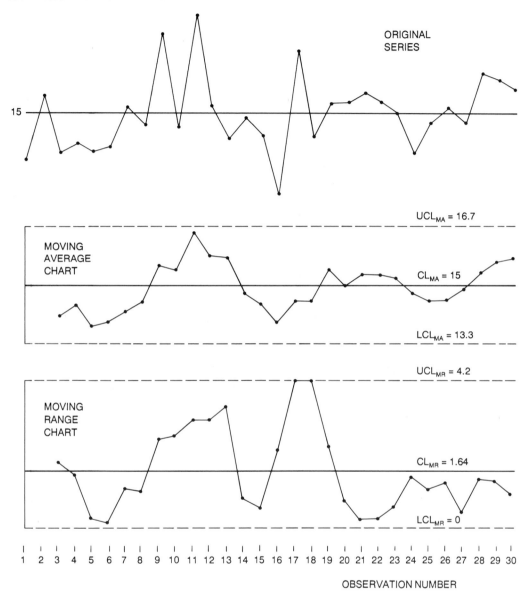

order to convey the basic ideas of control charting to those just beginning
to work with them.

Median charts are constructed much like mean charts except the
standard error of the sample median is used, which is accounted for by
a *different* tabulated factor. Although alternatives exist, the method we
shall use to find the center line is as the arithmetic mean of the sample
medians. The center line and control limits for a median chart based on
process data are found using the following formulas:

$$CL_{\tilde{x}} = \bar{\bar{X}}$$

$$= \frac{\Sigma \widetilde{X}}{k}$$

$$\text{UCL}_{\widetilde{x}} = \overline{\overline{X}} + 3\ \hat{\sigma}_{\widetilde{x}}$$

$$= \overline{\overline{X}} + A_6 \overline{R}$$

$$\text{LCL}_{\widetilde{x}} = \overline{\overline{X}} - 3\ \hat{\sigma}_{\widetilde{x}}$$

$$= \overline{\overline{X}} - A_6 \overline{R}$$

where $\widetilde{X}$ represents the median of an individual subgroup, $\overline{\overline{X}}$ is the arithmetic mean of the subgroup medians, and R is the arithmetic mean of the subgroup ranges. A_6 is a *tabulated factor* that accounts for the multiple of the estimated standard error of the sample median in terms of the subgroup range.

Selected values of A_6 are tabulated in **Exhibit 3** for subgroup sizes of 3, 5, and 7. Since use of the median merely is a computational convenience, the purpose is defeated if the sample either is large or an even number in size. Using odd numbers, one only has to select the median as the middle value, without calculating the mean of the middle two. Further, locating the median on a chart when all of the observations are plotted becomes automatic. The most convenient subgroup size is 3.

EXHIBIT 3
VALUES OF FACTOR FOR CALCULATING MEDIAN
CONTROL CHART LIMITS USING THE SAMPLE RANGE

Sample Size, n	A_6
3	1.187
5	.691
7	.509

Only the sample range is considered to measure variation in connection with medians since, obviously, if computational convenience is the goal, the range is nearly as simple to find as the median, especially for small samples.

The formulas given above assume that control limits are based on process data; that is the way they usually are presented. They can be adapted for use with targeted values of the process mean and standard deviation, μ_0 and σ_0. Under the usual assumption of a normally distributed process, μ_0 can be used as the center line owing to symmetry of the normal. The corresponding control limits for the sample median can be found using the formulas based on targets for sample means merely by multiplying the factor A in Table C-1 by 1.2533. This represents the multiple of the standard error of the sample mean which gives the standard error of the sample median based on a normal universe. The range chart based on σ_0 is the same as the one presented in Chapter 7.

Example 2   Reproduced in **Exhibit 4** are the data from the original
filling process problem first introduced in Chapter 7.
Accompanying the data are the sample medians and
sample ranges for each of the 18 subgroups. Use the
information given to construct a median control chart.

EXHIBIT 4
ORIGINAL FILLING PROCESS WEIGHTS IN GRAMS

Sample Number	Nozzle 1	Nozzle 2	Nozzle 3	Nozzle 4	Nozzle 5	$\tilde{X}$	R
1	448.3	449.1	447.9	445.0	448.1	448.1	4.1
2	449.6	446.6	448.4	449.5	445.4	448.4	4.2
3	440.5	439.9	435.8	433.9	439.1	439.1	6.6
4	435.4	440.6	437.7	438.1	438.7	438.1	5.2
5	440.5	438.6	439.7	432.9	438.3	438.6	7.6
6	435.2	438.1	436.8	435.6	437.9	436.8	2.9
7	445.4	445.7	445.0	439.9	444.8	445.0	5.5
8	441.8	445.9	443.3	442.6	443.9	443.3	4.1
9	439.5	439.7	440.4	434.0	438.7	439.5	6.4
10	435.8	437.5	437.2	436.6	437.0	437.0	1.7
11	437.8	438.4	438.0	432.0	437.4	437.8	6.4
12	433.6	435.3	436.1	434.8	437.8	435.3	3.2
13	439.2	440.1	440.6	434.0	438.9	439.2	6.6
14	435.5	440.0	437.5	437.1	438.6	437.5	4.5
15	439.9	439.6	440.7	434.5	439.2	439.6	6.2
16	435.8	439.6	437.5	437.4	438.4	437.5	3.8
17	440.9	441.5	442.2	434.1	439.9	440.9	8.1
18	437.2	439.7	439.3	437.6	440.0	439.3	2.8
Total						7921.0	89.9

$$CL_{\tilde{x}} = \bar{\bar{X}}$$

$$= \frac{\Sigma \tilde{X}}{k}$$

$$= \frac{7921}{18}$$

$$= 440$$

$$\bar{R} = \frac{\Sigma R}{k}$$

$$= \frac{89.9}{18}$$

$$= 5.0$$

$$UCL_{\tilde{x}} = \bar{\bar{X}} + A_6\bar{R}$$

$$= 440 + .691(5.0)$$

$$= 443.5$$

$$LCL_{\tilde{x}} = \bar{\bar{X}} - A_6\bar{R}$$

$$= 440 - .691(5.0)$$

$$= 436.5$$

The resulting control chart is presented in **Exhibit 5** When used in an actual application, it should be accompanied by a range chart.

Overall, the median chart provides the same type of information as the mean chart, although differences between the two charts will exist. By comparing the median chart with the mean chart appearing in Exhibit 5 in Chapter 7, we can see that sample points numbered 6 and 8 are in the control limits on the median chart, whereas they are not on the mean chart.

EXHIBIT 5
MEDIAN CHART FOR ORIGINAL FILLING PROBLEM BASED ON PROCESS DATA

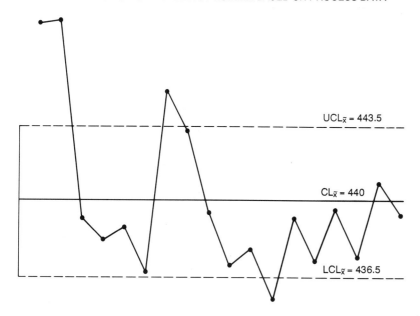

UCL$_{\tilde{x}}$ = 443.5

CL$_{\tilde{x}}$ = 440

LCL$_{\tilde{x}}$ = 436.5

The start-up pattern and wide swings and peaks are similar on both charts. The sudden and sustained shift downward on the mean chart is present on the median chart, but it appears more variable and jagged with a hint toward an upward trend. In general, the median is a less reliable estimator and as a consequence is subject to more random variation than the mean. This also becomes evident in terms of the wider control limits.

Appearing in **Exhibit 6** is the same median chart; however, it also presents the original subgroup observations superimposed on the chart. By plotting the original observations on the chart first, and using an odd subgroup size, one immediately can identify the middle value, or the median. When done in this way, the median is circled and these points are connected.

EXHIBIT 6
MEDIAN CHART FOR FILLING PROBLEM WITH INDIVIDUAL SUBGROUP DATA PLOTTED

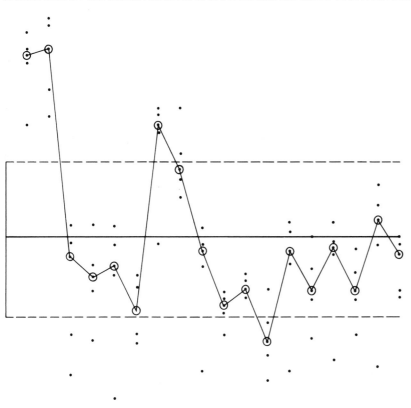

ADDITIONAL IDEAS

In this chapter we have introduced a moving average and moving range chart, and a median chart that would be accompanied by an ordinary range chart. The primary purpose of the chapter is to provide exposure to other types of control charts in addition to the basic more frequently used ones. Before closing, we shall briefly introduce some additional charts and ideas that can be pursued further if the need arises.

There is not very much to say about median charts since they represent a convenient yet not as efficient alternative to mean charts. The moving average chart is worthy of some additional attention since it has properties that are a little different from those exhibited by a simple mean chart.

A simple mean chart has no memory, so to speak, since observations within a particular subgroup do not appear in any other subgroup and, therefore, subgroups are considered independent. Depending on the size of the average or the number of observations used, moving averages have some *memory* in the sense that some of the observations in one subgroup are part of a fixed number of successive subgroups. This being the case, changes in a process that are occurring will be reflected in a subgroup, partly in terms of the past information as well as the present. Such a property is desirable in order for a chart to be more responsive to change. There is a tendency, however, for moving averages to display trend-cycle type patterns in a purely random series. Other charts are available that possess some of the better features of moving averages but do not suffer the limitations to the same degree.

Before surveying other types of control charts available, some important points about the use of control charts are worth noting. On the one hand, different charts may be useful for detecting different types of changes. For example, it has been demonstrated that the basic mean chart presented in Chapter 7 is best for detecting *large changes* from a target. Consequently, when beginning to set up a process control system, use of basic charts may be satisfactory. At the other extreme, when a process with two-sided dimensional specifications has been extensively improved, meaning that major causes have been eliminated such that it is mostly in control with respect to a targeted mean and the natural tolerances are far less than those specified, small changes do not have to be detected quickly since they do not affect specifications.

On the other hand, consider a situation similar to the filling process problem where it has been initially controlled and improved, where continued improvements lead to lower targets closer to the label claim limit in order to reduce overfill. When the process initially was out of control with a high mean, it was not necessary to worry about *small changes*. However, when the lower natural limit is close to the label claim, small changes in the process must be detected since these could lead to violations of the label claim. In other cases, such as of short runs, there may be a need for quick responsiveness and possible changes in the responsiveness of a chart. Different runs may require different degrees of responsiveness.

Another aspect of charting which has not been fully addressed is the use of fully automated data collection and processing systems for control purposes. With increased use of 100 percent automated inspection of the quality of final product and the eventual introduction of automated on-line charting systems, charts that have an adaptive or feedback capability provide the basis for quicker corrective action to changing processes. For this and other reasons cited, other types of control charts may be useful in order to meet specific process control needs.

Cumulative Sum Charts

An alternative to the basic, or Shewhart, control chart introduced earlier that has gained recognition is the cumulative sum, or CUSUM, chart. This is a chart in which the observations are successively cumulated, and each new sum is plotted at each point in time. In other words, the statistic

$$C = \sum^{t} X$$

is observed. As an alternative, deviations of the observations from a process target or some other quantity are used. That is, in the case of a targeted mean,

$$D = X - \mu_0$$

where the plotted quantity is the successive sum

$$\Sigma D = \overset{t}{\Sigma}(X - \mu_0)$$

Actually, this is the quantity that is plotted against time.

When the mean of the observations is the *same* as the target, the cumulative sum will behave randomly around *zero*. If the two means differ, then the sum should wander about this difference.

The control limits for cumulative sum charts are *not* parallel as in the simple mean chart. Instead, a so-called V-mask is used, which is a "V" placed on its side such that it is pointing in the direction of the passage of time. The sides of the V-mask are the control limits, which incrementally are moved with successive observations.

Inputs into the construction of the V-mask are the lead distance between the point of the mask and the last plotted point and the angle between the sides of the "V". These inputs account for the probability of a Type II error associated with an important shift in mean along with a chosen probability of a Type I error, which is considered but fixed with a Shewhart chart.

Unlike the Shewhart chart which has no memory and the moving average which considers the immediate history, the CUSUM chart remembers everything, so to speak, where all observations contained in the CUSUM are given equal weight. A CUSUM chart will pick up a process shift sooner than the ordinary control chart. An example of a CUSUM chart is presented in **Exhibit 7** along with the corresponding

EXHIBIT 7
EXAMPLE OF A BASIC SHEWHART CONTROL CHART
FOR MEANS AND CORRESPONDING CUSUM CHART

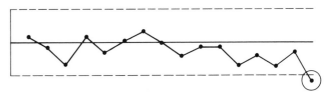

BASIC MEAN CHART

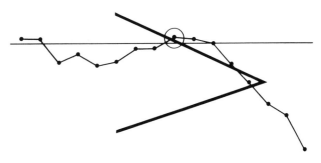

CUSUM CHART

Shewhart control chart for means. The main point to note about the exhibit is that an out-of-control indication, evidenced by the first point above the upper part of the "V" occurs long before out of control is indicated on the mean chart; seven more points must be observed before an out-of-control point occurs on the mean chart. Details regarding construction of a CUSUM chart can be found in Johnson and Leone [74] and additional considerations are discussed in Montgomery [12].

Exponentially Weighted Moving Average Charts

An exponentially weighted moving average, or EWMA, is a form of average that *weights* past data but applies a smaller and smaller weight to data as they get older. As such, a control chart based on an EWMA falls between a basic mean chart and a CUSUM chart, where these either have no memory or a complete memory. Exponentially weighted moving averages have long since been applied to economic problems and gained recognition in problems of inventory control for some time. With respect to economic type problems, they are used at the least to obtain one step ahead time series forecasts. Until recently less attention has been given to these in quality control applications.

An EWMA is based on a statistic of the form

$$EWMA = wX + (1 - w)\hat{X}$$

where X is an observed value of a variable measured at a given point in time and $\hat{X}$ is the EWMA calculated at the immediately preceding time point. It is written with a "hat" since the EWMA can be shown to be the forecast of that observation for the given point in time, and contains the history of the series.

The quantity w is a constant between zero and one that determines the degree of memory of the EWMA. The closer w is to one, the more weight is given to the most recent observation, and the closer w is to zero the less weight is given to a current observation. When w is close to one, the points plotted on a control chart appear similar to a Shewhart chart, and when near zero, the plotted points resemble a CUSUM chart.

Control limits for an EWMA can be calculated for given values of w for either individual observations or subgroup averages. The corresponding chart considers runs and other patterns discussed in Chapter 8 in a formal way. EWMA charts are useful for detecting small process shifts and can be enhanced to provide a quick initial response suited to adaptive systems noted earlier, which would appear to become a consideration in the not too distant future. A nice introduction to the EWMA is given in Hunter [30] and a detailed discussion regarding its properties and enhancements for control purposes can be found in Lucas and Saccucci [39].

Numerous other control charts and modifications of the ones discussed have been devised to meet special requirements. The interested reader is referred to the references appearing in the Appendix where he or she may find discussions of many of these.

CONCLUDING REMARKS

Unlike other books concerned with quality control, this book focuses on a highly directed theme. Instead of attempting to cover as many bases as possible, this book is not directed toward all aspects of the quality problem nor to all aspects of statistics as applied to the quality problem. Although many different perspectives are useful in order to deal with quality as it is being addressed today, it is virtually impossible to do it all in a digestible form in a single volume.

What this book does is focus on currently recognized statistical tools and methods that have unquestionable usefulness in controlling the quality of products and services that are produced or delivered on a repetitive basis. Not only does the book provide specific methods and guides for control but it also introduces a foundation on which one is able to build solutions to their own process problems by logically extending the fundamentals.

In order to summarize what we have done, a basic flowchart describing the control charting process appears on the following page. Highlighted in the chart are the four basic components of the control problem that have been emphasized throughout: measurement, control, capability, and improvement. By first introducing the nature of data in Module A and developing the underlying statistical basis for control charts in Module B, we were able to unite the four components in Module C in terms of variables data. Additional types of control charts that are based on the same principles are considered in the last module, D.

There are two key points about the flowchart that should be emphasized in addition to the *four main components*. First a starting position is indicated, but *no* stop. This has been done to reinforce the idea that process control and improvement should be never ending. Even in cases where a process is completely capable of meeting specifications, one may continually improve a process by *reducing variability*. The methods, however, to accomplish this are not as strictly defined as the

293

control charting procedure. In some cases, other methods of statistical analysis may be used based on carefully designed studies to further analyze the nature of a process. These methods may be used in conjunction with changes in a process or equipment aimed at further reductions in variation.

BASIC FLOWCHART DESCRIBING CONTROL CHARTING PROCESS

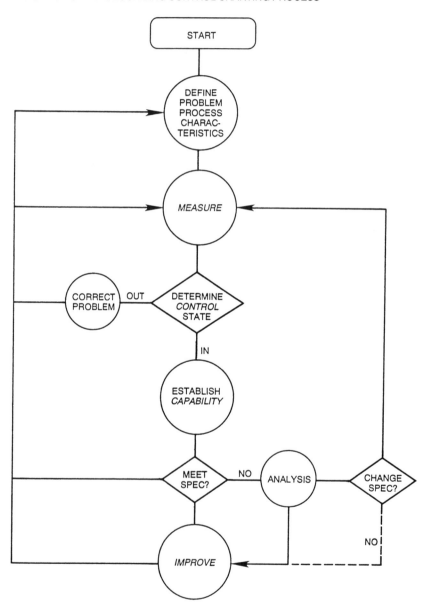

The second important point to note is that in all cases the control and improvement cycle continually feeds back not only to continued measurement but into continued *redefinition* of the process control problem and the characteristics used to measure process performance. In some cases, initial definitions may not be adequate but merely provide a starting point. In others, knowledge obtained from earlier attempts at control may lead to insights about the problem that require new or additional characteristics or a clearer definition of the process requiring control. Much of this activity is related to the particulars of a process within a specific organization. The cycle of control and continuous improvement, therefore, should involve continued redefinition in addition to the other four key components.

As stated, the ideas presented are applicable to *any repetitive process*. Although manufacturing applications were emphasized, the principles apply equally well to manufactured products or to many types of services. Such things as delivery time for hospital emergency services, invoice errors, and customer complaints can be charted as easily as the weights of ball bearings.

Because so much is being said nowadays about quality control and total quality management, there appears to be a possible misconception that it is necessary to put in place a total company-wide system before being able to implement the process control tools presented earlier. In reality, when beginning, it is advisable to begin with a highly visible and simply defined process, the control of which would lead to obvious improvements in terms of product quality, cost savings, or reductions in scrap and rework.

By interlocking the results and efforts associated with such component-type processes, it is possible to build the control effort to embrace a complete or extended company-wide process related to all or many of an organizations' activities over time. The chances for ultimate organization-wide success are increased by careful planning and management of variation across many different time horizons.

In addition to process control, a total company-wide quality effort must consider many things at all managerial, behavioral, and technical levels, which can include statistical methods to varying degrees. Ultimately, inputs from suppliers, customers, technical staff, and training should be an integral part of product and process initiation, design, and development, aimed at continued fulfillment of customers' needs, improved product quality, and greater productivity.

REFERENCES

Statistical Quality Control

[1] Bowker, A. H. and H. P. Goode: *Sampling Inspection by Variables*, McGraw-Hill Book Company, New York, 1952.

[2] Burr, I. W.: *Engineering Statistics and Quality Control*, McGraw-Hill Book Company, New York, 1953.

[3] Burr, I. W.: *Statistical Quality Control Methods*, Marcel Dekker, Inc., New York, 1976.

[4] Cowden, D. J.: *Statistical Methods in Quality Control*, Prentice Hall, Inc., Englewood Cliffs, New Jersey, 1957.

[5] DeBruyn, C. S. V.: *Cumulative Sum Tests: Theory and Practice*, Hafner Publishing Company, New York, 1968.

[6] Dodge, H. F.: "A General Procedure for Sampling Inspection by Attributes Based on the AQL Concept," Technical Report No. 10, The Statistics Center, Rutgers, The State University, New Brunswick, New Jersey, 1959

[7] Dodge, H. F. and H. G. Romig: *Sampling Inspection Tables Single and Double Sampling*, 2d ed., John Wiley & Sons, Inc., New York, 1959.

[8] Duncan, A. J.: *Quality Control and Industrial Statistics,* 5th ed., Richard D. Irwin, Inc., Homewood, Illinois, 1986.

[9] Enrick, N. L.: *Quality, Reliability, and Process Improvement,* 8th ed., The Industrial Press, New York, 1985.

[10] Gitlow, H., S. Gitlow, A. Oppenheim, and R. Oppenheim: *Tools and Methods for the Improvement of Quality*, Richard D Irwin, Inc., Homewood, Illinois, 1989.

[11] Grant, E. L. and R. S. Leavenworth: *Statistical Quality Control,* 6th ed., McGraw-Hill Book Company, New York, 1988.

[12] Montgomery, D. C.: *Introduction to Statistical Quality Control*, John Wiley & Sons, Inc., New York, 1985.

[13] Ott, Ellis R.: *Process Quality Control: Trouble-shooting and Interpretation of Data*, McGraw-Hill Book Company, New York, 1975.

[14] Shewhart, W. A.: *Economic Control of Quality of Manufactured Product*, D. Van Nostrand Company, Inc., Princeton, New Jersey, 1931, Reprinted by American Society for Quality Control, Milwaukee, Wisconsin.

[15] Shewhart, W. A. (edited by W. E. Deming): *Statistical Method from the Viewpoint of Quality Control*, The Graduate School, Department of Agriculture, Washington, DC, 1939.

[16] Western Electric Company: *Statistical Quality Control Handbook,* 2d ed., Western Electric Company, Inc., New York 1958.

General Quality Control

[17] Deming, W. E.: *Out of the Crisis*, Massachusetts Institute of Technology Center of Advanced Study, 1986.

[18] Deming, W. E.: *Quality, Productivity, and Competitive Position*, The MIT Press, Cambridge, Massachusetts, 1982.

[19] Feigenbaum, A. V.: *Total Quality Control—Engineering and Management*, 3d ed., McGraw-Hill Book Company, New York, 1983.

[20] Juran, J. W. and F. M. Gryna, Jr.: *Quality Planning and Analysis*, 2d ed., McGraw-Hill Book Company, New York, 1980.

Special Topics in Quality Control

[21] American Society for Quality Control, *Advanced Quality Planning*, Milwaukee; Wisconsin, (1988).

[22] Bazovsky, I.: *Reliability: Theory and Practice*, Prentice-Hall, Inc., Englewood Cliffs, New Jersey, 1961.

[23] Blanchard, B. S. and E. E. Lowery: *Maintainability*, McGraw-Hill Book Company, New York, 1969.

[24] Box, G. E. P. and N. R. Draper: *Evolutionary Operation*, John Wiley & Sons, Inc., New York, 1969.

[25] Chorafas, D. N.: *Statistical Processes and Reliability Engineering*, D. Van Nostrand Reinhold Company, Inc., Princeton, New Jersey, 1960.

[26] European Organization for Quality: *Glossary of Terms Used in Quality Control*, Rotterdam, 1965. (Includes terms in English, French, German, Dutch and Italian.)

[27] *Guide for Quality Control and Control Chart Method of Analyzing Data, ANSI Standards Z1.1-1975 and Z1.2-1975*, National American Standards Institute, New York, 1975.

[28] Haviland, R. P.: *Engineering Reliability and Long Life Design*, Van Nostrand Reinhold Company, Princeton, New Jersey, 1964.

[29] Hayes, G. E. and G. Romig: *Modern Quality Control*, Bruce Division of Benziger, Bruce, and Glencoe, Inc., Encino, California, 1982.

[30] Hunter, J. S.: "The Exponentially Weighted Moving Average," *Journal of Quality Technology*, Vol 18, No. 4, 203-210, 1986.

[31] Ireson, W. G. 9 eds.: *Reliability Handbook, 3d ed.*, McGraw-Hill Book Company, New York, 1988.

[32] Ireson, W. G and E. L. Grant (eds.): *Handbook of Industrial Engineering and Management, 2d ed.*, Prentice-Hall, Inc., Englewood Cliffs, New Jersey, 1971.

[33] Japanese Standards Association: *Standardization and Quality Control in Japan*, JSA Technical Report No. 1, Japanese Standards Association, Tokyo, Japan, 1963.

[34] Juran, J. M: *Management of Inspection and Quality Control*, Harper & Row, Publishers, Inc., New York, 1945.

[35] Juran, J. M. (ed): *Quality Control Handbook, 3d ed.*, McGraw-Hill Book Company, New York, 1974.

[36] Kalbfleisch, J. D. and R. L., Prentice: *The Statistical Analysis of Failure Time Data*, John Wiley & Sons, Inc., New York, 1980.

[37] Keats, J. B and N. F. Hubele, (eds.): *Statistical Process Control in Automated Manufacturing*, Marcel Dekker, Inc., New York, 1989.

[38] Ku, H. H. (ed.): *Precision Measurement and Calibration*, Superintendent of Documents, Government Printing Office, Washington, DC, 1969.

[39] Lucas, J. M. and H. S. Saccucci: "Exponentially Weighted Moving Average Control Schemes: Properties and Enhancements," *Technometrics*, Vol. 32, No. 1, 1-29, 1990.

[40] Military Standard 105D: *Sampling Procedures and Tables for Inspection by Attributes*, Superintendent of Documents, Government Printing Office, Washington, DC, 1963.

[41] Military Standard 414: *Sampling Procedures and Tables for Inspection by Variables for Percent Defective*, Superintendent of Documents, Government Printing Office, Washington, DC, 1957.

[42] *Supply and Logistics Handbook Inspection H 105, Administration of Sampling Procedures for Acceptance Inspection*, Superintendent of Documents, Government Printing Office, Washington, DC, 1954.

[43] Schmidt, S. R. and R. G. Launsby: *Understanding Industrial Designed Experiments*, Longmont: CQG Ltd. Printing, 1989.

[44] Taguchi, G.: *Introduction to Quality Engineering: Designing Quality into Products and Processes*, Kraus International Publications, White Plains, NY, 1986.

[45] Taguchi, G, and Y. Wu: *Introduction to Off-Line Quality Control*, Central Japan Quality Control Association, Nagoya, 1985.

Related Topics

[46] Drucker, P. E.: "The Emerging Theory of Manufacturing," *Harvard Business Review*, May-June, 1990.

[47] Garvin, D. A.: *Managing Quality: The Strategic and Competitive Edge*, Free Press, New York, 1988.

[48] Harris, D. H. and F. B. Chaney: *Human Factors in Quality Assistance*, John Wiley & Sons, Inc., New York, 1969.

[49] Hauser, J. R. and Clausing, D.: "The House of Quality," *Harvard Business Review*, May-June, 1988.

[50] Krantz, T. K.: "How Velcro Got Hooked on Quality," *Harvard Business Review*, Sep.-Oct., 1989.

[51] Whitney, D. E.: "Manufacturing By Design," *Harvard Business Review*, Jul.-Aug., 1988.

General Statistics and Data Analysis

[52] Brownlee, K. A.: *Statistical Theory and Methodology in Science and Engineering*, 2d ed., John Wiley & Sons, Inc., New York, 1965. Reprinted by R. E. Krieger Publishing Company, Inc., Melbourne, Florida, 1984.

[53] Dixon, W. J., and F. J. Massey, Jr.: *Introduction to Statistical Analysis*, 4th ed., McGraw-Hill Book Company, New York, 1983.

[54] Freeman, H. A.: *Introduction to Statistical Inference*, Addison-Wesley Book Company, Reading, Massachusets, 1963.

[55] Fry, T. C.: *Probability and Its Engineering Uses*, 2d ed., D. Van Nostrand Reinhold Company, Inc., Princeton, New Jersey, 1965.

[56] Guttman, Irwin and S. S. Wilks: *Introductory Engineering Statistics*, 3d ed., John Wiley & Sons, Inc., New York, 1982.

[57] Gulezian, R.: *Statistics For Decision Making*, W. B. Saunders Company, Philadelphia, 1979, Reprinted by QualityAlert Institute, New York, 1990.

[58] Hamburg, M.: *Statistical Analysis for Decision Making*, 4th ed., Harcourt Brace Jovanoivich, Inc., New York, 1987.

[59] Hoaglin, D. C., F. Mostellor, and J. Tukey, *Data Tables, Trends, and Shapes*, John Wiley & Sons, Inc., New York, 1985.

[60] Hoaglin, D. C.: *Understanding Robust and Exploratory Data Analysis*, John Wiley & Sons, Inc., New York, 1983.

[61] Hoel, P. G.: *Elementary Statistics*, 4th ed., John Wiley & Sons, Inc., New York, 1976.

[62] Hoel, P.G.: *Introduction to Mathematical Statistics*, 5th ed., John Wiley & Sons, Inc., New York, 1984.

Special Topics in Statistics

[63] Box, G. E. P., W. G. Hunter, and J. S. Hunter: *Statistics for Experimenters*, John Wiley & Sons, Inc., New York, 1978.

[64] Brown, R. G.: *Smoothing, Forecasting and Prediction of Discrete Time Series*, Prentice-Hall Inc., Englewood Cliffs, New Jersey, 1963.

[65] Cochran, W. G.: *Sampling Techniques*, 3d ed., John Wiley & Sons, Inc., New York, 1977.

[66] Cochran, W. G. and Cox, G. M.: *Experimental Design, 2d ed.*, John Wiley & Sons, Inc., New York, 1957.

[67] Cox, D. R.: *Planning of Experiments*, John Wiley & Sons, Inc., New York, 1958.

[68] Deming, W. E.: *Some Theory of Sampling*, John Wiley & Sons, Inc., New York, 1966.

[69] Draper, N. R., and H. Smith: *Applied Regression Analysis*, 2d ed., John Wiley & Sons., Inc., New York, 1981.

[70] Fleiss, J. L.: *Statistical Methods for Rates and Proportions*, John Wiley & Sons., Inc., New York, 1973.

[71] Gibbons, J. D.: *Nonparametric Methods for Quantitative Analysis*, American Science Press, New York, 1985.

[72] Gilchrist, W.: *Statistical Forecasting*, John Wiley & Sons, Inc., New York, 1976.

[73] Holman, J. P.: *Experimental Methods for Engineers*, 4th ed., McGraw-Hill Book Company, New York, 1983.

[74] Johnson, N. L. and Leone, F. C.: *Statistics and Experimental Design in Engineering and the Physical Sciences*, John Wiley & Sons, Inc., Vol. I and II, 2d ed., New York, 1977.

[75] Kish, L.: *Survey Sampling*, John Wiley & Sons, Inc., New York, 1965.

[76] Montgomery, D. C.: *Design and Analysis of Experiments*, John Wiley & Sons, Inc., 2d ed., New York, 1984.

[77] Morrison, D. F.: *Applied Linear Statistical Methods*, Prentice Hall, Inc., Englewood Cliffs, New Jersey, 1983.

APPENDIX

Areas of the Normal Curve

Table A-1 is used to find probabilities associated with normal variables in terms of areas under the standardized normal distribution

$$f(z) = \frac{1}{\sqrt{2\pi}}\, e^{-\frac{1}{2}z^2};\, -\infty < z < \infty$$

which is a normal curve with a mean of zero and variance of one. The area tabulated is shown in the following diagram.

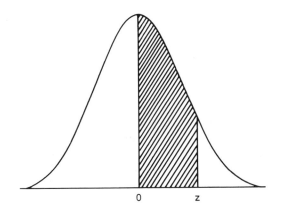

The area tabulated lies between the mean equal to zero and a value z of the standardized normal variate. Probabilities for any normal variate with mean μ and standard deviation σ are found in terms of z and the area provided in the table. The value of z is obtained with the formula

$$z = \frac{x - \mu}{\sigma}$$

The table is given in terms of *positive* values of z since the distribution is symmetric; however, probabilities in other regions of the curve also can be obtained.

Example A.1. To find the probability that x falls between the mean and 16.4 of a normal distribution with a mean of 10 and standard deviation of 5, we first compute z as

$$z = \frac{x - \mu}{\sigma} = \frac{16.4 - 10}{5} = \frac{6.4}{5} = 1.28$$

Values of z are given in the extreme left column of the table where the second decimal place is provided at the top row. Hence, we locate 1.2 in the left column and the second decimal place 0.08 at the top. The intersection of these values provides the required probability

$$= P(10 \leq x \leq 16.4) = .3997$$

All other probabilities are found in terms of positive values of z corresponding to areas between the mean and a given number.

Example A.2. The table can be used in reverse to find values of x when a probability is given. Using the same mean and standard deviation in the above example, let us find the value of x such that 30 percent of the area falls between x and the mean. This is found by substituting into the formula for z and solving for x:

$$z = \frac{x - \mu}{\sigma}$$

$$0.84 = \frac{x - 10}{5} \rightarrow x = 10 + (.84)(5)$$

$$= 14.2$$

The value of z is determined by finding the given probability in the body of the table and adding the corresponding values in the left column and top row. In this case 30 percent or 0.3000 is not included so the closest value of .2995 is used. It is not necessary to interpolate. When x is to the left of the mean, a minus sign must be attached to z.

Table A-1 AREAS OF THE NORMAL CURVE

Z	.00	.01	.02	.03	.04	.05	.06	.07	.08	.09
0.0	.0000	.0040	.0080	.0120	.0160	.0199	.0239	.0279	.0319	.0359
0.1	.0398	.0438	.0478	.0517	.0557	.0596	.0636	.0675	.0714	.0753
0.2	.0793	.0832	.0871	.0910	.0948	.0987	.1026	.1064	.1103	.1141
0.3	.1179	.1217	.1255	.1293	.1331	.1368	.1406	.1443	.1480	.1517
0.4	.1554	.1591	.1628	.1664	.1700	.1736	.1772	.1808	.1844	.1879
0.5	.1915	.1950	.1985	.2019	.2054	.2088	.2123	.2157	.2190	.2224
0.6	.2257	.2291	.2324	.2357	.2389	.2422	.2454	.2486	.2517	.2549
0.7	.2580	.2611	.2642	.2673	.2704	.2734	.2764	.2794	.2823	.2852
0.8	.2881	.2910	.2939	.2967	.2995	.3023	.3051	.3078	.3106	.3133
0.9	.3159	.3186	.3212	.3238	.3264	.3289	.3315	.3340	.3365	.3389
1.0	.3413	.3438	.3461	.3485	.3508	.3531	.3554	.3577	.3599	.3621
1.1	.3643	.3665	.3686	.3708	.3729	.3749	.3770	.3790	.3810	.3830
1.2	.3849	.3869	.3888	.3907	.3925	.3944	.3962	.3980	.3997	.4015
1.3	.4032	.4049	.4066	.4082	.4099	.4115	.4131	.4147	.4162	.4177
1.4	.4192	.4207	.4222	.4236	.4251	.4265	.4279	.4292	.4306	.4319
1.5	.4332	.4345	.4357	.4370	.4382	.4394	.4406	.4418	.4429	.4441
1.6	.4452	.4463	.4474	.4484	.4495	.4505	.4515	.4525	.4535	.4545
1.7	.4554	.4564	.4573	.4582	.4591	.4599	.4608	.4616	.4625	.4633
1.8	.4641	.4649	.4656	.4664	.4671	.4678	.4686	.4693	.4699	.4706
1.9	.4713	.4719	.4726	.4732	.4738	.4744	.4750	.4756	.4761	.4767
2.0	.4772	.4778	.4783	.4788	.4793	.4798	.4803	.4808	.4812	.4817
2.1	.4821	.4826	.4830	.4834	.4838	.4842	.4846	.4850	.4854	.4857
2.2	.4861	.4864	.4868	.4871	.4875	.4878	.4881	.4884	.4887	.4890
2.3	.4893	.4896	.4898	.4901	.4904	.4906	.4909	.4911	.4913	.4916
2.4	.4918	.4920	.4922	.4925	.4927	.4929	.4931	.4932	.4934	.4936
2.5	.4938	.4940	.4941	.4943	.4945	.4946	.4948	.4949	.4951	.4952
2.6	.4953	.4955	.4956	.4957	.4959	.4960	.4961	.4962	.4963	.4964
2.7	.4965	.4966	.4967	.4968	.4969	.4970	.4971	.4972	.4973	.4974
2.8	.4974	.4975	.4976	.4977	.4977	.4978	.4979	.4979	.4980	.4981
2.9	.4981	.4982	.4982	.4983	.4984	.4984	.4985	.4985	.4986	.4986
3.0	.4987	.4987	.4987	.4988	.4988	.4989	.4989	.4989	.4990	.4990
3.1	.4990	.4991	.4991	.4991	.4992	.4992	.4992	.4992	.4993	.4993
3.2	.4993	.4993	.4994	.4994	.4994	.4994	.4994	.4995	.4995	.4995
3.3	.4995	.4995	.4995	.4996	.4996	.4996	.4996	.4996	.4996	.4997
3.4	.4997	.4997	.4997	.4997	.4997	.4997	.4997	.4997	.4997	.4998

Binomial Probabilities

Table A-2 provides probabilities of f(x) for values of x based on selected values of n and P for the binomial distribution

$$f(x) = \begin{bmatrix} n \\ x \end{bmatrix} P^x(1-P)^{n-x}; x = 0, 1, ..., n$$

The table considers values of n between 1 and 20 and values of P between 0.05 and 0.50 in intervals of 0.05. When P exceeds 0.5, probabilities for values of x can be found by using n-x in place of x and 1-P in place of P.

Example A.3. In order to find the probability that x equals 5 given n equals 8 and P equals 0.35, locate n=8 in the left margin and P=.35 at the top. The probability is found below P across from x=5 associated with n.

$$P(x = 5 \mid n = 8, P = .35) = .0808$$

Example A.4. Suppose n equals 10 and P equals 0.8 and we want to find the probability that x equals 7. Since the table does not consider P greater than 0.5 directly, we must locate n-x = 10 -7 = 1-.8 = .2 corresponding to n = 8. That is

$$P(x = 7 \mid n = 10, P = .8) = P(n - x = 3 \mid n = 10, 1 - P = .2)$$

$$= .2013$$

Note. Some probabilities are recorded as 0.0000. This means that the actual probability is very small and results in zero when rounded to four decimal places.

Table A-2   BINOMIAL PROBABILITIES

n	x	.05	.10	.15	.20	.25	.30	.35	.40	.45	.50
1	0	.9500	.9000	.8500	.8000	.7500	.7000	.6500	.6000	.5500	.5000
	1	.0500	.1000	.1500	.2000	.2500	.3000	.3500	.4000	.4500	.5000
2	0	.9025	.8100	.7225	.6400	.5625	.4900	.4225	.3600	.3025	.2500
	1	.0950	.1800	.2550	.3200	.3750	.4200	.4550	.4800	.4950	.5000
	2	.0025	.0100	.0225	.0400	.0625	.0900	.1225	.1600	.2025	.2500
3	0	.8574	.7290	.6141	.5120	.4219	.3430	.2746	.2160	.1664	.1250
	1	.1354	.2430	.3251	.3840	.4219	.4410	.4436	.4320	.4084	.3750
	2	.0071	.0270	.0574	.0960	.1406	.1890	.2389	.2880	.3341	.3750
	3	.0001	.0010	.0034	.0080	.0156	.0270	.0429	.0640	.0911	.1250
4	0	.8145	.0561	.5220	.4096	.3164	.2401	.1785	.1296	.0915	.0625
	1	.1715	.2916	.3685	.4096	.4219	.4116	.3845	.3456	.2995	.2500
	2	.0135	.0486	.0975	.1536	.2109	.2646	.3105	.3456	.3675	.3750
	3	.0005	.0036	.0115	.0256	.0469	.0756	.1115	.1536	.2005	.2500
	4	.0000	.0001	.0005	.0016	.0039	.0081	.0150	.0256	.0410	.0625
5	0	.7738	.5905	.4437	.3277	.2373	.1681	.1160	.0778	.0503	.0312
	1	.2036	.3280	.3915	.4096	.3955	.3602	.3124	.2592	.2059	.1562
	2	.0214	.0729	.1382	.2048	.2637	.3087	.3364	.3456	.3369	.3125
	3	.0011	.0081	.0244	.0512	.0879	.1323	.1811	.2304	.2757	.3125
	4	.0000	.0004	.0022	.0064	.0146	.0284	.0488	.0768	.1128	.1562
	5	.0000	.0000	.0001	.0003	.0010	.0024	.0053	.0102	.0185	.0312
6	0	.7351	.5314	.3771	.2621	.1780	.1176	.0754	.0467	.0277	.0156
	1	.2321	.3543	.3993	.3932	.3560	.3025	.2437	.1866	.1359	.0938
	2	.0305	.0984	.1762	.2458	.2966	.3241	.3280	.3110	.2780	.2344
	3	.0021	.0146	.0415	.0819	.1318	.1852	.2355	.2765	.3032	.3125
	4	.0001	.0012	.0055	.0154	.0330	.0595	.0951	.1382	.1861	.2344
	5	.0000	.0001	.0004	.0015	.0044	.0102	.0205	.0369	.0609	.0938
	6	.0000	.0000	.0000	.0001	.0002	.0007	.0018	.0041	.0083	.0156
7	0	.6983	.4783	.3206	.2097	.1335	.0824	.0490	.0280	.0152	.0078
	1	.2573	.3720	.2960	.3670	.3115	.2471	.1848	.1306	.0872	.0547
	2	.0406	.1240	.2097	.2753	.3115	.3177	.2985	.2613	.2140	.1641
	3	.0036	.0230	.0617	.1147	.1730	.2269	.2679	.2903	.2918	.2734
	4	.0002	.0026	.0109	.0287	.0577	.0972	.1442	.1935	.2388	.2734
	5	.0000	.0002	.0012	.0043	.0115	.0250	.0466	.0774	.1172	.1641
	6	.0000	.0000	.0001	.0004	.0013	.0036	.0084	.0172	.0320	.0547
	7	.0000	.0000	.0000	.0000	.0001	.0002	.0006	.0016	.0037	.0078
8	0	.6634	.4305	.2725	.1678	.1001	.0576	.0319	.0168	.0084	.0039
	1	.2793	.3826	.3847	.3355	.2670	.1977	.1373	.0896	.0548	.0312
	2	.0515	.1488	.2376	.2936	.3115	.2965	.2587	.2090	.1569	.1094
	3	.0054	.0331	.0839	.1468	.2076	.2541	.2786	.2787	.2568	.2188
	4	.0004	.0046	.0185	.0459	.0865	.1361	.1875	.2322	.2627	.2734
	5	.0000	.0004	.0026	.0092	.0231	.0467	.0808	.1239	.1719	.2188
	6	.0000	.0000	.0002	.0011	.0038	.0100	.0217	.0413	.0703	.1094
	7	.0000	.0000	.0000	.0001	.0004	.0012	.0033	.0079	.0164	.0312
	8	.0000	.0000	.0000	.0000	.0000	.0001	.0002	.0007	.0017	.0039

Table A-2 BINOMIAL PROBABILITIES

n	x	.05	.10	.15	.20	.25	.30	.35	.40	.45	.50
9	0	.6302	.3874	.2316	.1342	.0751	.0404	.0207	.0101	.0046	.0020
	1	.2985	.3874	.3679	.3020	.2253	.1556	.1004	.0605	.0339	.0176
	2	.0629	.1722	.2597	.3020	.3003	.2668	.2162	.1612	.1110	.0703
	3	.0077	.0446	.1069	.1762	.2336	.2668	.2716	.2508	.2119	.1641
	4	.0006	.0074	.0283	.0661	.1168	.1715	.2194	.2508	.2600	.2461
	5	.0000	.0008	.0050	.0165	.0389	.0735	.1181	.1672	.2128	.2461
	6	.0000	.0001	.0006	.0028	.0087	.0210	.0424	.0743	.1160	.1641
	7	.0000	.0000	.0000	.0003	.0012	.0039	.0098	.0212	.0407	.0703
	8	.0000	.0000	.0000	.0000	.0001	.0004	.0013	.0035	.0083	.0176
	9	.0000	.0000	.0000	.0000	.0000	.0000	.0001	.0003	.0008	.0020
10	0	.5987	.3487	.1969	.1074	.0563	.0282	.0135	.0060	.0025	.0010
	1	.3151	.3874	.3474	.2684	.1877	.1211	.0725	.0403	.0207	.0098
	2	.0746	.1937	.2759	.3020	.2816	.2335	.1757	.1209	.0763	.0439
	3	.0105	.0574	.1298	.2013	.2503	.2668	.2522	.2150	.1665	.1172
	4	.0010	.0112	.0401	.0881	.1460	.2001	.2377	.2508	.2384	.2051
	5	.0001	.0015	.0085	.0264	.0584	.1029	.1536	.2007	.2340	.2461
	6	.0000	.0001	.0012	.0055	.0162	.0368	.0689	.1115	.1596	.2051
	7	.0000	.0000	.0001	.0008	.0031	.0090	.0212	.0425	.0746	.1172
	8	.0000	.0000	.0000	.0001	.0004	.0014	.0043	.0106	.0229	.0439
	9	.0000	.0000	.0000	.0000	.0000	.0001	.0005	.0016	.0042	.0098
	10	.0000	.0000	.0000	.0000	.0000	.0000	.0000	.0001	.0003	.0010
11	0	.5688	.3138	.1673	.0859	.0422	.0198	.0088	.0036	.0014	.0005
	1	.3293	.3835	.3248	.2362	.1549	.0932	.0518	.0266	.0125	.0054
	2	.0867	.2131	.2866	.2953	.2581	.1998	.1395	.0887	.0513	.0269
	3	.0137	.0710	.1517	.2215	.2581	.2568	.2254	.1774	.1259	.0806
	4	.0014	.0158	.0536	.1107	.1721	.2201	.2428	.2365	.2060	.1611
	5	.0001	.0025	.0132	.0388	.0803	.1321	.1830	.2207	.2360	.2256
	6	.0000	.0003	.0023	.0097	.0268	.0566	.0985	.1471	.1931	.2256
	7	.0000	.0000	.0003	.0017	.0064	.0173	.0379	.0701	.1128	.1611
	8	.0000	.0000	.0000	.0002	.0011	.0037	.0102	.0234	.0462	.0806
	9	.0000	.0000	.0000	.0000	.0001	.0005	.0018	.0052	.0126	.0269
	10	.0000	.0000	.0000	.0000	.0000	.0000	.0002	.0007	.0021	.0054
	11	.0000	.0000	.0000	.0000	.0000	.0000	.0000	.0000	.0002	.0005
12	0	.5404	.2824	.1422	.0687	.0317	.0138	.0057	.0022	.0008	.0002
	1	.3413	.3766	.3012	.2062	.1267	.0712	.0368	.0174	.0075	.0029
	2	.0988	.2301	.2924	.2835	.2323	.1678	.1088	.0639	.0339	.0161
	3	.0173	.0852	.1720	.2362	.2581	.2397	.1954	.1419	.0923	.0537
	4	.0021	.0213	.0683	.1329	.1936	.2311	.2367	.2128	.1700	.1208
	5	.0002	.0038	.0193	.0532	.1032	.1585	.2039	.2270	.2225	.1934
	6	.0000	.0005	.0040	.0155	.0401	.0792	.1281	.1766	.2124	.2256
	7	.0000	.0000	.0006	.0033	.0115	.0291	.0591	.1009	.1489	.1934
	8	.0000	.0000	.0001	.0005	.0024	.0078	.0199	.0420	.0762	.1208
	9	.0000	.0000	.0000	.0001	.0004	.0015	.0048	.0125	.0277	.0537
	10	.0000	.0000	.0000	.0000	.0000	.0002	.0008	.0025	.0068	.0161
	11	.0000	.0000	.0000	.0000	.0000	.0000	.0001	.0003	.0010	.0029
	12	.0000	.0000	.0000	.0000	.0000	.0000	.0000	.0000	.0001	.0002

Table A-2   BINOMIAL PROBABILITIES

n	x	P .05	.10	.15	.20	.25	.30	.35	.40	.45	.50
13	0	.5133	.2542	.1209	.0550	.0238	.0097	.0037	.0013	.0004	.0001
	1	.3512	.3672	.2774	.1787	.1029	.0540	.0259	.0113	.0045	.0016
	2	.1109	.2448	.2937	.2680	.2059	.1388	.0836	.0453	.0220	.0095
	3	.0214	.0997	.1900	.2457	.2517	.2181	.1651	.1107	.0660	.0349
	4	.0028	.0277	.0838	.1535	.2097	.2337	.2222	.1845	.1350	.0873
	5	.0003	.0055	.0266	.0691	.1258	.1803	.2154	.2214	.1989	.1571
	6	.0000	.0008	.0063	.0230	.0559	.1030	.1546	.1968	.2169	.2095
	7	.0000	.0001	.0011	.0058	.0186	.0442	.0833	.1312	.1775	.2095
	8	.0000	.0000	.0001	.0011	.0047	.0142	.0336	.0656	.1089	.1571
	9	.0000	.0000	.0000	.0001	.0009	.0034	.0101	.0243	.0495	.0873
	10	.0000	.0000	.0000	.0000	.0001	.0006	.0022	.0065	.0162	.0349
	11	.0000	.0000	.0000	.0000	.0000	.0001	.0003	.0012	.0036	.0095
	12	.0000	.0000	.0000	.0000	.0000	.0000	.0000	.0001	.0005	.0016
	13	.0000	.0000	.0000	.0000	.0000	.0000	.0000	.0000	.0000	.0001
14	0	.4877	.2288	.1028	.0440	.0178	.0068	.0024	.0008	.0002	.0001
	1	.3593	.3559	.2539	.1539	.0832	.0407	.0181	.0073	.0027	.0009
	2	.1229	.2570	.2912	.2501	.1802	.1134	.0634	.0317	.0141	.0056
	3	.0259	.1142	.2056	.2501	.2402	.1943	.1366	.0845	.0462	.0222
	4	.0037	.0349	.0998	.1720	.2202	.2290	.2022	.1549	.1040	.0661
	5	.0004	.0078	.0352	.0860	.1468	.1963	.2178	.2066	.1701	.1222
	6	.0000	.0013	.0093	.0322	.0734	.1262	.1759	.2066	.2088	.1833
	7	.0000	.0002	.0019	.0092	.0280	.0618	.1082	.1574	.1952	.2095
	8	.0000	.0000	.0003	.0020	.0082	.0232	.0510	.0918	.1398	.1833
	9	.0000	.0000	.0000	.0003	.0018	.0066	.0183	.0408	.0762	.1222
	10	.0000	.0000	.0000	.0000	.0003	.0014	.0049	.0136	.0312	.0611
	11	.0000	.0000	.0000	.0000	.0000	.0002	.0010	.0033	.0093	.0222
	12	.0000	.0000	.0000	.0000	.0000	.0000	.0001	.0005	.0019	.0056
	13	.0000	.0000	.0000	.0000	.0000	.0000	.0000	.0001	.0002	.0009
	14	.0000	.0000	.0000	.0000	.0000	.0000	.0000	.0000	.0000	.0001
15	0	.4633	.2059	.0874	.0352	.0134	.0047	.0016	.0005	.0001	.0000
	1	.3658	.3432	.2312	.1319	.0668	.0305	.0126	.0047	.0016	.0005
	2	.1348	.2669	.2856	.2309	.1559	.0916	.0476	.0219	.0090	.0032
	3	.0307	.1285	.2184	.2501	.2252	.1700	.1110	.0634	.0318	.0139
	4	.0049	.0428	.1156	.1876	.2252	.2186	.1792	.1268	.0780	.0417
	5	.0006	.0105	.0449	.1032	.1651	.2061	.2123	.1859	.1404	.0916
	6	.0000	.0019	.0132	.0430	.0917	.1472	.1906	.2066	.1914	.1527
	7	.0000	.0003	.0030	.0138	.0393	.0811	.1319	.1771	.2013	.1964
	8	.0000	.0000	.0005	.0035	.0131	.0348	.0710	.1181	.1647	.1964
	9	.0000	.0000	.0001	.0007	.0034	.0116	.0298	.0612	.1048	.1527
	10	.0000	.0000	.0000	.0001	.0007	.0030	.0096	.0245	.0515	.0916
	11	.0000	.0000	.0000	.0000	.0001	.0006	.0024	.0074	.0191	.0417
	12	.0000	.0000	.0000	.0000	.0000	.0001	.0004	.0016	.0052	.0139
	13	.0000	.0000	.0000	.0000	.0000	.0000	.0001	.0003	.0010	.0032
	14	.0000	.0000	.0000	.0000	.0000	.0000	.0000	.0000	.0001	.0005
	15	.0000	.0000	.0000	.0000	.0000	.0000	.0000	.0000	.0000	.0000

Table A-2   BINOMIAL PROBABILITIES

n	x	.05	.10	.15	.20	.25	.30	.35	.40	.45	.50
16	0	.4401	.1853	.0743	.0281	.0100	.0033	.0010	.0003	.0001	.0000
	1	.3706	.3294	.2097	.1126	.0535	.0228	.0087	.0030	.0009	.0002
	2	.1463	.2745	.2775	.2111	.1336	.0732	.0353	.0150	.0056	.0018
	3	.0359	.1423	.2285	.2463	.2079	.1465	.0888	.0468	.0215	.0085
	4	.0061	.0514	.1311	.2001	.2252	.2040	.1553	.1014	.0572	.0278
	5	.0008	.0137	.0555	.1201	.1802	.2099	.2008	.1623	.1123	.0667
	6	.0001	.0028	.0180	.0550	.1101	.1649	.1982	.1983	.1684	.1222
	7	.0000	.0004	.0045	.0197	.0524	.1010	.1524	.1889	.1969	.1746
	8	.0000	.0001	.0009	.0055	.0197	.0487	.0923	.1417	.1812	.1964
	9	.0000	.0000	.0001	.0012	.0058	.0185	.0442	.0840	.1318	.1746
	10	.0000	.0000	.0000	.0002	.0014	.0056	.0167	.0392	.0755	.1222
	11	.0000	.0000	.0000	.0000	.0002	.0013	.0049	.0142	.0337	.0667
	12	.0000	.0000	.0000	.0000	.0000	.0002	.0011	.0040	.0115	.0278
	13	.0000	.0000	.0000	.0000	.0000	.0000	.0002	.0008	.0029	.0085
	14	.0000	.0000	.0000	.0000	.0000	.0000	.0000	.0001	.0005	.0018
	15	.0000	.0000	.0000	.0000	.0000	.0000	.0000	.0000	.0001	.0002
	16	.0000	.0000	.0000	.0000	.0000	.0000	.0000	.0000	.0000	.0000
17	0	.4181	.1668	.0631	.0225	.0075	.0023	.0007	.0002	.0000	.0000
	1	.3741	.3150	.1893	.0957	.0426	.0169	.0060	.0019	.0005	.0001
	2	.1575	.2800	.2673	.1914	.1136	.0581	.0260	.0102	.0035	.0010
	3	.0415	.1556	.2359	.2393	.1893	.1245	.0701	.0341	.0144	.0052
	4	.0076	.0605	.1457	.2093	.2209	.1868	.1320	.0796	.0411	.0182
	5	.0010	.0175	.0668	.1361	.1914	.2081	.1849	.1379	.0875	.0472
	6	.0001	.0039	.0236	.0680	.1276	.1784	.1991	.1839	.1432	.0944
	7	.0000	.0007	.0065	.0267	.0668	.1201	.1685	.1927	.1841	.1484
	8	.0000	.0001	.0014	.0084	.0279	.0644	.1134	.1606	.1883	.1855
	9	.0000	.0000	.0003	.0021	.0093	.0276	.0611	.1070	.1540	.1855
	10	.0000	.0000	.0000	.0004	.0025	.0095	.0263	.0571	.1008	.1484
	11	.0000	.0000	.0000	.0001	.0005	.0026	.0090	.0242	.0525	.0944
	12	.0000	.0000	.0000	.0000	.0001	.0006	.0024	.0081	.0215	.0472
	13	.0000	.0000	.0000	.0000	.0000	.0001	.0005	.0021	.0068	.0182
	14	.0000	.0000	.0000	.0000	.0000	.0000	.0001	.0004	.0016	.0052
	15	.0000	.0000	.0000	.0000	.0000	.0000	.0000	.0001	.0003	.0010
	16	.0000	.0000	.0000	.0000	.0000	.0000	.0000	.0000	.0000	.0001
	17	.0000	.0000	.0000	.0000	.0000	.0000	.0000	.0000	.0000	.0000
18	0	.3972	.1501	.0536	.0180	.0056	.0016	.0004	.0001	.0000	.0000
	1	.3763	.3002	.1704	.0811	.0338	.0126	.0042	.0012	.0003	.0001
	2	.1683	.2835	.2556	.1723	.0958	.0458	.0190	.0069	.0022	.0006
	3	.0473	.1680	.2406	.2297	.1704	.1046	.0547	.0246	.0095	.0031
	4	.0093	.0700	.1592	.2153	.2130	.1681	.1104	.0614	.0291	.0117
	5	.0014	.0218	.0787	.1507	.1988	.2017	.1664	.1146	.0666	.0327
	6	.0002	.0052	.0301	.0816	.1436	.1873	.1941	.1655	.1181	.0708
	7	.0000	.0010	.0091	.0350	.0820	.1376	.1792	.1892	.1657	.1214
	8	.0000	.0002	.0022	.0120	.0376	.0811	.1327	.1734	.1864	.1669
	9	.0000	.0000	.0004	.0033	.0139	.0386	.0794	.1284	.1694	.1855

Table A-2 BINOMIAL PROBABILITIES

n	x	.05	.10	.15	.20	.25	.30	.35	.40	.45	.50
18	10	.0000	.0000	.0001	.0008	.0042	.0149	.0385	.0771	.1248	.1669
	11	.0000	.0000	.0000	.0001	.0010	.0046	.0151	.0374	.0742	.1214
	12	.0000	.0000	.0000	.0000	.0002	.0012	.0047	.0145	.0354	.0708
	13	.0000	.0000	.0000	.0000	.0000	.0002	.0012	.0045	.0134	.0327
	14	.0000	.0000	.0000	.0000	.0000	.0000	.0002	.0011	.0039	.0117
	15	.0000	.0000	.0000	.0000	.0000	.0000	.0000	.0002	.0009	.0031
	16	.0000	.0000	.0000	.0000	.0000	.0000	.0000	.0000	.0001	.0006
	17	.0000	.0000	.0000	.0000	.0000	.0000	.0000	.0000	.0000	.0001
	18	.0000	.0000	.0000	.0000	.0000	.0000	.0000	.0000	.0000	.0000
19	0	.3774	.1351	.0456	.0144	.0042	.0011	.0003	.0001	.0000	.0000
	1	.3774	.2852	.1529	.0685	.0268	.0093	.0029	.0008	.0002	.0000
	2	.1787	.2852	.2428	.1540	.0803	.0358	.0138	.0046	.0013	.0003
	3	.0533	.1796	.2428	.2182	.1517	.0869	.0422	.0175	.0062	.0018
	4	.0112	.0798	.1714	.2182	.2023	.1491	.0909	.0467	.0203	.0074
	5	.0018	.0266	.0907	.1636	.2023	.1916	.1468	.0933	.0497	.0222
	6	.0002	.0069	.0374	.0955	.1574	.1916	.1844	.1451	.0949	.0518
	7	.0000	.0014	.0122	.0443	.0974	.1525	.1844	.1797	.1443	.0961
	8	.0000	.0002	.0032	.0166	.0487	.0981	.1489	.1797	.1771	.1442
	9	.0000	.0000	.0007	.0051	.0198	.0514	.0980	.1464	.1771	.1762
	10	.0000	.0000	.0001	.0013	.0066	.0220	.0528	.0976	.1449	.1762
	11	.0000	.0000	.0000	.0003	.0018	.0077	.0233	.0532	.0970	.1442
	12	.0000	.0000	.0000	.0000	.0004	.0022	.0083	.0237	.0529	.0961
	13	.0000	.0000	.0000	.0000	.0001	.0005	.0024	.0085	.0233	.0518
	14	.0000	.0000	.0000	.0000	.0000	.0001	.0006	.0024	.0082	.0222
	15	.0000	.0000	.0000	.0000	.0000	.0000	.0001	.0005	.0022	.0074
	16	.0000	.0000	.0000	.0000	.0000	.0000	.0000	.0001	.0005	.0018
	17	.0000	.0000	.0000	.0000	.0000	.0000	.0000	.0000	.0001	.0003
	18	.0000	.0000	.0000	.0000	.0000	.0000	.0000	.0000	.0000	.0000
	19	.0000	.0000	.0000	.0000	.0000	.0000	.0000	.0000	.0000	.0000
20	0	.3585	.1216	.0388	.0115	.0032	.0008	.0002	.0000	.0000	.0000
	1	.3774	.2702	.1368	.0576	.0211	.0068	.0020	.0005	.0001	.0000
	2	.1887	.2852	.2293	.1369	.0669	.0278	.0100	.0031	.0008	.0002
	3	.0596	.1901	.2428	.2054	.1339	.0716	.0323	.0123	.0040	.0011
	4	.0133	.0898	.1821	.2182	.1897	.1304	.0738	.0350	.0139	.0046
	5	.0022	.0319	.1028	.1746	.2023	.1789	.1272	.0746	.0365	.0148
	6	.0003	.0089	.0454	.1091	.1686	.1916	.1712	.1244	.0746	.0370
	7	.0000	.0020	.0160	.0545	.1124	.1643	.1844	.1659	.1221	.0739
	8	.0000	.0004	.0046	.0222	.0609	.1144	.1614	.1797	.1623	.1201
	9	.0000	.0001	.0011	.0074	.0271	.0654	.1158	.1597	.1771	.1602
	10	.0000	.0000	.0002	.0020	.0099	.0308	.0686	.1171	.1593	.1762
	11	.0000	.0000	.0000	.0005	.0030	.0120	.0336	.0710	.1185	.1602
	12	.0000	.0000	.0000	.0001	.0008	.0039	.0136	.0355	.0727	.1201
	13	.0000	.0000	.0000	.0000	.0002	.0010	.0045	.0146	.0366	.0739
	14	.0000	.0000	.0000	.0000	.0000	.0002	.0012	.0049	.0150	.0370
	15	.0000	.0000	.0000	.0000	.0000	.0000	.0003	.0013	.0049	.0148
	16	.0000	.0000	.0000	.0000	.0000	.0000	.0000	.0003	.0013	.0046
	17	.0000	.0000	.0000	.0000	.0000	.0000	.0000	.0000	.0002	.0011
	18	.0000	.0000	.0000	.0000	.0000	.0000	.0000	.0000	.0000	.0002
	19	.0000	.0000	.0000	.0000	.0000	.0000	.0000	.0000	.0000	.0000
	20	.0000	.0000	.0000	.0000	.0000	.0000	.0000	.0000	.0000	.0000

Poisson Probabilities

Table A-3 provides probabilities or f(x) for values of x based on selected values of μ for the Poison distribution:

$$f(x) = \frac{\mu^x e^{-\mu}}{x!}; x = 0, 1, 2, ...$$

The table includes values of μ between 0.1 and 20 in units of 0.1.

Example A.5. In order to find the probability that x equals 4 when μ equals 3.5, locate 3.5 in the center of the table and look down the column until you reach the row corresponding to x on the left equal to 4. The probability is found as

$$P(x = 4 | \mu = 3.5) = .1888$$

Note. Some probabilities are recorded as 0.0000. This means that the actual probability is very small and results in zero when rounded to four decimal places.

Table A-3 POISSON PROBABILITIES

x	0.1	0.2	0.3	0.4	0.5	μ 0.6	0.7	0.8	0.9	1.0
0	.9048	.8187	.7408	.6703	.6065	.5488	.4966	.4493	.4066	.3679
1	.0905	.1637	.2222	.2681	.3033	.3293	.3476	.3595	.3659	.3679
2	.0045	.0164	.0333	.0536	.0758	.0988	.1217	.1438	.1647	.1839
3	.0002	.0011	.0033	.0072	.0126	.0198	.0284	.0383	.0494	.0613
4	.0000	.0001	.0002	.0007	.0016	.0030	.0050	.0077	.0111	.0153
5	.0000	.0000	.0000	.0001	.0002	.0004	.0007	.0012	.0020	.0031
6	.0000	.0000	.0000	.0000	.0000	.0000	.0001	.0002	.0003	.0005
7	.0000	.0000	.0000	.0000	.0000	.0000	.0000	.0000	.0000	.0001

x	1.1	1.2	1.3	1.4	1.5	μ 1.6	1.7	1.8	1.9	2.0
0	.3329	.3012	.2725	.2466	.2231	.2019	.1827	.1653	.1496	.1353
1	.3662	.3614	.3543	.3452	.3347	.3230	.3106	.2975	.2842	.2707
2	.2014	.2169	.2303	.2417	.2510	.2584	.2640	.2678	.2700	.2707
3	.0738	.0867	.0998	.1128	.1255	.1378	.1496	.1607	.1710	.1804
4	.0203	.0260	.0324	.0395	.0471	.0551	.0636	.0723	.0812	.0902
5	.0045	.0062	.0084	.0111	.0141	.0176	.0216	.0260	.0309	.0361
6	.0008	.0012	.0018	.0026	.0035	.0047	.0061	.0078	.0098	.0120
7	.0001	.0002	.0003	.0005	.0008	.0011	.0015	.0020	.0027	.0034
8	.0000	.0000	.0001	.0001	.0001	.0002	.0003	.0005	.0006	.0009
9	.0000	.0000	.0000	.0000	.0000	.0000	.0001	.0001	.0001	.0002

x	2.1	2.2	2.3	2.4	2.5	μ 2.6	2.7	2.8	2.9	3.0
0	.1225	.1108	.1003	.0907	.0821	.0743	.0672	.0608	.0550	.0498
1	.2572	.2438	.2306	.2177	.2052	.1931	.1815	.1703	.1596	.1494
2	.2700	.2681	.2652	.2613	.2565	.2510	.2450	.2384	.2314	.2240
3	.1890	.1966	.2033	.2090	.2138	.2176	.2205	.2225	.2237	.2240
4	.0992	.1082	.1169	.1254	.1336	.1414	.1488	.1557	.1622	.1680
5	.0417	.0476	.0538	.0602	.0668	.0735	.0804	.0872	.0940	.1008
6	.2146	.0174	.0206	.0241	.0278	.0319	.0362	.0407	.0455	.0504
7	.0044	.0055	.0068	.0083	.0099	.0118	.0139	.0163	.0188	.0216
8	.0011	.0015	.0019	.0025	.0031	.0038	.0047	.0057	.0068	.0081
9	.0003	.0004	.0005	.0007	.0009	.0011	.0014	.0018	.0022	.0027
10	.0001	.0001	.0001	.0002	.0002	.0003	.0004	.0005	.0006	.0008
11	.0000	.0000	.0000	.0000	.0000	.0001	.0001	.0001	.0002	.0002
12	.0000	.0000	.0000	.0000	.0000	.0000	.0000	.0000	.0000	.0001

x	3.1	3.2	3.3	3.4	3.5	μ 3.6	3.7	3.8	3.9	4.0
0	.0450	.0408	.0369	.0334	.0302	.0273	.0247	.0224	.0202	.0183
1	.1397	.1304	.1217	.1135	.1057	.0984	.0915	.0850	.0789	.0733
2	.2165	.2087	.2008	.1929	.1850	.1771	.1692	.1615	.1539	.1465
3	.2237	.2226	.2209	.2186	.2158	.2125	.2087	.2046	.2001	.1954
4	.1734	.1781	.1823	.1858	.1888	.1912	.1931	.1944	.1951	.1954
5	.1075	.1140	.1203	.1264	.1322	.1377	.1429	.1477	.1522	.1563
6	.0555	.0608	.0662	.0716	.0771	.0826	.0881	.0936	.0989	.1042
7	.0246	.0278	.0312	.0348	.0385	.0425	.0466	.0508	.0551	.0595
8	.0095	.0111	.0129	.0148	.0169	.0191	.0215	.0241	.0269	.0298
9	.0033	.0040	.0047	.0056	.0066	.0076	.0089	.0102	.0116	.0132

Table A-3   POISSON PROBABILITIES

x	3.1	3.2	3.3	3.4	3.5	μ 3.6	3.7	3.8	3.9	4.0
10	.0010	.0013	.0016	.0019	.0023	.0028	.0033	.0039	.0045	.0053
11	.0003	.0004	.0005	.0006	.0007	.0009	.0011	.0013	.0016	.0019
12	.0001	.0001	.0001	.0002	.0002	.0003	.0003	.0004	.0005	.0006
13	.0000	.0000	.0000	.0000	.0001	.0001	.0001	.0001	.0002	.0002
14	.0000	.0000	.0000	.0000	.0000	.0000	.0000	.0000	.0000	.0001

x	4.1	4.2	4.3	4.4	4.5	μ 4.6	4.7	4.8	4.9	5.0
0	.0166	.0150	.0136	.0123	.0111	.0101	.0091	.0082	.0074	.0067
1	.0679	.0630	.0583	.0540	.0500	.0462	.0427	.0395	.0365	.0337
2	.1393	.1323	.1254	.1188	.1125	.1063	.1005	.0948	.0894	.0842
3	.1904	.1852	.1798	.1743	.1687	.1631	.1574	.1517	.1460	.1404
4	.1951	.1944	.1933	.1917	.1898	.1875	.1849	.1820	.1789	.1755
5	.1600	.1633	.1662	.1687	.1708	.1725	.1738	.1747	.1753	.1755
6	.1093	.1143	.1191	.1237	.1281	.1323	.1362	.1398	.1432	.1462
7	.0640	.0686	.0732	.0778	.0824	.0869	.0914	.0959	.1002	.1044
8	.0328	.0360	.0393	.0428	.0463	.0500	.0537	.0575	.0614	.0653
9	.0150	.0168	.0188	.0209	.0232	.0255	.0280	.0307	.0334	.0363
10	.0061	.0071	.0081	.0092	.0104	.0118	.0132	.0147	.0164	.0181
11	.0023	.0027	.0032	.0037	.0043	.0049	.0056	.0064	.0073	.0082
12	.0008	.0009	.0011	.0014	.0016	.0019	.0022	.0026	.0030	.0034
13	.0002	.0003	.0004	.0005	.0006	.0007	.0008	.0009	.0011	.0013
14	.0001	.0001	.0001	.0001	.0002	.0002	.0003	.0003	.0004	.0005
15	.0000	.0000	.0000	.0000	.0001	.0001	.0001	.0001	.0001	.0002

x	5.1	5.2	5.3	5.4	5.5	μ 5.6	5.7	5.8	5.9	6.0
0	.0061	.0055	.0050	.0045	.0041	.0037	.0033	.0030	.0027	.0025
1	.0311	.0287	.0265	.0244	.0225	.0207	.0191	.0176	.0162	.0149
2	.0793	.0746	.0701	.0659	.0618	.0580	.0544	.0509	.0477	.0446
3	.1348	.1293	.1239	.1185	.1133	.1082	.1033	.0985	.0938	.0892
4	.1719	.1681	.1641	.1600	.1558	.1515	.1472	.1428	.1383	.1339
5	.1753	.1748	.1740	.1728	.1714	.1697	.1678	.1656	.1632	.1606
6	.1490	.1515	.1537	.1555	.1571	.1584	.1594	.1601	.1605	.1606
7	.1086	.1125	.1163	.1200	.1234	.1267	.1298	.1326	.1353	.1377
8	.0692	.0731	.0771	.0810	.0849	.0887	.0925	.0962	.0998	.1033
9	.0392	.0423	.0454	.0486	.0519	.0552	.0586	.0620	.0654	.0688
10	.0200	.0220	.0241	.0262	.0285	.0309	.0334	.0359	.0386	.0413
11	.0093	.0104	.0116	.0129	.0143	.0157	.0173	.0190	.0207	.0225
12	.0039	.0045	.0051	.0058	.0065	.0073	.0082	.0092	.0102	.0113
13	.0015	.0018	.0021	.0024	.0028	.0032	.0036	.0041	.0046	.0052
14	.0006	.0007	.0008	.0009	.0011	.0013	.0015	.0017	.0019	.0022
15	.0002	.0002	.0003	.0003	.0004	.0005	.0006	.0007	.0008	.0009
16	.0001	.0001	.0001	.0001	.0001	.0002	.0002	.0002	.0003	.0003
17	.0000	.0000	.0000	.0000	.0000	.0001	.0001	.0001	.0001	.0001

Table A-3 POISSON PROBABILITIES

x	6.1	6.2	6.3	6.4	6.5	μ 6.6	6.7	6.8	6.9	7.0
0	.0022	.0020	.0018	.0017	.0015	.0014	.0012	.0011	.0010	.0009
1	.0137	.0126	.0116	.0106	.0098	.0090	.0082	.0076	.0070	.0064
2	.0417	.0390	.0364	.0340	.0318	.0296	.0276	.0258	.0240	.0223
3	.0848	.0806	.0765	.0726	.0688	.0652	.0617	.0584	.0552	.0521
4	.1294	.1249	.1205	.1162	.1118	.1076	.1034	.0992	.0952	.0912
5	.1579	.1549	.1519	.1487	.1454	.1420	.1385	.1349	.1314	.1277
6	.1605	.1601	.1595	.1586	.1575	.1562	.1546	.1529	.1511	.1490
7	.1399	.1418	.1435	.1450	.1462	.1472	.1480	.1486	.1489	.1490
8	.1066	.1099	.1130	.1160	.1188	.1215	.1240	.1263	.1284	.1304
9	.0723	.0757	.0791	.0825	.0858	.0891	.0923	.0954	.0985	.1014
10	.0441	.0469	.0498	.0528	.0558	.0588	.0618	.0649	.0679	.0710
11	.0245	.0265	.0285	.0307	.0330	.0353	.0377	.0401	.0426	.0452
12	.0124	.0137	.0150	.0164	.0179	.0194	.0210	.0227	.0245	.0264
13	.0058	.0065	.0073	.0081	.0089	.0098	.0108	.0119	.0130	.0142
14	.0025	.0029	.0033	.0037	.0041	.0046	.0052	.0058	.0064	.0071
15	.0010	.0012	.0014	.0016	.0018	.0020	.0023	.0026	.0029	.0033
16	.0004	.0005	.0005	.0006	.0007	.0008	.0010	.0011	.0013	.0014
17	.0001	.0002	.0002	.0002	.0003	.0003	.0004	.0004	.0005	.0006
18	.0000	.0001	.0001	.0001	.0001	.0001	.0001	.0002	.0002	.0002
19	.0000	.0000	.0000	.0000	.0000	.0000	.0000	.0001	.0001	.0001

x	7.1	7.2	7.3	7.4	7.5	μ 7.6	7.7	7.8	7.9	8.0
0	.0008	.0007	.0007	.0006	.0006	.0005	.0005	.0004	.0004	.0003
1	.0059	.0054	.0049	.0045	.0041	.0038	.0035	.0032	.0029	.0027
2	.0208	.0194	.0180	.0167	.0156	.0145	.0134	.0125	.0116	.0107
3	.0492	.0464	.0438	.0413	.0389	.0366	.0345	.0324	.0305	.0286
4	.0874	.0836	.0799	.0764	.0729	.0696	.0663	.0632	.0602	.0573
5	.1241	.1204	.1167	.1130	.1094	.1057	.1021	.0986	.0951	.0916
6	.1468	.1445	.1420	.1394	.1367	.1339	.1311	.1282	.1252	.1221
7	.1489	.1486	.1481	.1474	.1465	.1454	.1442	.1428	.1413	.1396
8	.1321	.1337	.1351	.1363	.1373	.1382	.1388	.1392	.1395	.1396
9	.1042	.1070	.1096	.1121	.1144	.1167	.1187	.1207	.1224	.1241
10	.0740	.0770	.0800	.0829	.0858	.0887	.0914	.0941	.0967	.0993
11	.0478	.0504	.0531	.0558	.0585	.0613	.0640	.0667	.0695	.0722
12	.0283	.0303	.0323	.0344	.0366	.0388	.0411	.0434	.0457	.0481
13	.0154	.0168	.0181	.0196	.0211	.0227	.0243	.0260	.0278	.0296
14	.0078	.0086	.0095	.0104	.0113	.0123	.0134	.0145	.0157	.0169
15	.0037	.0041	.0046	.0051	.0057	.0062	.0069	.0075	.0083	.0090
16	.0016	.0019	.0021	.0024	.0026	.0030	.0033	.0037	.0041	.0045
17	.0007	.0008	.0009	.0010	.0012	.0013	.0015	.0017	.0019	.0021
18	.0003	.0003	.0004	.0004	.0005	.0006	.0006	.0007	.0008	.0009
19	.0001	.0001	.0001	.0002	.0002	.0002	.0003	.0003	.0003	.0004
20	.0000	.0000	.0001	.0001	.0001	.0001	.0001	.0001	.0001	.0002
21	.0000	.0000	.0000	.0000	.0000	.0000	.0000	.0000	.0001	.0001

Table A-3   POISSON PROBABILITIES

x	8.1	8.2	8.3	8.4	8.5 μ	8.6	8.7	8.8	8.9	9.0
0	.0003	.0003	.0002	.0002	.0002	.0002	.0002	.0002	.0001	.0001
1	.0025	.0023	.0021	.0019	.0017	.0016	.0014	.0013	.0012	.0011
2	.0100	.0092	.0086	.0079	.0074	.0068	.0063	.0058	.0054	.0050
3	.0269	.0252	.0237	.0222	.0208	.0195	.0183	.0171	.0160	.0150
4	.0544	.0517	.0491	.0466	.0443	.0420	.0398	.0377	.0357	.0337
5	.0882	.0849	.0816	.0784	.0752	.0722	.0692	.0663	.0635	.0607
6	.1191	.1160	.1128	.1097	.1066	.1034	.1003	.0972	.0941	.0911
7	.1378	.1358	.1338	.1317	.1294	.1271	.1247	.1222	.1197	.1171
8	.1395	.1392	.1388	.1382	.1375	.1366	.1356	.1344	.1332	.1318
9	.1256	.1269	.1280	.1290	.1299	.1306	.1311	.1315	.1317	.1318
10	.1017	.1040	.1063	.1084	.1104	.1123	.1140	.1157	.1172	.1186
11	.0749	.0776	.0802	.0828	.0853	.0878	.0902	.0925	.0948	.0970
12	.0505	.0530	.0555	.0579	.0604	.0629	.0654	.0679	.0703	.0728
13	.0315	.0334	.0354	.0374	.0395	.0416	.0438	.0459	.0481	.0504
14	.0182	.0196	.0210	.0225	.0240	.0256	.0272	.0289	.0306	.0324
15	.0098	.0107	.0116	.0126	.0136	.0147	.0158	.0169	.0182	.0194
16	.0050	.0055	.0060	.0066	.0072	.0079	.0086	.0093	.0101	.0109
17	.0024	.0026	.0029	.0033	.0036	.0040	.0044	.0048	.0053	.0058
18	.0011	.0012	.0014	.0015	.0017	.0019	.0021	.0024	.0026	.0029
19	.0005	.0005	.0006	.0007	.0008	.0009	.0010	.0011	.0012	.0014
20	.0002	.0002	.0002	.0003	.0003	.0004	.0004	.0005	.0005	.0006
21	.0001	.0001	.0001	.0001	.0001	.0002	.0002	.0002	.0002	.0003
22	.0000	.0000	.0000	.0000	.0001	.0001	.0001	.0001	.0001	.0001

x	9.1	9.2	9.3	9.4	9.5 μ	9.6	9.7	9.8	9.9	10.0
0	.0001	.0001	.0001	.0001	.0001	.0001	.0001	.0001	.0001	.0000
1	.0010	.0009	.0009	.0008	.0007	.0007	.0006	.0005	.0005	.0005
2	.0046	.0043	.0040	.0037	.0034	.0031	.0029	.0027	.0025	.0023
3	.0140	.0131	.0123	.0115	.0107	.0100	.0093	.0087	.0081	.0076
4	.0319	.0302	.0285	.0269	.0254	.0240	.0226	.0213	.0201	.0189
5	.0581	.0555	.0530	.0506	.0483	.0460	.0439	.0418	.0398	.0378
6	.0881	.0851	.0822	.0793	.0764	.0736	.0709	.0682	.0656	.0631
7	.1145	.1118	.1091	.1064	.1037	.1010	.0982	.0955	.0928	.0901
8	.1302	.1286	.1269	.1251	.1232	.1212	.1191	.1170	.1148	.1126
9	.1317	.1315	.1311	.1306	.1300	.1293	.1284	.1274	.1263	.1251
10	.1198	.1210	.1219	.1228	.1235	.1241	.1245	.1249	.1250	.1251
11	.0991	.1012	.1031	.1049	.1067	.1083	.1098	.1112	.1125	.1137
12	.0752	.0776	.0799	.0822	.0844	.0866	.0888	.0908	.0928	.0948
13	.0526	.0549	.0572	.0594	.0617	.0640	.0662	.0685	.0707	.0729
14	.0342	.0361	.0380	.0399	.0419	.0439	.0459	.0479	.0500	.0521
15	.0208	.0221	.0235	.0250	.0265	.0281	.0297	.0313	.0330	.0347
16	.0118	.0127	.0137	.0147	.0157	.0168	.0180	.0192	.0204	.0217
17	.0063	.0069	.0075	.0081	.0088	.0095	.0103	.0111	.0119	.0128
18	.0032	.0035	.0039	.0042	.0046	.0051	.0055	.0060	.0065	.0071
19	.0015	.0017	.0019	.0021	.0023	.0026	.0028	.0031	.0034	.0037
20	.0007	.0008	.0009	.0010	.0011	.0012	.0014	.0015	.0017	.0019
21	.0003	.0003	.0004	.0004	.0005	.0006	.0006	.0007	.0008	.0009
22	.0001	.0001	.0002	.0002	.0002	.0002	.0003	.0003	.0004	.0004
23	.0000	.0001	.0001	.0001	.0001	.0001	.0001	.0001	.0002	.0002
24	.0000	.0000	.0000	.0000	.0000	.0000	.0000	.0001	.0001	.0001

Table A-3   POISSON PROBABILITIES

x	11	12	13	14	15	μ 16	17	18	19	20
0	.0000	.0000	.0000	.0000	.0000	.0000	.0000	.0000	.0000	.0000
1	.0002	.0001	.0000	.0000	.0000	.0000	.0000	.0000	.0000	.0000
2	.0010	.0004	.0002	.0001	.0000	.0000	.0000	.0000	.0000	.0000
3	.0037	.0018	.0008	.0004	.0002	.0001	.0000	.0000	.0000	.0000
4	.0102	.0053	.0027	.0013	.0006	.0003	.0001	.0001	.0000	.0000
5	.0224	.0127	.0070	.0037	.0019	.0010	.0005	.0002	.0001	.0001
6	.0411	.0255	.0152	.0087	.0048	.0026	.0014	.0007	.0004	.0002
7	.0646	.0437	.0281	.0174	.0104	.0060	.0034	.0018	.0010	.0005
8	.0888	.0655	.0457	.0304	.0194	.0120	.0072	.0042	.0024	.0013
9	.1085	.0874	.0661	.0473	.0324	.0213	.0135	.0083	.0050	.0029
10	.1194	.1048	.0859	.0663	.0486	.0341	.0230	.0150	.0095	.0058
11	.1194	.1144	.1015	.0844	.0663	.0496	.0355	.0245	.0164	.0106
12	.1094	.1144	.1099	.0984	.0829	.0661	.0504	.0368	.0259	.0176
13	.0926	.1056	.1099	.1060	.0956	.0814	.0658	.0509	.0378	.0271
14	.0728	.0905	.1021	.1060	.1024	.0930	.0800	.0655	.0514	.0387
15	.0534	.0724	.0885	.0989	.1024	.0992	.0906	.0786	.0650	.0516
16	.0367	.0543	.0719	.0866	.0960	.0992	.0963	.0884	.0772	.0646
17	.0237	.0383	.0550	.0713	.0847	.0934	.0963	.0936	.0863	.0760
18	.0145	.0256	.0397	.0554	.0706	.0830	.0909	.0936	.0911	.0844
19	.0084	.0161	.0272	.0409	.0557	.0699	.0814	.0887	.0911	.0888
20	.0046	.0097	.0177	.0286	.0418	.0559	.0692	.0798	.0866	.0888
21	.0024	.0055	.0109	.0191	.0299	.0426	.0560	.0684	.0783	.0846
22	.0012	.0030	.0065	.0121	.0204	.0310	.0433	.0560	.0676	.0769
23	.0006	.0016	.0037	.0074	.0133	.0216	.0320	.0438	.0559	.0669
24	.0003	.0008	.0020	.0043	.0083	.0144	.0226	.0328	.0442	.0557
25	.0001	.0004	.0010	.0024	.0050	.0092	.0154	.0237	.0336	.0446
26	.0000	.0002	.0005	.0013	.0029	.0057	.0101	.0164	.0246	.0343
27	.0000	.0001	.0002	.0007	.0016	.0034	.0063	.0109	.0173	.0254
28	.0000	.0000	.0001	.0003	.0009	.0019	.0038	.0070	.0117	.0181
29	.0000	.0000	.0001	.0002	.0004	.0011	.0023	.0044	.0077	.0125
30	.0000	.0000	.0000	.0001	.0002	.0006	.0013	.0026	.0049	.0083
31	.0000	.0000	.0000	.0000	.0001	.0003	.0007	.0015	.0030	.0054
32	.0000	.0000	.0000	.0000	.0001	.0001	.0004	.0009	.0018	.0034
33	.0000	.0000	.0000	.0000	.0000	.0001	.0002	.0005	.0010	.0020
34	.0000	.0000	.0000	.0000	.0000	.0000	.0001	.0002	.0006	.0012
35	.0000	.0000	.0000	.0000	.0000	.0000	.0000	.0001	.0003	.0007
36	.0000	.0000	.0000	.0000	.0000	.0000	.0000	.0001	.0002	.0004
37	.0000	.0000	.0000	.0000	.0000	.0000	.0000	.0000	.0001	.0002
38	.0000	.0000	.0000	.0000	.0000	.0000	.0000	.0000	.0000	.0001
39	.0000	.0000	.0000	.0000	.0000	.0000	.0000	.0000	.0000	.0001

Values of Student's t-Distribution

Table A-4 is used to find values of Student's t-statistic corresponding to a limited selection of probability values needed in problems in inference. Positive values of t are given that correspond to probabilities comprising both tails of the distribution, which are shown in the following diagram:

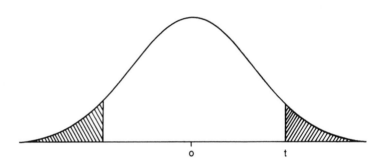

The probabilities in the table represent the sum of the shaded areas and are divided equally between the tails. Since the distribution is symmetric, only *positive* values of t are given for the right half of the curve.

The t-distribution changes with the number of degrees of freedom, DF, which is determined on the basis of a particular problem. In order to determine t, it is necessary to know the degrees of freedom, the probability level, and whether this is one-tailed or two-tailed. Values of t corresponding to one tail are found by locating the two-tailed probability equal to twice the required one-tailed probability.

Example A.6. Suppose it is of interest to find the values of t such that 2.5 percent of the area lies in each tail (or corresponding to 95 percent in the center of the distribution) where the number of degrees of freedom equals 16. The degrees of freedom are given in the left-most column. Consequently, we locate the two-tailed probability of 0.05 in the top column and find t equal to 2.120 across from 16 degrees of freedom. This is the value of t in the upper tail. The value in the lower tail is the same except a minus sign is attached: –2.120.

Note. For degrees of freedom not considered in the table the standardized normal variate, z, from Table 2 can be used to approximate t. The values on the bottom row (DF = ∞) in Table 4 are the same as values of z; however, they correspond to a limited number of probability levels.

Table A-4   STUDENT'S T-DISTRIBUTION

DF	Two-Tailed Probabilities					
	0.2	0.1	0.05	0.02	0.01	0.001
1	3.078	6.314	12.706	31.821	63.657	636.619
2	1.886	2.920	4.303	6.965	9.925	31.598
3	1.638	2.353	3.182	4.541	5.841	12.924
4	1.533	2.132	2.776	3.747	4.604	8.610
5	1.476	2.015	2.571	3.365	4.032	6.869
6	1.440	1.943	2.447	3.143	3.707	5.959
7	1.415	1.895	2.365	2.998	3.499	5.408
8	1.397	1.860	2.306	2.896	3.355	5.041
9	1.383	1.833	2.262	2.821	3.250	4.781
10	1.372	1.812	2.228	2.764	3.169	4.587
11	1.363	1.796	2.201	2.718	3.106	4.437
12	1.356	1.782	2.179	2.681	3.055	4.318
13	1.350	1.771	2.160	2.650	3.012	4.221
14	1.345	1.761	2.145	2.624	2.977	4.140
15	1.341	1.753	2.131	2.602	2.947	4.073
16	1.337	1.746	2.120	2.583	2.921	4.015
17	1.333	1.740	2.110	2.567	2.898	3.965
18	1.330	1.734	2.101	2.552	2.878	3.922
19	1.328	1.729	2.093	2.539	2.861	3.883
20	1.325	1.725	2.086	2.528	2.845	3.850
21	1.323	1.721	2.080	2.518	2.831	3.819
22	1.321	1.717	2.074	2.508	2.819	3.792
23	1.319	1.714	2.069	2.500	2.807	3.767
24	1.318	1.711	2.064	2.492	2.797	3.745
25	1.316	1.708	2.060	2.485	2.787	3.725
26	1.315	1.706	2.056	2.479	2.779	3.707
27	1.314	1.703	2.052	2.473	2.771	3.690
28	1.313	1.701	2.048	2.467	2.763	3.674
29	1.311	1.699	2.045	2.462	2.756	3.659
30	1.310	1.697	2.042	2.457	2.750	3.646
40	1.303	1.684	2.021	2.423	2.704	3.551
60	1.296	1.671	2.000	2.390	2.660	3.460
120	1.289	1.658	1.980	2.358	2.617	3.373
∞	1.282	1.645	1.960	2.326	2.576	3.291

Values of the Chi-Square Distribution

Table A-5 is used to find values of the chi-square statistic, χ^2, corresponding to selected probabilities required in problems in inference. The values of chi-square correspond to these probabilities that are associated with the right tail of the distribution as shown in the following diagram:

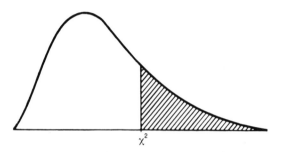

Although probabilities in the table are associated with the right tail, values of χ^2 corresponding to probabilities in the left tail can be obtained by subtracting the tail area from one and locating this probability at the top row on the left half of the table.

The chi-square distribution changes with the number of degrees of freedom, DF, which is determined on the basis of a particular problem. In order to determine χ^2, it is necessary to know the degrees of freedom, the probability level, and to which tail or tails the probability applies.

Example A.7. Consider finding the value of χ^2 corresponding to a probability of 0.05 or 5 percent in the upper or right tail when the degrees of freedom equals 14. This is obtained by locating 0.05 on the top row and 14 in the left-most column labeled DF. The corresponding value of chi-square is

$$\chi^2 = \chi^2_{.05}$$

$$= 23.685$$

The subscript is used to indicate the area above the value of χ^2.

Example A.8. Suppose we want to find the values of χ^2 corresponding to 0.05 or 5 percent of the area split equally in both tails when the degrees of freedom equals 14. Here we are to find two values of χ^2 corresponding to 0.025 in each tail. The value in the left tail is obtained by locating 0.025 at the top of the table directly and is

$$\chi^2 = \chi^2_{.025}$$

$$= 26.119$$

In order to find the value in the lower or left tail we subtract 0.025 from 1, or .975, and locate the result at the top of the table:

$$\chi^2 = \chi^2_{1-.025}$$

$$= \chi^2_{.975} = 5.629$$

The subscript indicates the area to the right or *above* the value of χ^2, which is equivalent to this value subtracted from 1 in the lower or left tail.

Table A-5   THE CHI-SQUARE DISTRIBUTION

DF	\.995	\.99	\.975	\.95	\.05	\.025	\.01	\.005
				Right-Tail Probabilities				
1	.0000	.0002	.001	.004	3.841	5.024	6.635	7.879
2	.010	.020	.051	.103	5.991	7.378	9.210	10.597
3	.072	.115	.216	.352	7.815	9.348	11.345	12.838
4	.207	.297	.484	.711	9.488	11.143	13.277	14.860
5	.412	.554	.831	1.145	11.070	12.832	15.086	16.750
6	.676	.872	1.237	1.635	12.592	14.449	16.812	18.548
7	.989	1.239	1.690	2.167	14.067	16.013	18.475	20.278
8	1.344	1.646	2.180	2.733	15.507	17.535	20.090	21.955
9	1.735	2.088	2.700	3.325	16.919	19.023	21.666	23.589
10	2.156	2.558	3.247	3.940	18.307	20.483	23.209	25.188
11	2.603	3.053	3.816	4.575	19.675	21.920	24.725	26.757
12	3.074	3.571	4.404	5.226	21.026	23.337	26.217	28.300
13	3.565	4.107	5.009	5.892	22.362	24.736	27.688	29.819
14	4.075	4.660	5.629	6.571	23.685	26.119	29.141	31.319
15	4.601	5.229	6.262	7.261	24.996	27.488	30.578	32.801
16	5.142	5.812	6.908	7.962	26.296	28.845	32.000	34.267
17	5.697	6.408	7.564	8.672	27.587	30.191	33.409	35.718
18	6.265	7.015	8.231	9.390	28.869	31.526	34.805	37.156
19	6.884	7.633	8.907	10.117	30.144	32.852	36.191	38.582
20	7.434	8.260	9.591	10.851	31.410	34.170	37.566	39.997
21	8.034	8.897	10.283	11.591	32.671	35.479	38.932	41.401
22	8.643	9.542	10.982	12.338	33.924	36.781	40.289	42.796
23	9.260	10.196	11.689	13.091	35.172	38.076	41.638	44.181
24	9.886	10.856	12.401	13.848	36.415	39.364	42.980	45.558
25	10.520	11.524	13.120	14.611	37.652	40.646	44.314	46.928
26	11.160	12.198	13.844	15.379	38.885	41.923	45.642	48.290
27	11.808	12.879	14.573	16.151	40.113	43.194	46.963	49.645
28	12.461	13.565	15.308	16.928	41.337	44.461	48.278	50.993
29	13.121	14.256	16.047	17.708	42.557	45.722	49.588	52.336
30	13.787	14.953	16.791	18.493	43.773	46.979	50.892	53.672

Values of the F-Distribution
5 Percent Level

Table A-6(a) is used to find values of the F-statistic corresponding only to one probability level equal to 0.05 in the upper or right tail of the F-distribution. Hence, values of F in Table 6(a) correspond to the area shown in the following diagram:

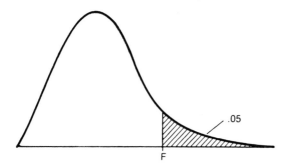

The F-distribution changes with two quantities or parameters, the numerator degrees of freedom, DF_1, and the denominator degrees of freedom, DF_2.

Example A.9. Suppose we are to find the value of F corresponding to 0.05 in the upper or right tail where the numerator degrees of freedom equals 9 and the denominator degrees of freedom equals 20. This is found by locating $DF_1 = 9$ at the top row and $DF_2=20$ at the left-most column. The corresponding value of F equals 2.39.

Table A-6(a)   THE F-DISTRIBUTION

Denominator Degrees of Freedom (DF_2)

DF_2	1	2	3	4	5	6	7	8	9	10	12	15	20	24	30	40	60	120	∞
1	161	200	216	225	230	234	237	239	241	242	244	246	248	249	250	251	252	253	254
2	18.5	19.0	19.2	19.2	19.3	19.3	19.4	19.4	19.4	19.4	19.4	19.4	19.4	19.5	19.5	19.5	19.5	19.5	19.5
3	10.1	9.55	9.28	9.12	9.01	8.94	8.89	8.85	8.81	8.79	8.74	8.70	8.66	8.64	8.62	8.59	8.57	8.55	8.53
4	7.71	6.94	6.59	6.39	6.26	6.16	6.09	6.04	6.00	5.96	5.91	5.86	5.80	5.77	5.75	5.72	5.69	5.66	5.63
5	6.61	5.79	5.41	5.19	5.05	4.95	4.88	4.82	4.77	4.74	4.68	4.62	4.56	4.53	4.50	4.46	4.43	4.40	4.37
6	5.99	5.14	4.76	4.53	4.39	4.28	4.21	4.15	4.10	4.06	4.00	3.94	3.87	3.84	3.81	3.77	3.74	3.70	3.67
7	5.59	4.74	4.35	4.12	3.97	3.87	3.79	3.73	3.68	3.64	3.57	3.51	3.44	3.41	3.38	3.34	3.30	3.27	3.23
8	5.32	4.46	4.07	3.84	3.69	3.58	3.50	3.44	3.39	3.35	3.28	3.22	3.15	3.12	3.08	3.04	3.01	2.97	2.93
9	5.12	4.26	3.86	3.63	3.48	3.37	3.29	3.23	3.18	3.14	3.07	3.01	2.94	2.90	2.86	2.83	2.79	2.75	2.71
10	4.96	4.10	3.71	3.48	3.33	3.22	3.14	3.07	3.02	2.98	2.91	2.85	2.77	2.74	2.70	2.66	2.62	2.58	2.54
11	4.84	3.98	3.59	3.36	3.20	3.09	3.01	2.95	2.90	2.85	2.79	2.72	2.65	2.61	2.57	2.53	2.49	2.45	2.40
12	4.75	3.38	3.49	3.26	3.11	3.00	2.91	2.85	2.80	2.75	2.69	2.62	2.54	2.51	2.47	2.43	2.38	2.34	2.30
13	4.67	3.81	3.41	3.18	3.03	2.92	2.83	2.77	2.71	2.67	2.60	2.53	2.46	2.42	2.38	2.34	2.30	2.25	2.21
14	4.60	3.74	3.34	3.11	2.96	2.85	2.76	2.70	2.65	2.60	2.53	2.46	2.39	2.35	2.31	2.27	2.22	2.18	2.13
15	4.54	3.68	3.29	3.06	2.90	2.79	2.71	2.64	2.59	2.54	2.48	2.40	2.33	2.29	2.25	2.20	2.16	2.11	2.07
16	4.49	3.63	3.24	3.01	2.85	2.74	2.66	2.59	2.54	2.49	2.42	2.35	2.28	2.24	2.19	2.15	2.11	2.06	2.01
17	4.45	3.59	3.20	2.96	2.81	2.70	2.61	2.55	2.49	2.45	2.38	2.31	2.23	2.19	2.15	2.10	2.06	2.01	1.96
18	4.41	3.55	3.16	2.93	2.77	2.66	2.58	2.51	2.46	2.41	2.34	2.27	2.19	2.15	2.11	2.06	2.02	1.97	1.92
19	4.38	3.52	3.13	2.90	2.74	2.63	2.54	2.48	2.42	2.38	2.31	2.23	2.16	2.11	2.07	2.03	1.98	1.93	1.88
20	4.35	3.49	3.10	2.87	2.71	2.60	2.51	2.45	2.39	2.35	2.28	2.20	2.12	2.08	2.04	1.99	1.95	1.90	1.84
21	4.32	3.47	3.07	2.84	2.68	2.57	2.49	2.42	2.37	2.32	2.25	2.18	2.10	2.05	2.01	1.96	1.92	1.87	1.81
22	4.30	3.44	3.05	2.82	2.66	2.55	2.46	2.40	2.34	2.30	2.23	2.15	2.07	2.03	1.98	1.94	1.89	1.84	1.78
23	4.28	3.42	3.03	2.80	2.64	2.53	2.44	2.37	2.32	2.27	2.20	2.13	2.05	2.01	1.96	1.91	1.86	1.81	1.76
24	4.26	3.40	3.01	2.78	2.62	2.51	2.42	2.36	2.30	2.25	2.18	2.11	2.03	1.98	1.94	1.89	1.84	1.79	1.73
25	4.24	3.39	2.99	2.76	2.60	2.49	2.40	2.34	2.28	2.24	2.16	2.09	2.01	1.96	1.92	1.87	1.82	1.77	1.71
30	4.17	3.32	2.92	2.69	2.53	2.42	2.33	2.27	2.21	2.16	2.09	2.01	1.93	1.89	1.84	1.79	1.74	1.68	1.62
40	4.08	3.23	2.84	2.61	2.45	2.34	2.25	2.18	2.12	2.08	2.00	1.92	1.84	1.79	1.74	1.69	1.64	1.58	1.51
60	4.00	3.15	2.76	2.53	2.37	2.25	2.17	2.10	2.04	1.99	1.92	1.84	1.75	1.70	1.65	1.59	1.53	1.47	1.39
120	3.92	3.07	2.68	2.45	2.29	2.18	2.09	2.02	1.96	1.91	1.83	1.75	1.66	1.61	1.55	1.50	1.43	1.35	1.25
∞	3.84	3.00	2.60	2.37	2.21	2.10	2.01	1.94	1.88	1.83	1.75	1.67	1.57	1.52	1.46	1.39	1.32	1.22	1.00

Numerator Degrees of Freedom (DF_1)

A. 23

Values of the F-Distribution
1 Percent Level

Table A-6(b) is used to find values of the F-statistic corresponding only to one level equal to 0.01 in the upper or right tail of the F-distribution. Hence, values of F in Table 6(b) correspond to the area shown in the following diagram:

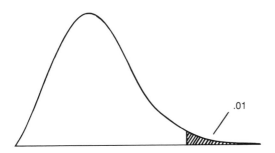

The F-distribution changes with two quantities or parameters, the numerator degrees of freedom, DF_1, and the denominator degrees of freedom, DF_2.

Example A.10. Suppose we are to find the value of F corresponding to 0.01 in the upper right tail where the numerator degrees of freedom equals 4 and the denominator degrees of freedom equals 13. This is found by locating $DF_1 = 4$ at the top row and $DF_2 = 13$ at the left-most column. The corresponding F equals 5.21.

Note. Values of F for a probability of 0.05 in the upper tail appear in Table 6(a).

Table A-6(b)   THE F-DISTRIBUTION

		Numerator Degrees of Freedom (DF_1)																	
	1	2	3	4	5	6	7	8	9	10	12	15	20	24	30	40	60	120	∞
1	4052	5000	5403	5625	5764	5859	5928	5982	6023	6056	6106	6157	6209	6235	6261	6287	6313	6339	6366
2	98.5	99.0	99.2	99.2	99.3	99.3	99.4	99.4	99.4	99.4	99.4	99.4	99.4	99.5	99.5	99.5	99.5	99.5	99.5
3	34.1	30.8	29.5	28.7	28.2	27.9	27.7	27.5	27.3	27.2	27.1	26.9	26.7	26.6	26.5	26.4	26.3	26.2	26.1
4	21.2	18.0	16.7	16.0	15.5	15.2	15.0	14.8	14.7	14.5	14.4	14.2	14.0	13.9	13.8	13.7	13.7	13.6	13.5
5	16.3	13.3	12.1	11.4	11.0	10.7	10.5	10.3	10.2	10.1	9.89	9.72	9.55	9.47	9.38	9.29	9.20	9.11	9.02
6	13.7	10.9	9.78	9.15	8.75	8.47	8.26	8.10	7.98	7.87	7.72	7.56	7.40	7.31	7.23	7.14	7.06	6.97	6.88
7	12.2	9.55	8.45	7.85	7.46	7.19	6.99	6.84	6.72	6.62	6.47	6.31	6.16	6.07	5.99	5.91	5.82	5.74	5.65
8	11.3	8.65	7.59	7.01	6.63	6.37	6.18	6.03	5.91	5.81	5.67	5.52	5.36	5.28	5.20	5.12	5.03	4.95	4.86
9	10.6	8.02	6.99	6.42	6.06	5.80	5.61	5.47	5.35	5.26	5.11	4.96	4.81	4.73	4.65	4.57	4.48	4.40	4.31
10	10.0	7.56	6.55	5.99	5.64	5.39	5.20	5.06	4.94	4.85	4.71	4.56	4.41	4.33	4.25	4.17	4.08	4.00	3.91
11	9.65	7.21	6.22	5.67	5.32	5.07	4.89	4.74	4.63	4.54	4.40	4.25	4.10	4.02	3.94	3.86	3.78	3.69	3.60
12	9.33	6.93	5.95	5.41	5.06	4.82	4.64	4.50	4.39	4.30	4.16	4.01	3.86	3.78	3.70	3.62	3.54	3.45	3.36
13	9.07	6.70	5.74	5.21	4.86	4.62	4.44	4.30	4.19	4.10	3.96	3.82	3.66	3.59	3.51	3.43	3.34	3.25	3.17
14	8.86	6.51	5.56	5.04	4.70	4.46	4.28	4.14	4.03	3.94	3.80	3.66	3.51	3.43	3.35	3.27	3.18	3.09	3.00
15	8.68	6.36	5.42	4.89	4.56	4.32	4.14	4.00	3.89	3.80	3.67	3.52	3.37	3.29	3.21	3.13	3.05	2.96	2.87
16	8.53	6.23	5.29	4.77	4.44	4.20	4.03	3.89	3.78	3.69	3.55	3.41	3.26	3.18	3.10	3.02	2.93	2.84	2.75
17	8.40	6.11	5.19	4.67	4.34	4.10	3.93	3.79	3.68	3.59	3.46	3.31	3.16	3.08	3.00	2.92	2.83	2.75	2.65
18	8.29	6.01	5.09	4.58	4.25	4.01	3.84	3.71	3.60	3.51	3.37	3.23	3.08	3.00	2.92	2.84	2.75	2.66	2.57
19	8.19	5.93	5.01	4.50	4.17	3.94	3.77	3.63	3.52	3.43	3.30	3.15	3.00	2.92	2.84	2.76	2.67	2.58	2.49
20	8.10	5.85	4.94	4.43	4.10	3.87	3.70	3.56	3.46	3.37	3.23	3.09	2.94	2.86	2.78	2.69	2.61	2.52	2.42
21	8.02	5.78	4.87	4.37	4.04	3.81	3.64	3.51	3.40	3.31	3.17	3.03	2.68	2.80	2.72	2.64	2.55	2.46	2.36
22	7.95	5.72	4.82	4.31	3.99	3.76	3.59	3.45	3.35	3.26	3.12	2.98	2.83	2.75	2.67	2.58	2.50	2.40	2.31
23	7.88	5.66	4.76	4.26	3.94	3.71	3.54	3.41	3.30	3.21	3.07	2.93	2.78	2.70	2.62	2.54	2.45	2.35	2.26
24	7.82	5.61	4.72	4.22	3.90	3.67	3.50	3.36	3.26	3.17	3.03	2.89	2.74	2.66	2.58	2.49	2.40	2.31	2.21
25	7.77	5.57	4.68	4.18	3.86	3.63	3.46	3.32	3.22	3.13	2.99	2.85	2.70	2.62	2.53	2.45	2.36	2.27	2.17
30	7.56	5.39	4.51	4.02	3.70	3.47	3.30	3.17	3.07	2.98	2.84	2.70	2.55	2.47	2.39	2.30	2.21	2.11	2.01
40	7.31	5.18	4.31	3.83	3.51	3.29	3.12	2.99	2.89	2.80	2.66	2.52	2.37	2.29	2.20	2.11	2.02	1.92	1.80
60	7.08	4.98	4.13	3.65	3.34	3.12	2.95	2.82	2.72	2.63	2.50	2.35	2.20	2.12	2.03	1.94	1.84	1.73	1.60
120	6.85	4.79	3.95	3.48	3.17	2.96	2.79	2.66	2.56	2.47	2.34	2.19	2.03	1.95	1.86	1.76	1.66	1.53	1.38
∞	6.63	4.61	3.78	3.32	3.02	2.80	2.64	2.51	2.41	2.32	2.18	2.04	1.88	1.79	1.70	1.59	1.47	1.32	1.00

Denominator Degrees of Freedom (DF_2)

Random Digits

This table provides a sequence of integers between 0 and 9 that are generated at random with equal probabilities. Consequently, each of the digits between 0 and 9 has the same chance of occupying each position in the table. The order in which the digits appear in the table is important. The table can be used to generate sequences of values at random and to select random samples. It is possible to begin generating a sequence at any point in the table. Once a starting place is chosen, however, values should be selected consecutively. It does not matter whether the table is used vertically or horizontally.

Example B.1. Suppose we are interested in generating 8 values representing the outcomes of 8 rolls of a balanced die. There are 6 possible outcomes, 1 through 6, each with an equal probability of occurrence. Starting at the beginning of the table working horizontally across the page, the sequence is

$$3 \quad 2 \quad 3 \quad 3 \quad 3 \quad 1 \quad 5 \quad 6$$

This is obtained by recording values between 1 and 6 when they appear in sequence and ignoring values not falling within the prescribed range.

Example B.2. Suppose we are to select a random sample of 5 units from a universe of 10,000. Each unit is assigned a number between 1 and 10,000. Starting at the beginning of the table working vertically, the units in the sample are the ones with the following numbers:

$$2610 \quad 7633 \quad 1384 \quad 3290 \quad 4905$$

In order to obtain this sample, columns of 5-digit numbers must be considered since the maximum number of units equal to 10,000 contains 5 digits. By examining each number in the first column of five digits, we see that 02610 is the first that falls within the range 1-10000. The zero on the left does not contribute to the value. Hence, the first unit in the sample is the one numbered 2610. Remaining numbers are obtained by working successively down the table ignoring those not falling within the desired range. If the first column is exhausted before obtaining the full sample, the next adjacent column of 5 digits is examined, and so on, until the required number in the sample is achieved.

Note. When a number that already has been chosen appears again, it should be ignored if selection or sampling is done without replacement.

Table B-1   UNIFORM RANDOM DIGITS

38233	37175	96866	39089	94736	61380	90458	05453	95172	93666	16125	67117
43130	40343	75278	89548	44005	58983	39772	16209	79469	06890	74283	27617
84903	43656	46791	28316	55508	09819	79956	97381	73127	72270	65239	84032
44696	43241	46124	94993	97925	03283	80883	02641	43499	60871	50818	84164
56914	28546	02030	17028	97370	83488	36378	75833	05283	78546	10998	91909
30074	14617	18690	48142	92617	04706	12193	04771	84140	48038	26780	85503
20186	41895	73296	19029	00017	17641	81484	64558	83465	67338	46930	46501
81313	63535	95939	18038	73134	74555	54081	98885	51761	42759	04733	33750
51911	73119	08690	19557	86258	71465	32769	63188	61768	34779	33360	88759
35822	38244	43411	03660	49072	29825	06263	23218	26840	04273	81301	88710
21200	07495	16847	20245	32743	74882	68751	31859	33004	29646	19058	62772
17417	12765	58163	20019	30924	47313	25911	37882	00192	96388	05688	71174
02610	46858	91521	49759	53071	88082	41613	23758	59294	71824	92870	54956
76214	63470	74371	86432	02076	35952	57370	92730	67174	35663	44502	19696
95452	79732	96859	95538	68381	73400	47510	58040	96700	97139	84996	12714
07633	63263	24574	26378	63566	83696	11401	70684	91342	84727	10146	23469
72042	00641	81750	78377	33868	28119	01150	17941	04988	11946	48036	97893
55017	29963	95676	10470	69049	35296	54500	17695	31023	54065	07728	27436
79373	74634	74692	04302	07451	02186	45172	83942	16003	08011	47475	78049
51264	44022	55247	95868	59779	68605	87917	19981	60743	22596	55917	89792
27403	76388	26201	99006	55602	18836	32165	55814	14742	87484	09982	85667
60098	12629	01946	85056	89691	28961	47774	90510	03251	35537	03149	02248
92874	29696	32607	41121	86929	17624	49885	06170	22467	89743	38396	86827
01384	00359	13967	74694	96676	29890	63198	21550	10503	53028	30119	45589
65038	15306	14625	97516	28773	92337	55423	23585	71623	98086	40452	28854
34178	00711	64194	28906	82587	68647	46919	06417	64283	40403	72294	39812
43857	60911	27444	39676	86000	73531	27995	70736	01360	07415	93227	17228
22682	72727	84258	05945	38458	08768	25120	76183	40021	10244	43034	66001
16118	67929	90658	20236	13861	97215	79892	75693	68930	98897	53307	33544
45477	71051	85585	61148	52906	91412	79272	32953	67456	02582	53693	87317
62687	93308	41231	21079	52014	85361	17064	99581	23395	11456	43531	85369
95291	57926	31321	89373	20798	67556	95510	19295	76399	25342	34494	84500
92790	81492	62506	74903	08086	99645	46788	72726	90034	41450	53635	22932
36075	73713	84367	10676	80894	88053	12374	74655	65485	81645	86568	23826
98531	46896	53996	81931	31164	46261	63045	86942	41462	96916	85889	40630
03290	68482	67799	53067	63728	59944	30196	32109	96288	65025	04731	53148
78103	04530	16513	99222	43653	37343	22291	47259	34643	38773	82291	07756
66372	30789	05371	49879	92604	59276	73829	14774	39016	01309	26079	16144
27992	03779	43576	97076	27486	89095	29984	91535	58244	16233	04470	96479
04905	59642	30039	20427	86340	20760	73340	78022	54618	99610	61963	88360
45256	83348	29872	43056	25182	08510	96386	93812	22892	25673	07773	69867
35618	29606	95609	48843	25573	46922	03378	91542	22102	91005	06780	84021
03465	74932	99918	57257	10335	74548	11548	74584	59835	27434	85148	35165
85648	19233	32351	35552	76958	19304	60048	02535	74722	95768	42093	50214
55909	81536	28816	26003	39209	87926	96746	89508	03941	08420	51650	88066
39831	63766	71470	34381	87796	45784	50744	32883	93233	20975	88307	33284
45153	96739	66624	25320	13057	51228	91404	88157	25574	33774	78890	16494
57331	98914	46826	88573	81418	88299	83879	06773	92077	26997	81559	77863
11014	16511	31929	45930	30669	35548	60438	62509	67195	88450	74572	18858
55067	94103	12842	57212	00579	42647	01978	95511	02975	10540	72882	46012

Table B-1   UNIFORM RANDOM DIGITS

80228	94650	98827	06716	87059	73187	65518	87934	08550	59656	83690	38975
11628	38997	97586	17935	97437	54708	13404	11193	84853	77658	58313	55704
43420	51029	99009	83003	47328	99892	18167	81569	83295	31972	02170	39666
76561	37244	12465	17759	85036	45863	14640	79790	89190	20710	51329	56776
74957	08291	47782	13230	34106	21974	77973	81218	18775	08881	05113	82676
17857	53018	97004	76189	38481	33208	81353	31089	21716	40228	12298	44686
74209	72305	84513	89882	71566	30761	42122	91567	55794	26241	84936	36160
06176	24713	97697	39673	59473	91209	66041	01213	48396	36481	97767	95281
82957	72585	35475	41563	45263	14207	74938	75034	41973	43327	32136	68343
28228	66034	12211	11844	01217	34302	48485	07762	59084	33426	54929	28011
75919	33487	11989	60556	60113	22576	93990	19218	05551	22933	83481	95931
20842	60990	31592	50027	64811	34888	77983	42946	39834	11439	43635	15042
72056	49717	06979	69247	74524	12800	45286	43799	42544	39434	88354	07696
83900	61343	87267	52372	37856	75464	63123	04188	95258	26525	67631	69272
41926	00734	02848	41214	69593	77667	37071	62873	28348	34055	61767	29547
97901	38306	46821	84480	81685	23016	82173	16436	66247	40469	49584	57456
90322	43071	35187	00860	76653	57091	95909	50875	21026	56531	44297	33361
44128	53542	68835	15958	76449	77345	01697	57285	68631	76385	01655	28788
81720	36056	41871	91299	79629	80832	10982	48432	77144	31235	54595	78635
62256	54387	64865	40942	99468	47925	90785	83794	61495	83373	68138	64344
09037	63822	57470	15697	17281	35058	35878	93637	46550	44815	76902	05397
02973	05968	03677	76158	81972	21463	82953	26812	18909	03341	34154	99904
11475	26812	56902	74434	33760	48934	82864	90109	27106	25841	34312	33383
04506	14722	85609	76330	75980	77913	34274	37106	72182	90220	43418	11832
48027	51541	76252	86682	24490	33832	64295	61186	04560	09276	88982	27419
78120	74800	16316	53910	22373	60069	29176	18712	92241	64558	74629	98190
07704	00578	47435	12270	09210	13455	27487	58044	74194	22157	63786	14118
70490	04712	14274	35419	59534	68114	36753	00056	57203	32586	44801	49757
13256	37320	48708	81848	12740	20222	59529	87733	97509	13314	59894	69263
10357	10777	03801	23868	03059	18363	41766	24135	80203	81121	79301	81680
92558	26337	75235	14392	18976	18443	71538	18253	69296	92136	58600	82617
73835	50639	70723	66490	30539	46388	19351	32391	72274	12025	86124	77430
94221	89718	07488	70786	91259	73006	89240	32329	16459	60913	62818	08058
01635	60086	61853	72973	78432	51237	72938	97573	57152	73065	62558	18645
38476	85892	69540	18513	65122	41428	64928	70054	31090	53622	79981	83737
63561	04605	36213	06086	35035	82470	42488	20855	17410	41883	05123	35179
74798	15466	08644	84800	02514	77952	16826	65657	08263	62803	99395	79936
98232	41347	80149	25949	34773	74012	08134	15469	37418	22953	42447	90327
73801	44228	35735	99481	46230	53734	65386	56253	62061	54645	75542	47052
88685	30803	50191	35266	81908	54935	81996	00244	83251	78318	44869	12451
50918	39339	04620	45761	98565	99832	65390	90342	52824	11916	74527	38278
76972	63515	27344	79145	70830	75215	56903	39838	02384	36844	41336	25291
44798	33099	77580	90821	85738	71605	45953	68589	24634	75604	18616	25007
43090	53160	68401	76922	37911	58696	73453	68064	00312	36254	47019	81014
13184	69776	35497	31620	25270	62956	10519	61186	54264	84923	60330	22623
53643	99337	83120	34317	14515	51128	43477	58882	04263	07096	69400	87031
90719	95160	87829	61810	49374	56180	52461	50929	64841	75886	04665	80557
46470	82358	44671	73821	09842	32049	01303	75933	21142	35346	50014	71633
65219	53520	39176	69765	56080	90261	08420	40769	19363	62689	35658	81457
73177	77210	64760	93880	58426	42365	53378	66232	46636	72268	08995	42642

Table B-1 UNIFORM RANDOM DIGITS

33255	77597	82636	01540	41252	23256	57104	88153	73494	92847	06685	43760
52110	36679	33180	57528	88574	21752	64514	36838	29592	32613	43824	44821
49724	02434	79764	42128	01297	46861	75309	50919	84548	12363	51724	55064
63218	28093	06185	94336	42090	92688	20351	87406	08412	92490	04955	42539
20820	81591	10937	10811	22028	45262	15007	78704	81433	45672	11319	91354
38154	69894	76979	74498	26156	98314	63302	64068	28585	36104	54369	21737
04433	14304	52877	96866	41630	13202	15856	99431	34958	17891	47984	86766
73661	21572	30879	55989	93324	71338	58043	08107	42947	92527	37639	60202
18078	00594	27325	94334	90476	68364	12247	77629	31686	58280	67311	41421
73464	68195	53634	06160	44590	91060	06245	28800	90885	75519	26913	63938
78663	54499	58401	23231	51736	91994	09902	64231	18132	54202	37154	49943
64222	96579	76956	97402	11638	25796	52965	10354	29830	90838	93788	84782
79326	34374	78108	64987	43089	88539	76989	10736	59897	64524	84026	16148
85463	05580	69330	02129	82967	05645	05836	67999	45477	94024	98069	26415
03023	96022	62898	60543	37511	18443	92349	94910	97598	83368	00259	42382
21027	99681	35894	61017	34863	64252	17228	72684	98362	67117	52234	06031
11955	75760	34998	17682	35782	46381	08534	38225	82042	89214	97958	51242
56909	36853	16058	85018	56805	00943	95238	58988	38904	32031	26230	92545
29334	33862	00108	23064	44791	25110	58014	16598	83511	33264	40138	54568
10765	00871	07801	60791	78961	58905	32766	13376	33672	08941	03639	93893
24081	69803	82886	91452	20943	73239	70125	42409	12304	76701	18121	10054
04661	47837	08117	73998	12579	82671	65503	80482	89384	81226	71224	25205
13499	42688	42336	60110	21880	87308	44181	90390	42075	82315	67003	22505
62292	42075	53448	22982	51526	17664	24960	17915	57696	17864	37774	64038
91584	59594	36326	61174	55769	64704	32465	42273	40119	64547	06342	12300
56194	24129	96414	81266	30130	84285	00510	88250	34893	71038	63743	49160
70743	21519	09591	13219	71653	69303	23400	57601	23804	14610	53925	87459
57004	25184	35576	75356	14207	53792	41855	99772	63044	56179	37786	25528
30272	19702	52128	99543	94812	77339	25477	82053	09093	48374	93788	47385
64508	52255	64062	40984	66757	40352	21913	56532	04744	24825	02216	21643
03741	26783	98474	27392	27728	32479	55520	16504	31591	84996	44965	19757
64419	01135	66313	58441	99798	88595	60532	62720	58679	40748	04217	15748
26355	48041	80211	28669	90666	62506	46309	47703	30024	49206	44138	93655
76886	30111	93914	82466	78801	17595	26728	55906	19092	93410	47246	80913
69096	47867	69082	01328	01931	34262	49708	88103	56873	45574	19293	85930
13994	55387	12445	23890	46451	02407	99000	83494	42941	61500	29393	81854
50921	86611	61002	31440	50808	61446	97848	04977	58034	59372	20380	85719
30520	71749	86337	03639	48603	00872	42854	30109	13380	95533	64354	69056
94577	99083	44517	72310	81184	54709	70004	85790	53722	66252	55694	27951
00553	67018	93443	52360	06795	59784	52416	43796	25508	78186	86017	51589
83953	87879	22522	03411	38367	96669	13698	36016	75546	60892	74119	94520
14956	72005	15476	23294	38440	48492	23094	16872	54839	82670	27900	49870
03385	75508	71168	77587	71621	72714	59627	00661	40315	91222	59099	56509
06587	21658	50227	48412	31230	95724	08929	90411	05927	60230	14430	39148
07980	32848	64258	46088	51701	52489	62201	33796	21999	90674	27873	60081
04543	65204	48934	32547	78900	78021	02897	87963	32579	19179	28079	81952
30886	06211	94863	87216	11682	06179	82187	35036	05644	61011	19105	99404
25140	88666	13234	35261	83960	20072	60178	63242	90632	46467	16281	51051
37588	13740	53497	36387	52724	94148	86686	74844	82596	88876	24891	84966
55943	62642	48795	16085	64427	33439	58415	59700	36889	20459	96007	53980

FACTORS FOR VARIABLES
CONTROL CHARTS

Summary of Formulas for Variables Control Charts

$CL_{\bar{x}} = \mu_0$

$CL_{\bar{x}} = \bar{\bar{X}}$

$UCL_{\bar{x}} = \mu_0 + A\sigma_0$

$UCL_{\bar{x}} = \bar{\bar{X}} + A_3\bar{S}$

$LCL_{\bar{x}} = \mu_0 - A\sigma_0$

$LCL_{\bar{x}} = \bar{\bar{X}} - A_3\bar{S}$

$CL_s = c_4\sigma_0$

$CL_s = \bar{S}$

$UCL_s = B_6\sigma_0$

$UCL_s = B_4\bar{S}$

$LCL_s = B_5\sigma_0$

$LCL_s = B_3\bar{S}$

$CL_{\bar{x}} = \bar{\bar{X}}$

$UCL_{\bar{x}} = \bar{\bar{X}} + A_2\bar{R}$

$LCL_{\bar{x}} = \bar{\bar{X}} - A_2\bar{R}$

$CLR_R = d_2\sigma_0$

$CL_R = \bar{R}$

$UCL_R = D_2\sigma_0$

$UCL_R = D_4\bar{R}$

$LCL_R = D_1\sigma_0$

$LCL_R = D_3\bar{R}$

$$\hat{\sigma} = \frac{\bar{S}}{c_4}$$

$$\hat{\sigma} = \frac{\bar{R}}{d_2}$$

Table C-1 FACTORS FOR $\overline{X}$ AND S Charts

n	A	A_1	A_3	B_1	B_2	B_3	B_4	B_5	B_6	c_2	c_4
2	2.12	3.76	2.66	0	1.84	0	3.27	0	2.61	0.5642	0.7979
3	1.73	2.39	1.95	0	1.86	0	2.57	0	2.28	0.7236	0.8862
4	1.50	1.88	1.63	0	1.81	0	2.27	0	2.09	0.7979	0.9213
5	1.34	1.60	1.43	0	1.76	0	2.09	0	1.96	0.8407	0.9400
6	1.22	1.41	1.29	0.03	1.71	0.03	1.97	0.03	1.87	0.8686	0.9515
7	1.13	1.28	1.18	0.10	1.67	0.12	1.88	0.11	1.81	0.8882	0.9594
8	1.06	1.17	1.10	0.17	1.64	0.19	1.81	0.18	1.75	0.9027	0.9650
9	1.00	1.09	1.03	0.22	1.61	0.24	1.76	0.23	1.71	0.9139	0.9693
10	0.95	1.03	0.98	0.26	1.58	0.28	1.72	0.28	1.67	0.9227	0.9727
11	0.90	0.97	0.93	0.30	1.56	0.32	1.68	0.31	1.64	0.9300	0.9754
12	0.87	0.93	0.89	0.33	1.54	0.35	1.65	0.35	1.61	0.9359	0.9776
13	0.83	0.88	0.85	0.36	1.52	0.38	1.62	0.37	1.59	0.9410	0.9794
14	0.80	0.85	0.82	0.38	1.51	0.41	1.59	0.40	1.56	0.9453	0.9810
15	0.77	0.82	0.79	0.41	1.49	0.43	1.57	0.42	1.54	0.9490	0.9823
16	0.75	0.79	0.76	0.43	1.48	0.45	1.55	0.44	1.53	0.9523	0.9835
17	0.73	0.76	0.74	0.44	1.47	0.47	1.53	0.46	1.51	0.9551	0.9845
18	0.71	0.74	0.72	0.46	1.45	0.48	1.52	0.48	1.50	0.9576	0.9854
19	0.69	0.72	0.70	0.48	1.44	0.50	1.50	0.49	1.48	0.9599	0.9862
20	0.67	0.70	0.68	0.49	1.43	0.51	1.49	0.50	1.47	0.9619	0.9869

Table C-2 FACTORS FOR $\overline{X}$ AND R CHARTS

n	A_2	D_1	D_2	D_3	D_4	d_2	d_3
2	1.88	0	3.69	0	3.27	1.128	0.8525
3	1.02	0	4.36	0	2.57	1.693	0.8884
4	0.73	0	4.70	0	2.28	2.059	0.8798
5	0.58	0	4.92	0	2.11	2.326	0.8641
6	0.48	0	5.08	0	2.00	2.534	0.8480
7	0.42	0.20	5.20	0.08	1.92	2.704	0.8332
8	0.37	0.39	5.31	0.14	1.86	2.847	0.8198
9	0.34	0.55	5.39	0.18	1.82	2.970	0.8078
10	0.31	0.69	5.47	0.21	1.78	3.078	0.7971
11	0.29	0.81	5.53	0.26	1.74	3.173	0.7873
12	0.27	0.92	5.59	0.28	1.72	3.258	0.7785
13	0.25	1.03	5.65	0.31	1.69	3.336	0.7704
14	0.24	1.12	5.69	0.33	1.67	3.407	0.7630
15	0.22	1.21	5.74	0.35	1.65	3.472	0.7562
16	0.21	1.28	5.78	0.36	1.64	3.532	0.7499
17	0.20	1.36	5.82	0.38	1.62	3.588	0.7441
18	0.19	1.43	5.85	0.39	1.61	3.640	0.7386
19	0.19	1.49	5.89	0.40	1.60	3.689	0.7335
20	0.18	1.55	5.92	0.41	1.59	3.735	0.7287

FLOWCHART SYMBOLS

Appearing below are a collection of commonly used flowchart symbols. Some of the symbols are taken from information processing while others apply to manufacturing. The three symbols at the end are specially applicable to services.

Operation	◯	Scrap	
Transportation	⇒	Rework	
Inspection	▭	Pack	
Storage	▽	Document	
Delay	◗	Make ready/Setup	⬡
Start/Stop	⬭	Productive Operation	⬤
Decision	◇	Non-productive Operation	◯
Flowline	→	Required but Non-productive or Review	◑

CAUSE-AND-EFFECT DIAGRAMS

Whenever a problem is solved using statistical methods, the problem should be carefully defined and the nature of the approach used to solve it should be carefully delineated. Although recently attention is given to this in the statistical literature, most of what statisticians do is embedded within proper application of research methods or the scientific method as part of a total solution.

Currently within the area of quality control there have been many influences besides statistics which are directed toward formalizing the way in which quality problems are solved. Some of these have emanated from behavioral areas, and are being incorporated into quality control activities.

Part of any decision-making or problem solving effort involves thought provoking activity for which there are few rules or formal ways to reach a conclusion, or the next step toward which to proceed. Especially in process control problems before a process has been improved, many factors may have to be considered in order to track down a problem associated with an indication of an assignable cause.

A graphic device for delineating problems and hunting down the root of a problem is a cause-and-effect diagram, also referred to as a fishbone diagram or an Ishikawa diagram, the latter being named after its developer.

A basic diagram depicting the structure of a cause-and-effect diagram is presented in **Exhibit E-1**. At the extreme right of the diagram appears an effect, or problem requiring attention or in need of a solution. For example, it could represent discarded product, poor sales, machine misalignment, or sources of variation, to name a few.

The remainder of the diagram is comprised of causes or reasons for the effect or problem of interest. Major or primary causes are associated with the diagonal branches on this part of the diagram. For quality control purposes, it is sometimes convenient to identify these as materials, work methods, personnel, equipment, measurement, and environment.

The horizontal branches emanating from the primary branches correspond to secondary contributors to cause, or sub-causes. More detail can be added to the diagram by including tertiary causes, which can be represented as vertical or diagonal lines emanating from the horizontal secondary causes. The idea is to get as many reasons or causes on the chart as possible and properly ordered in order to provide input into the solution of the problem or determining the reasons for the effort of concern.

Exhibit E-2 presents an example of a completed cause-and-effect diagram. The diagram represents what could result from an attempt to determine the next step *before* deciding how to install control charts in

EXHIBIT E-1
BASIC STRUCTURE OF A CAUSE-AND-EFFECT DIAGRAM

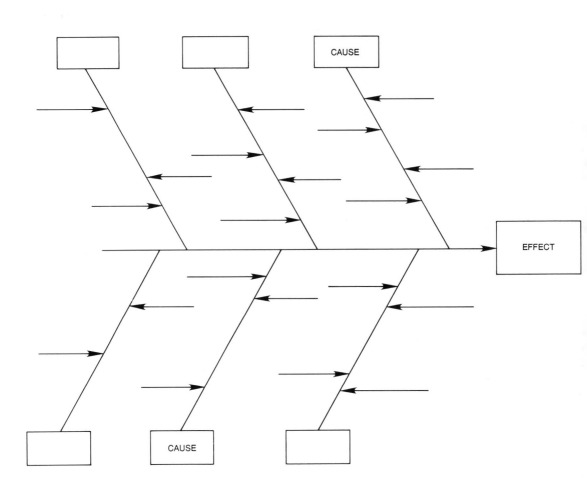

order to control a printing process. Such a chart is useful also *after* charting has been instituted in order to find problems in the process. Examination of such a chart may suggest that other more complicated forms of analysis may be required.

The utility of the approach lies in providing a structured delineation and presentation of the relevant factors relating to a particular problem. As such, it provides a means of communicating, at the very least, those factors to all parties concerned in a single place, or together. By doing so, it provides a basis for deciding on the possible approaches that may be considered in the next step in the solution of a particular problem. The example of a cause-and-effect diagram presented serves also to illustrate that a gap exists between the delineation of a problem and its solution. The combination of steps and methods used will depend on the nature of the problem and the process and the individuals involved in the solution.

E. 3

EXHIBIT E-2
EXAMPLE OF A CAUSE-AND-EFFECT DIAGRAM

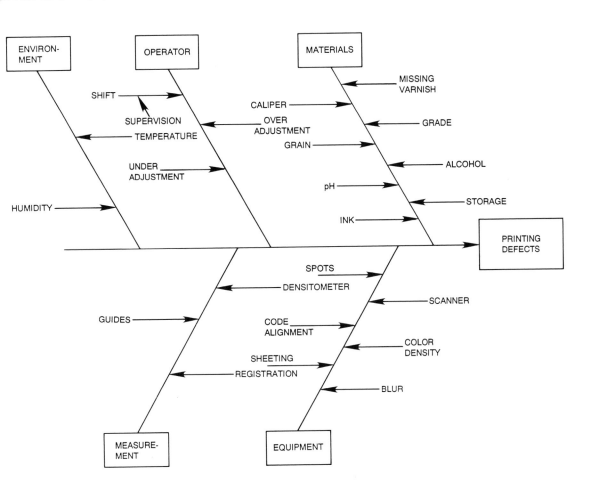

REVIEW OF BASIC
MATHEMATICS

This appendix is included in order to serve as a guide for under-standing basic mathematical concepts that are used in the main text which you may have forgotten or may not have seen before. Many of the concepts should not be completely new. Consequently, the material presented here should not be considered as a complete treatment of basic mathematics but as a review that highlights concepts used in the book and that may require some additional understanding.

USING SYMBOLS TO REPRESENT NUMBERS

We use numbers to represent many things. These can be charac-terized in different ways depending on the nature of a particular prob-lem. The numbers we use are part of the real number system with which we all are familiar. Different parts are used at varying levels of refine-ment in different problems.

Measurements are used to represent such things as dimensions, quantities, or distances and are considered over the entire scale of positive and negative real numbers. *Measurements* are expressed in terms of a whole number and a decimal, or fractional part. The part of the real numbers that are expressed strictly as a whole number without a fractional part are referred to as integers. Non-negative integers that are used for the purpose of enumeration are referred to as *counts*, or frequencies.

In cases where measurements or counts can assume different values, we refer to them in general terms as a variable. When dealing with variables it is convenient to represent them in terms of symbols. For example, a concise way of representing the real numbers is as follows:

$$-\infty < X < \infty$$

The symbol "X" stands for any value that falls between the limits of plus-and-minus infinity, where minus-infinity is given by the symbol $-\infty$. We use the term infinity to represent a quantity that is limitlessly large, or one that has no limit to its size. The symbol "<" stands for "less than" and the symbol ">" stands for "greater than." Hence, we can say that a real number is any value that is greater than minus infinity or is less than plus infinity.

The expression given above provides a way of expressing the real numbers in a general way without specifying a particular number. Another way that this can be accomplished is in terms of absolute values. An *absolute value* of a number X, written as $|X|$, indicates the size or magnitude of X without regard to its sign. For example, $|4|=4$ and $|-4|=4$: the absolute value of 4 equals 4 and the absolute value of -4 equals 4. Using the absolute value notation, the real numbers can be expressed as

$$0 \leq |X| < \infty$$

In other words, a real number is one such that its absolute magnitude equals any value between zero and infinity.

Notice that in the last expression, two kinds of *inequalities* are used. The symbol "$\leq$" means "less than or equal to," which in the example indicates that the absolute value of X either can equal zero or be greater. If the equal sign does not accompany the inequality, the specified value cannot assume the lower limit specified. Both "$<$" and "$>$" are referred to as inequalities; however, "$<$" is referred to as a strict inequality since it is not accompanied by an equal sign. In a similar way "$>$" and "$\geq$" are inequalities indicating "greater than" and "greater than or equal to," where "$>$" also is a strict inequality. A strict equality is an expression that contains only an equal sign and no inequalities. For example, if we want to write that X equals a specific number "a", it is expressed as

$$X = a$$

which is a strict equality.

Using some of the ideas given above, let us apply them to specific segments or intervals of real numbers. If, for example, we are interested in representing all numbers that can assume values between but not including 17 and 25, we may write this as

$$17 < X < 25$$

Numbers such as 19, 23.4567, 17,000002, and 24.999999 correspond to values specified in the above expression whereas 14, 15.37924, and 29.3 do not. Further, since strict inequalities are used, the endpoints or limits of 17 and 25 do not correspond to the specified values. Such an interval is referred to as *open* or open-ended since the limits are not considered as part of the interval. If, however, the interval were specified as

$$17 \leq X \leq 25$$

the values of 17 and 25 are considered part of the interval since equal signs accompany the inequalities. Intervals like this where the limits are considered as part of the interval are referred to as *closed*.

Sometimes it is useful to specify intervals in terms of general limits that represent numbers but are in terms of symbols. For example, if we let a and b represent any two numbers, we could write the above expressions as

$$a < X < b$$

$$a \leq X \leq b$$

The first of these is interpreted to mean that the variable X can assume a value that is greater than "a" but less than , "b," whereas the second states that X can equal a or b or any value between the two.

Intervals like the ones considered above are defined as *continuous* since values everywhere within the interval are considered. Let us now represent numbers that are not continuous in the form of symbols. For example, assume we are interested in the *integers* ranging from 1 to 10. This can be written as

$$X = 1,2,3,4,5,6,7,8,9,10$$

which means that X is a variable that can assume any integer between 1 and 10. The same thing can be accomplished with inequalities such as

$$1 \leq X \leq 10; X \text{ is an integer}$$

Notice an accompanying label is necessary to indicate that only integers and not all values in the interval are considered.

The above set of values can be represented in another way:

$$X = 1,2,.....,10$$

Here, all the values are not specified and intervening values are indicated with dots. When represented in this way one assumes that the same pattern of integer values continues up to 10.

The set of integers given in the last example is an illustration of a variable that is referred to as discrete. A variable is *discrete* if there are a finite number of values in the interval. This means that one actually can count the number of values in the interval.

Suppose we are interested in all positive integers. These can be represented in the following manner.

$$X = 1,2,3,4,.......$$

Recognize that the positive integers comprise a set of values that is infinite since one can continue to write out integers in an endless sequence. However, the set of all positive integers *also* is considered discrete since any subinterval contains a finite or countable number of values. By way of contrast, the set of positive real numbers is continuous since any subinterval of reals contains an infinite number of values that exist everywhere in any chosen subinterval.

SOME BASIC OPERATIONS AND RULES

Just as we can represent numbers in the form of symbols, we can use symbols to describe various operations that are performed on numbers in order to obtain different results. Associated with these operations are certain rules and conventions. In this section we shall review ones that are more commonly used. Suppose we have two variables denoted with the symbols X and Y, such that each variable can assume certain specified values. We can write the *sum* of the two variables as X+Y, which means that the value of Y is added to the value of X. If, for example, X equals 44 and Y equals 10, the sum is written as

$$X + Y = 44 + 10$$

$$= 54$$

In a similar manner, the *difference* between X and Y is written as X – Y and is given as

$$X - Y = 44 - 10$$

$$= 34$$

The two operations just illustrated are addition and subtraction which are familiar to us all.

Now suppose Y is *negative* and we want the sum of X and Y. This is written as

$$X + (-Y) = X - Y$$

Parentheses are used to enclose –Y in order to indicate that negative Y is added to X. The result is equivalent to the difference between positive values of both X and Y. On the other hand, if Y is negative and it is to be subtracted from X we have

$$X - (-Y) = X + Y$$

Hence, the difference between X and –Y is the sum of the absolute values of the two quantities whereas the sum of a positive and negative value is the difference between the absolute values.

In order to illustrate these concepts, let X equal 44 and Y equal –10. Then,

$$X + Y = 44 + (-10)$$

$$= 44 - 10$$

$$= 34$$

$$X - Y = 44 - (-10)$$

$$= 44 + 10$$

$$= 54$$

Consider the same example; however, let X equal –44 and Y equal 10. In this case we have

$$X + Y = -44 + 10$$

$$= 10 - 44$$

$$= -34$$

$$X - Y = -44 - 10$$

$$= -44 - 10$$

$$= -54$$

In the previous two examples the results are positive whereas in these two cases the results are negative. What we commonly think of as addition and subtraction *algebraically* are considered as addition of positive and negative numbers. When finding the algebraic sum of a positive and a negative number, the values are subtracted and the result retains the sign of the value with the larger absolute value. When both values have the same sign, the values are added and the result has the sign possessed by both values: if both are positive the result is positive and if both are negative the result is negative.

Plus-Minus

A combination of addition and subtraction in a single rule or formula frequently is used in problems in statistics. This involves the addition and subtraction of the same quantity to another number. In symbols, the problem is written as

$$X \pm k$$

This expression tells us to add and subtract the value of k to and from X in order to obtain two values or limits. The symbol ± is read as "plus and minus." The entire expression, therefore, is read as "X-plus-and-minus-k."

As an example, consider X equal to 25 and k equal to 8. Therefore, we can write

$$X \pm k$$

$$25 \pm 8$$

$$(25 - 8) - (25 + 8)$$

$$17 - 33$$

Hence, the result of 25 plus and minus 8 are the limits 17 and 33.

The Reciprocal

A quantity that is used indirectly in many operations is the reciprocal. For a variable X that can assume a certain set of values, the reciprocal is found as X divided into one, or "one over X." For example the reciprocal of 5 is given as

$$\frac{1}{X} = \frac{1}{5}$$

$$= 0.2$$

By taking the reciprocal of the reciprocal of a number, one obtains the original number. In terms of the above example we have

$$\frac{1}{1/X} = X$$

$$= \frac{1}{0.2}$$

$$= 5$$

The reciprocal of a number represents the value that when multiplied by the original number equals 1. For X equal to 5,

$$X \times \frac{1}{X} = 5 \times \frac{1}{5}$$

$$= 1$$

Multiplication and Division

Although the concepts of multiplication and division are quite basic, let us review these operations in terms of the different symbols that can be used and illustrate the fundamental algebraic properties of these operations in terms of positive and negative values.

In terms of two variables X and Y that may assume particular numerical values, the product of X and Y can be written in a number of ways. Alternative possibilities are

$$X \times Y$$

$$X \cdot Y$$

$$X(Y)$$

$$(X)(Y)$$

$$XY$$

Multiplication usually is indicated with a "times sign" ($\times$) which is provided as the first alternative. A dot ($\cdot$) sometimes is used in its place but is not as common. When multiplication is indicated in terms of variables represented by symbols, the sign usually is dropped (XY) unless each of the terms is a more complicated expression. When this occurs parentheses or brackets are used to separate the quantities multiplied. *Parentheses* also are useful when actual numbers are substituted in order to separate the values multiplied in a clearer manner.

Suppose X equals 7 and Y equals 5. When multiplied their product can be written in the following way.

$$XY = 7 \times 5$$

$$= 7 \cdot 5$$

$$= 7(5)$$

$$= (7)(5)$$

$$= 35$$

In this example, both X and Y are positive, and therefore the result is positive. Suppose, however, that X is positive and Y is negative. In terms of symbols the product is written as

$$X(-Y) = -XY$$

Parentheses are needed on the left in order to associate the minus sign with Y. The term on the right indicates that the result is negative.

Using the same numbers for X and Y above but changing the sign of the Y-value to minus, we can write the product as

$$XY = 7(-5)$$

$$= -35$$

On the other hand, if both X and Y are negative, we have

$$-X(-Y) = (-X)(-Y)$$

$$= XY$$

Using numbers, the result is

$$XY = -7(-5)$$

$$= (-7)(-5)$$

$$= 35$$

The above results can be generalized and indicate that the product of two positive or two negative numbers is positive whereas the product of a positive and a negative number is negative. If more than two numbers are multiplied, an even number of minus signs provides a positive result and an odd number of minus signs yields a negative result, regardless of the number of plus signs.

When a quantity denoted as X is divided by another quantity Y, we can write symbolically

$$X \div Y$$

$$\frac{X}{Y}$$

$$X/Y$$

The divide symbol (÷) is the one commonly used; however, values separated by a horizontal or diagonal line are used more frequently in mathematics. Regardless of the way division is represented each of the expressions is read as "X is divided by Y." X is referred to as the numerator and Y is the denominator. The result is referred to as the quotient. The quotient of two numbers also is referred to as a *ratio*. By writing the quotient as X/Y we can see that division can be thought of as multiplication where the numerator is multiplied by the reciprocal of the denominator. Hence

$$\frac{X}{Y} = X\left[\frac{1}{Y}\right]$$

Consequently, the rules regarding plus and minus signs in division are similar to those in multiplication. The quotient of two positive or of two negative numbers is positive whereas the quotient of a positive and a negative number is negative.

In order to illustrate these concepts consider X equal to 24 and Y equal to 8, where both are first considered as positive, then Y is considered negative, and then both are considered negative. We have,

$$\frac{X}{Y} = \frac{24}{8}$$

$$= 3$$

$$\frac{X}{Y} = \frac{24}{-8}$$

$$= -3$$

$$\frac{X}{Y} = \frac{-24}{-8}$$

$$= 3$$

In cases where division actually is not performed, two out of three signs in the expression for a quotient can be changed without altering its value. For example,

$$\frac{-X}{Y} = -\frac{X}{Y}$$

$$= \frac{X}{-Y}$$

In other words, the minus sign of X can be changed to plus and placed in front of the entire expression or it can be attached to Y without altering the result, which is negative. Similarly, we can write

$$\frac{X}{Y} = \frac{-X}{-Y}$$

$$= -\frac{-X}{Y}$$

$$= -\frac{X}{-Y}$$

In this case, the ratio of two positive quantities can be expressed as the ratio of their negatives or as the negative of the ratio of one positive and one negative quantity.

The numerator and denominator of the ratio of two quantities can be multiplied or divided by the same *constant* value without altering the value of the ratio. In other words, for some constant A the following holds.

$$\frac{X}{Y} = \frac{AX}{AY}$$

$$= \frac{X/A}{Y/A}$$

Furthermore, if the denominator of the ratio is divided by a constant, the result is the same as the numerator multiplied by a constant or the entire ratio multiplied by the constant. That is,

$$\frac{X}{Y/A} = \frac{AX}{Y}$$

$$= A\left[\frac{X}{Y}\right]$$

As an example, let X equal 44, Y equal 10, and A equal 6. The ratio of X to Y is given as

$$\frac{X}{Y} = \frac{44}{10}$$

$$= 4.4$$

If we multiply the numerator and denominator by A, we have

$$\frac{AX}{AY} = \frac{6(44)}{6(10)}$$

$$= \frac{264}{60}$$

$$= 4.4$$

Dividing both numerator and denominator by A yields

$$\frac{X/A}{Y/A} = \frac{44/6}{10/6}$$

$$= \frac{7.33333333}{1.66666667}$$

$$= 4.4$$

By dividing the denominator by A we get

$$\frac{X}{Y/A} = \frac{44}{10/6}$$

$$= \frac{44}{1.66666667}$$

$$= 26.4$$

which is the same as the numerator multiplied by A.

$$\frac{AX}{Y} = \frac{6(44)}{10}$$

$$= \frac{264}{10}$$

$$= 26.4$$

The same result is obtained when the entire ratio is multiplied by A equal to 6.

Parentheses and Order of Operations

Earlier we used parentheses to identify a negative number that was part of a product of two numbers. Parentheses are used quite commonly in mathematics and in statistics to separate more complicated quantities and to act as a convenient aid in identifying the way various operations are to be performed. For example, consider the expression

$$3[(5X - 4Y)(2X + 7Y)]$$

If we assign values to X and Y we can evaluate the expression and obtain a single numerical result. The *order* in which the various operations indicated in the expression are performed is important. In general, multiplications (and divisions) within parentheses or brackets should be performed *before* addition and subtraction.

Suppose X equals 6 and Y equals 8. Owing to the products inside the parentheses in the above expression, we shall need more brackets to separate the operations. Based on the given values of X and Y we can re-express the above expression as

$$3\{[5(6) - 4(8)][2(6) + 7(8)]\}$$

In order to evaluate this expression we first must multiply the terms within the square brackets and then subtract and add accordingly. Once this is done the quantities within the square brackets can be multiplied together and multiplied by 3 in any order. The result of this procedure is

$$3\{[30 - 32][12 + 56]\} = 3\{[-2][68]\}$$

$$= 3(-2 \times 68)$$

$$= 3(-136)$$

$$= -408$$

Consider a more complicated expression as another example.

$$23 + \frac{2 + 3\{[5/X + 4Y][2X + 7Y] - [3X - 4Y]\}}{2[6X - Y/4]}$$

Assuming the same values of X and Y, we first must substitute these into the above expression. We then perform all multiplications and divisions within each bracket, first working from the *inside out*. Next we perform the addition in the numerator, divided by the quantity in the denominator, and then add 23 last.

$$23 + \frac{2 + 3\{[5/6 + 4(8)][2(6) + 7(8)] - [3(6) - 4(8)]\}}{2[6(6) - 8/4]}$$

$$= 23 + \frac{2 + 3\{[.833 + 32][12 + 56] - [18 - 32]\}}{2[36 - 2]}$$

$$= 23 + \frac{2 + 3\{[32.833][68] - [-14]\}}{2[34]}$$

$$= 23 + \frac{2 + 3\{2232.644 + 14\}}{68}$$

$$= 23 + \frac{2 + 3(2246.644)}{68}$$

$$= 23 + \frac{2(6739.932)}{68}$$

$$= 23 + \frac{13479.864}{68}$$

$$= 23 + 198.233$$

$$= 221.233$$

In this section we illustrated the order in which operations involv
ing brackets and parentheses are performed in order to evaluate mor
complicated expressions. The expressions involved addition, subtrac
tion, multiplication, and division only. Another example is provided i
the next main section that involves exponents. This is presented afte
the concept of exponents are discussed in some detail.

Equalities

We introduced the idea of a strict equality where we expressed th
variable X in terms of some constant, A. When equalities are expresse
in terms of relationships among variables the relationship is referred t
as an *equation*. By performing various operations on equations it i
possible to solve for an unknown value of a variable in terms of the othe
quantities based on a specified relationship. Although there are equa
tions that are very complex and require complicated operations an
methods to reach a solution, simpler equations that we generally dea
with can be handled with four basic operations expressed in terms of th
following *rules*.

(1) Adding the same quantity to both sides of an equation does not alte
the equality.

(2) Subtracting the same quantity from both sides of an equation doe
not alter the equality.

(3) Multiplying both sides of an equation by the same quantity doe
not alter the equality.

(4) Dividing both sides of an equation by the same quantity does no
alter the equality.

Assume the simplest case where X = 4 and a constant value of 2 in orde
to illustrate the four rules.

(1) $X + 2 = 4 + 2 = 6$

(2) $X - 2 = 4 - 2 = 2$

$$(3)\ 2X = 2(4) = 8$$

$$(4)\ X/2 = 4/2 = 2$$

In each case illustrated, the equality holds.

Now consider a slightly more complicated equation where we apply the rules in order to isolate X to determine its value. In the following equation we are interested in "solving" for X.

$$\frac{1}{4}(X - 3) = 10$$

By multiplying both sides of the equation by 4 we can eliminate the ¼ from the left side of the equation since 4⁄4 equals 1.

$$4\left[\frac{1}{4}\right](X - 3) = 4(10)$$

$$\frac{4}{4}(X - 3) = 40$$

$$X - 3 = 40$$

Adding 3 to both sides eliminates the –3 on the left since the algebraic sum of 3 and –3 equals zero.

$$X - 3 + 3 = 40 + 3$$

$$X = 43$$

Therefore, the value of X that satisfies the initial equation is 43. This can be *checked* by substituting 43 into the equation and simplifying.

$$\frac{1}{4}(43 - 3) = 10$$

$$\frac{1}{4}(40) = 10$$

$$\frac{40}{4} = 10$$

$$10 = 10$$

Since both sides of the expression are equal, 43 is the value of X satisfying the relationship. The same procedures can be applied to expressions where X appears more than once on both sides of the equation.

Inequalities

The operations given above in the form of rules apply to inequalitie with one exception. When both sides of the inequality are multiplied divided by a negative number, the *direction* of the inequality is reverse For example, obviously 10 is greater than 4. This can be written as

$$10 > 4$$

If we add 2 to both sides of the inequality, we obtain

$$10 + 2 > 4 + 2$$

$$12 > 6$$

The inequality still holds. The same would be true if we subtracted 1 from each side.

$$10 - 12 > 4 - 12$$

$$-2 > -8$$

Although both numbers are negative, –2 is greater than –8 since –2 i closer to zero.

If we multiply the original expression by 2 we get

$$2(10) > 2(4)$$

$$20 > 8$$

Since 20 is greater than 8, multiplying by the positive constant of 2 doe not alter the direction of the inequality. However, if we multiply by –2

$$10 > 4$$

$$-2(10) < -2(4)$$

$$-20 < -8$$

we can see that the inequality must be reversed since –20 is less thar –8 since –20 is further from zero in the negative direction.

POWERS AND EXPONENTS

When the same quantity is multiplied by itself repeatedly, the product is referred to as the *power* of that quantity. For example, a quantity X multiplied by itself equals X times X, or X × X. In this case we say that X is raised to the second power, which commonly is written as X^2 and is read as "X-square." Similarly, the symbol X^3 represents the quantity X multiplied by itself three times and is read as X to the third power or "X cube."

In general, we can represent a quantity raised to any power as X^n which is read as "X to the nth power." In this case, X is referred to as the base and n is an exponent. A base raised to the first power or an exponent of 1 equals the value of the base. By definition a base raised to the "zeroth power," or an exponent of zero, equals 1; ie., $X^0 = 1$.

As an example, consider a base of 2, or X=2, raised to various powers.

$$2^0 = 1$$

$$2^1 = 2$$

$$2^2 = 2 \times 2$$

$$2^3 = 2 \times 2 \times 2 = 8$$

$$2^4 = 2 \times 2 \times 2 \times 2 = 16$$

$$2^5 = 2 \times 2 \times 2 \times 2 \times 2 = 32$$

Consider the last expression in the above example, 2^5. This represents the product of five 2's which equals 32.

Raising numbers to higher powers directly becomes tedious and time consuming. Nowadays, it is a simple matter to evaluate powers of numbers with calculators equipped with an appropriate function key. The remaining parts of this section present some common operations associated with exponents.

Negative Exponents

Exponents considered above are positive; however, exponents also can be negative. For example, X^{-n} is read as "X to the minus nth power." Exponents that are negative indicate that the quantity equals the reciprocal with a change in the sign of the exponent. Hence,

$$X^{-n} = \frac{1}{X^n}$$

$$\frac{1}{X^{-n}} = X^n$$

In other words, switching a power to a numerator or to a denominator changes the sign of the exponent. As an example, suppose we want to evaluate 3^{-4}. This can be written as

$$3^{-4} = \frac{1}{3^4}$$

$$= \frac{1}{3 \times 3 \times 3 \times 3}$$

$$= 81$$

$$= 0.012345679$$

It is interesting to note that the reciprocal of any quantity can be represented as the quantity raised to the "minus one" power. That is,

$$\frac{1}{X} = X^{-1}$$

In general, this form of the reciprocal is referred to as the *inverse* of X.

Operations

Three basic operations regarding powers are important to consider. These are listed as follows:

(1) The product of the same base raised to separate powers equals the base raised to the sum of the exponents, or $X^n X^m = X^{n+m}$.

(2) The quotient of the same base raised to separate powers equals the base raised to the difference of the exponents, or $X^n / X^m = X^{n-m}$.

(3) A power of a quantity raised to a power equals the base raised to the product of the exponents, or $(X^n)^m = X^{nm}$.

In order to illustrate these rules, let X equal 5, n equal 7, and m equal 4. Then,

(1) $X^n X^m = (5)^7 (5)^4 = 5^{7+4} = 5^{11}$

(2) $X^n / X^m = 5^7 / 5^4 = 5^{7-4} = 5^3$

(3) $(X^n)^m = (5^7)^4 = 5^{7 \times 4} = 5^{28}$

When there are negative exponents, the same rules apply where the rules of algebraic addition and multiplication apply to the exponents.

Fractional Base

In general, the quotient of two powers with different bases must be evaluated by expanding the numerator and the denominator separately and then dividing the results. For example, consider the following

$$\frac{5^2}{4^3} = \frac{5 \times 5}{4 \times 4 \times 4}$$

$$= \frac{25}{64}$$

$$= 0.390625$$

In cases where the bases are different but the exponents are the same, division can be performed first and then the result can be raised to the common power. In general terms, this is written as

$$\frac{X^n}{Y^n} = \left[\frac{X}{Y}\right]^n$$

In other words, X to the nth power divided by Y to the nth power equals X divided by Y to the nth power. We can illustrate this by letting X equal 24, Y equal 4, and n equal 2. We have

$$\frac{X^n}{Y^n} = \frac{24^2}{4^2}$$

$$= \left[\frac{24}{4}\right]^2$$

$$= 6^2 = 36$$

In words, the square of 24 divided by the square of 4 equals 6 squared, or 36.

Scientific Notation

There are problems where it is convenient to express numbers in terms of powers of 10 so that they are easier to work with. Numbers expressed in this way are said to be represented in scientific notation. Scientific calculators and computers use this notation in order to present and process results that are too large relative to the capacity of the machine. In general, a number X expressed in scientific notation usually takes the form

$$X = A \cdot 10^n$$

where A represents a number between one and 10 (ie., $1 < A < 10$) and n is the power of 10. Hence, a number represented in scientific notation is expressed as a product of a number between 1 and 10 and a power of 10.

In order to understand the concept, consider the number 250. In scientific notation this is written as

$$250 = 2.5 \cdot 10^2$$

Notice that 10^2 equals 100 which when multiplied by 2.5 provides the original value of 250. Examples of other numbers expressed in scientific notation are given as follows:

$$25000 = 2.5 \times 10^4$$

$$.25 = 2.5 \times 10^{-1}$$

$$345 = 3.45 \times 10^2$$

$$579463 = 5.79463 \times 10^5$$

$$3000000 = 3 \times 10^6$$

Except for the second case, the other examples should be clear. In the second case, the power of 10 is negative which means that 2.5 is *divided* by 10 raised to the positive value of that power (ie., $2.5/10 = .25$).

In general, the power of 10 equals the number of decimal places that the decimal point is moved. If the decimal point is moved to the left the power of 10 is positive and if moved to the right the power of 10 is negative. The process is reversed in order to obtain the original number based on the one expressed in scientific notation. A positive exponent indicates that the multiplier is multiplied by that power of 10 where the power represents the number of decimal places shifted to the right. A negative exponent indicates that the multiplier is divided by that power of 10 where the power represents the number of decimal places shifted to the left.

Operations can be performed on numbers expressed in terms of scientific notation. These involve a combination of rules for ordinary algebraic operations and those for exponents. For example, suppose we want to multiply 6.5×10^2 by 3.7×10^3. This is performed in the following way.

$$[6.5 \times 10^2][3.7 \times 10^3] = (6.5)(3.7)(10^{2+3})$$

$$= 24.05 \times 10^5$$

$$= 2.405 \times 10^6$$

When multiplying, the multipliers of the powers of 10 are multiplied together and the powers of 10 are multiplied by adding the exponents. The final result is adjusted so that the multiplier lies between 1 and 10 which in the example raises the power of 10 in the result from 5 to 6. Division is performed similarly except the multipliers are divided, and the powers of 10 are subtracted.

Addition of numbers in scientific notation is a little trickier since it is not done directly like multiplication or division. The numbers first must be converted to the same power of 10. Then the multipliers are added. For example, consider the sum of 6.5×10^2 and 3.7×10^3. We have,

$$6.5 \times 10^2 + 3.7 \times 10^3 = 6.5 \times 10^2 + 37 \times 10^2$$

$$= (6.5 + 37)(10^2)$$

$$= 43.5 \times 10^2$$

$$= 4.35 \times 10^3$$

Parentheses and Order of Operations

Earlier we introduced two examples where we showed that operations within parentheses would be performed first, and that multiplication and division precede addition and subtraction. When dealing with complicated expressions that also involve exponents the same rules apply however numbers should be raised to a power *before* multiplication and division. For example, consider the following expression

$$17 + \frac{\{[2X + Y^2][5X^3 - Y]\}^2}{9(X + Y)}$$

If we let X equal 3 and Y equal 4, the above expression can be evaluated in the following way.

$$17 + \frac{\{[2(3) + (4^2)][5(3)^3 - 4]\}^2}{9(3 + 4)}$$

$$= 17 + \frac{\{[6 + 16][5(27) - 4]^2\}}{9(7)}$$

$$= 17 + \frac{[22(135 - 4)]^2}{63}$$

$$= 17 + \frac{[22(131)]^2}{63}$$

$$= 17 + \frac{[2882]^2}{63} = 17 + \frac{8305924}{63}$$

$$= 17 + 131840.063492 = 131857.063492$$

In order to evaluate the expression, powers within each inside bracket are found first, then products, and sums and differences last. The product within the curled brackets, { }, is computed before squaring the result (each component of the product could have been squared before multiplying; however, this involves more work). Division by the term 9(X+Y), or 63 is performed before adding 17.

ROOTS

In the previous section we dealt with exponents that were integers. If we consider the quantity X raised to the nth power, X^n, where n is a positive integer, then the nth root of X^n is X. In other words, the nth root of a number represents the value which when multiplied by itself n times equals that number. In this sense a root can be considered as the *inverse* of taking a power and is given by an exponent equal to the reciprocal of the power. Hence, in terms of exponents we can write

$$(X^n)^{1/n} = X^{n/n}$$

$$= X^1$$

$$= X$$

In other words, the nth root (ie., $1/n$) of X to the nth power equals X.

Another way of expressing a root is in the form of the "radical." That is,

$$\sqrt[n]{X^n} = X$$

where the root appears in the "v-like" portion of the radical sign.

Consider the number 27 and let us find the third root, or the cube root. This is written as

$$(27)^{1/3} = \sqrt[3]{27}$$

$$= \sqrt[3]{3 \times 3 \times 3}$$

$$= \sqrt[3]{3^3} = 3 = (3^3)^{1/3}$$

Notice that this is easy to do in this case since 27 is the product of three 3's and we are to find the third root. Hence, 3 is the value which when multiplied by itself 3 times equals 27. The same is true of the cube root of 1000 which equals 10 since $10^3 = 10 \times 10 \times 10 = 1000$. All roots, however, are easily determined with a scientific calculator equipped with an appropriate function key.

The most common root and one that frequently is used in statistics is the *square root*. The square root of a number is the value such that when multiplied by itself the result equals the original number. In general terms, the square root can be written in two ways.

$$\text{Square root of } X = X^{1/2}$$

$$= \sqrt{X}$$

In other words, the square root can be expressed as X to the "one-half" power or in terms of a radical. When a radical is used, the "2" indicating the root is omitted in the case of a square root only.

In order to understand the meaning of a square root, consider a simple example where we want to find the square root of 25. We all know that "5 times 5" equals 25, so that the answer is 5. This is written as

$$\sqrt{25} = \sqrt{5 \times 5}$$

$$= \sqrt{5^2} = 5$$

In this case the result is an integer and represents part of our basic multiplication table so that there is no difficulty in establishing the square root. For larger numbers or in cases where the result is not an integer one cannot automatically determine the square root in the same way. Frequently, when the result is not an integer it is in the form of an approximation that is rounded since many decimal places are required for an exact answer.

There are a number of ways to determine the square root of a number. The simplest, of course is to use a calculator with a square root key. The accuracy of the answer when using a calculator depends on the magnitude of the result and the number of display positions in the calculator readout. Square roots of very large numbers can be found on scientific calculators in terms of scientific notation, or powers of 10. Without a calculator, square roots can be found using a table.

Some Simple Operations

Since roots can be expressed as exponents, the rules for exponents given earlier apply to roots in general. Because square roots mainly are used in the text, we shall discuss some basic operations in terms of the square root. Two basic rules of importance apply to square roots:

(1) The square root of the product of two values equals the product of their square roots, or $\sqrt{XY} = \sqrt{X} \cdot \sqrt{Y}$.

(2) The square root of the quotient of two values equals the quotient of their square roots, or $\sqrt{X/Y} = \sqrt{X}/\sqrt{Y}$.

In order to illustrate these rules, consider the following example.

$$\sqrt{25 \times 49} = \sqrt{25} \times \sqrt{49}$$

$$= 5 \times 7 = 35$$

Similarly,

$$\sqrt{\frac{25}{49}} = \frac{\sqrt{25}}{\sqrt{49}}$$

$$= \frac{5}{7} = 0.71429$$

In both cases illustrated, perfect squares were used to make the point more clearly; however, the rules apply to any numbers in general.

SUBSCRIPTS AND SUMMATIONS

We have seen how to represent numbers in terms of symbols. Frequently, when dealing with a group of numbers or observations symbols accompanied by a subscript are used to identify individual observations. In some cases, this is useful in order to determine which values are involved in different operations specified by particular formulas.

As an example of the use of subscripts, consider the following values that could represent a set of measurements.

24 13 7 33 19 22 8 16

If we want to represent the measurements in terms of symbols, we can use X to stand for the quantity being measured in general terms and identify each value in terms of specific X's using a numerical *subscript* in the following way.

$$X_1 = 24$$

$$X_2 = 13$$

$$X_3 = 7$$

$$X_4 = 33$$

$$X_5 = 19$$

$$X_6 = 22$$

$$X_7 = 8$$

$$X_8 = 16$$

Instead of using different letters to represent each number, we use the same letter X to indicate that the values vary and a numerical subscript to distinguish among the values in general terms. Consider the last value $X_8 = 16$. In words, this is read as "X-sub-eight equals sixteen," where the subscript equal to 8 indicates the eighth observation. The remaining values are read in the same way except the value of the subscript differs.

If we want to represent the values completely in general terms using symbols, we can represent them in the following way.

$$X_i \; ; \; i = 1, 2, \ldots, 8$$

This is read as the variable "X-sub-i", where i "goes" from 1 to 8. By using dots to represent intervening values it is assumed that these values are integers. Here we can see that the subscript is designated with a symbol that varies also. In cases where we want to represent any number of observations in completely general terms we can write

$$X_i \; ; \; i = 1, 2, \ldots, n$$

This tells us that X is a variable where i goes from 1 to n, which indicates that there are "n" values of X.

Sometimes we deal with two sets of observations at the same time. If these are to be represented generally, different letters can be used for the variable and the subscript. That is,

$$Y_j \; ; \; j = 1, 2, \ldots, m$$

This tells us that Y is a variable that assumes m values, where j goes from 1 to m. X, say, could represent the incomes of a group of n people, and Y could represent the incomes of another group of m people. In cases where two observations are taken from the same people, for example income and mortgage payment, then both variables would have the same subscript and would be written as

$$X_i, Y_i; i = 1, 2, \ldots, n$$

In such a case, the subscripts indicate that a *pair* of observations are obtained from the same n subjects.

Summations

An operation that is used quite frequently in statistical work is summation. When formulas are given in general terms which indicate that a summation is necessary, a special symbol is used. This is the upper case Greek letter Σ, which is read as "sigma". Anything that follows this symbol is summed.

As an example, consider n values of a variable X that are to be added and we want to represent this in the form of symbols. This can be written in terms of sigma as

$$\sum_{i=1}^{n} X_i$$

This expression tells us to add the n values of X. We know that n values are to be added since the "index of summation," i, attached to the summation symbol is specified between 1 and n. When it is clear that all the observations are to be added the index of summation can be omitted and the sum simply can be represented as ΣX.

Suppose we want to find the sum of the eight values given above. This can be written as

$$\sum_{i=1}^{8} X_i = \Sigma X$$

$$= 24 + 13 + 7 + 33 + 19 + 22 + 8 + 16$$

$$= 142$$

There are problems when a *partial* sum is required, which means that only a portion of the values are to be added. In such a case it is necessary to retain the index of summation. Suppose we want to find the sum of the first "k" values in our example where k equals 3. This can be written as

$$\sum_{i=1}^{k} X_i = \sum_{i=1}^{3} X_i$$

$$= 24 + 13 + 7$$

$$= 44$$

If, instead, we wanted the sum of the last three values we could write this as

$$\sum_{i=6}^{n} X_i = \sum_{i=6}^{8} X_i$$

$$= 22 + 8 + 16$$

$$= 46$$

Notice the index of summation tells us to sum the values between 6 and the last, n, equal to 8.

In many cases it is necessary to perform various operations before a summation is performed. One common operation is to multiply each value by a constant before summing. This can be simplified, however, since the sum of a constant times a variable is the constant times the sum of the variable. That is

$$\Sigma CX = C\Sigma X$$

where C represents a constant. For example, consider our original eight values again and suppose we want to find the following quantity:

$$\sum_{i=1}^{8} 13X_i$$

This can be simplified as follows:

$$\Sigma 13X = 13\Sigma X$$

$$= 13(142)$$

$$= 1846$$

Another common operation is to square each value before summing. This is written symbolically as

$$\Sigma X^2$$

This instruction indicates that each value must be squared and then the squares must be added. In order to illustrate this operation, consider the values

$$3 \quad 5 \quad 2 \quad 4 \quad 6$$

The sum of the squares is found as

$$\sum_{i=1}^{5} X_i^2 = \Sigma X^2$$

$$= 3^2 + 5^2 + 2^2 + 4^2 + 6^2$$

$$= 9 + 25 + 4 + 16 + 36$$

$$= 90$$

Frequently, the operations that are to be performed before summing are more complicated and it is easy to get a wrong answer because the instructions provided in the formula are not clearly understood. Let us consider three common situations in order to understand this point.

Suppose you are given the instruction, $(\Sigma X)^2$. This tells us to find the sum of the X values and *then* square the sum; the answer is not the same as the sum of the squares, ΣX^2. This can be seen in terms of the last five values used above. We have,

$$\left(\sum_{i=1}^{5} X_i \right)^2 = (\Sigma X)^2$$

$$= (3 + 5 + 2 + 4 + 6)^2$$

$$= (20)^2$$

$$= 20 \times 20 = 400$$

Another commonly encountered problem is to subtract a constant from a set of values and then find the sum of the squares of these differences, or $\Sigma(X - C)$. In such a case the constant must be subtracted from *each* value first before squaring. For example, let us use the same values and let C equal 2.

$$\sum_{i=1}^{5}(X_i - C)^2 = \Sigma(X - C)^2$$

$$= (3 - 2)^2 + (5 - 2)^2 + (2 - 2)^2 + (4 - 2)^2 + (6 - 2)^2$$

$$= (1)^2 + (3)^2 + (0)^2 + (2)^2 + (4)^2$$

$$= 1 + 9 + 0 + 4 + 16$$

$$= 30$$

If the square in the above example were not present, it would not be necessary to subtract the constant from each value. Instead, we could add the X-values and subtract the constant *times* the number of observations from the total. In general terms, this can be written as

$$\sum_{i=1}^{n}(X_i - C) = \sum_{i=1}^{n}X_i - \sum_{i=1}^{n}C$$

$$= \sum_{i=1}^{n}X_i - nC$$

In terms of the above example, we have

$$\sum_{i=1}^{5}(X - 2) = \Sigma X - 5(2)$$

$$= (3 + 5 + 2 + 4 + 6) - 10$$

$$= 20 - 10 = 10$$

Cumulative Crossproducts

Earlier in this section we mentioned that a *pair* of observations can be associated with a similar element. In such cases it is common to find the sum of the products of the paired values, XY, which is referred to as a cumulative crossproduct. For example, consider the following set of observations of two variables that are associated with the same element.

i	1	2	3	4	5
X_i	9	6	7	12	11
Y_i	13	8	14	11	10

Here, we have five elements (i=1,2,3,4,5) where an X and a Y-value are associated with each. Based on these values, let us find the cumulative crossproduct, ΣXY.

$$\sum_{i=1}^{5} X_i Y_i = \Sigma XY$$

$$= (9 \times 13) + (6 \times 8) + (7 \times 14) + (12 \times 11) + (11 \times 10)$$

$$= 117 + 48 + 98 + 132 + 110$$

$$= 505$$

In many problems, summations such as the ones illustrated here are part of more complex calculations. The rules associated with the order of operations presented earlier also apply to problems involving summations.

Double Summation

Consider the following table of values.

		Characteristic A				
		j=1	j=2	j=3	j=4	j=5	Total
	i=1	X_{11}=3	X_{12}=9	X_{13}=6	X_{14}=8	X_{15}=11	37
Characteristic B	i=2	X_{21}=7	X_{22}=2	X_{23}=13	X_{24}=5	X_{25}=4	31
	Total	10	11	19	13	15	68

Normally, a table such as this is presented just in terms of numbers without the symbols; however, it has been done here in order to illustrate another concept involving summations. The table provides X-values associated with *two* characteristics, A and B, simultaneously. A is indexed by j and B is indexed by i. The j's correspond to columns and the i's correspond to rows. If we let the number of rows equal r and the number of columns equal c, then in general terms we can write

$$i = 1,2,....r$$

$$j = 1,2,......c$$

In the example, r equals 2 and c equals 5.

Each of the X-values has two subscripts in order to identify the row and column to which each value belongs. The first subscript indicates the row and the second subscript indicates the column. For example, X_{23} represents the value in the second row and the third column, and equals 13.

We can generalize this idea in terms of symbols. In other words, we can write X_{ij} to stand for the value in the "ith row" and the "jth column." By representing the X's in this way, we are able to write a general expression for the different totals appearing in the table. For example, suppose we want to represent the sum of the first *row* symbolically. This is written as

$$\sum_{j=1}^{5} X_{ij} = 37$$

Notice the index of summation is for j between 1 and 5 which means that the sum is over all *columns*. Since the first subscript attached to X equals 1 and is fixed, the sum is performed for all columns and is restricted to the first row. If the sum were performed, it would be done as follows:

$$\sum_{j=1}^{5} X_{1j} = X_{11} + X_{12} + X_{13} + X_{14} + X_{15}$$

$$= 3 + 9 + 6 + 8 + 11$$

$$= 37$$

Similar sums can be defined for any of the columns. For example, the sum of the values in the second column is given as

$$\sum_{i=1}^{2} X_{i2} = X_{12} + X_{22}$$

$$= 9 + 2$$

$$= 11$$

Here, the column number is held fixed and the summation is taken over all rows.

Now suppose we are interested in the sum of all of the X-values in the table. This can be represented symbolically in general terms as

$$\sum_{i=1}^{r} \sum_{j=1}^{c} X_{ij}$$

This expression is referred to as a *double* summation indicated by the two sigma signs. The double summation specifies that the total of all observations is obtained by first summing over all columns and then summing over all rows. Double summations generally are performed from the "inside out", which means perform the sum on the inside first and then the one on the outside. In our example, the order does not matter.

With respect to the numerical example, the sum of all of the observations is represented as

$$\sum_{i=1}^{2} \sum_{j=1}^{5} X_{ij}$$

The inner sum is made of two totals, one corresponding to the first row and the other to the second row. The totals are summed to obtain the overall total. Hence:

$$\sum_{i=1}^{2} \sum_{j=1}^{5} X_{ij} = (3 + 9 + 6 + 8 + 11) + (7 + 2 + 13 + 5 + 4)$$

$$= 37 + 31$$

$$= 68$$

The example of a double summation given here is very simple and is used in order to provide a basic understanding of the concept. Double sums can, however, be more complicated. For example, they may involve two variables where each sum is associated with a different variable. In such cases, the sum with respect to the first variable is performed where the second variable is treated as a constant and then the sum is performed with respect to the second variable.

ROUNDING

When working with large numbers or ones with decimals carried to a large number of places, it frequently becomes necessary to round results when doing computations by hand or with calculating devices of limited capacity. Otherwise, the calculations may be unmanageable. When values in a set of calculations are rounded, the result obtained is an approximation which varies to different degrees depending on the size of the numbers involved and the way in which rounding is performed. If a result is rounded at the *end* of a set of calculations there is no real problem, however rounding at *intermediate* stages of a complicated set of calculations may provide a result that is far from the correct value. Consequently, it is important to consider rounding problems in your calculations. In some cases, you may perform the operations correctly but obtain the wrong result due to the way rounding is performed.

In order to understand the problem, consider a simple example where two numbers are multiplied.

$$33 \times 34 = 1122$$

When 33 and 34 are multiplied directly, the exact result equals 1122. If, in a particular problem, the requirement is that the final answer should be accurate to the nearest ten, the value of 1122 can be rounded to 1120. A general rule for rounding is that at the point where rounding occurs the value remains the same if the next digit is less than 5 and is increased by one if the next digit is 5 or greater; the next digit and those to the right are changed to zero. Variations to this rule exist; however, this is the one used in this book.

Now suppose we round the original two numbers in our example to the nearest ten before we perform the multiplication. The result is

$$30 \times 30 = 900$$

The result differs substantially from the exact value and also differs from the value based on the requirement of accuracy used above.

The example just presented is somewhat exaggerated in order to make the initial point that rounding can affect an answer differently depending on how it is done. Let us consider another example that on the surface does not seem so extreme but does have important implications when combined with other calculations. Suppose we seek a result, call it A, that is the product of two decimals.

$$A = .3696 \times .5238$$

$$= .19359648$$

The exact answer is given to eight decimal places and equals 0.19359648. If we rounded the answer to the second decimal place we would obtain 0.19, and to the first decimal place the answer would be 0.20.

Now suppose we find A in terms of the product of the original two numbers rounded to the first decimal place. We have,

$$A \doteq .4 \times .5$$

$$= .20$$

In this case, we can see that intermediate rounding before multiplication is performed yields a result that seems close to the result that is rounded at the end. The symbol "$\doteq$" is used to mean "approximately equal to."

Consider another quantity, B, that is given by the following expression

$$B = 35 - 180A$$

where the value of A is necessary to find B. Assume that the value of B must fall between 0 and 1 and that it is necessary to obtain B accurate to three decimal places. The exact value of B is found as

$$B = 35 - 180(.19359648)$$

$$= 35 - 34.8473664$$

$$= .1526336 \text{ or } .153 \text{ (rounded)}$$

In terms of the way in which A is calculated, we can consider two problems relating to the accuracy of B: (1) the effect of rounding of the initial values in order to determine A, and (2) the effect of rounding the end result when calculating A. Results associated with the first problem are presented in the following table.

Initial Values	A	B
.4 and .5	.2	-1
.37 and .52	.1924	.368
.370 and .524	.19388	.1016

The initial values in the table are rounded to the three possible levels. Notice that in each case the values of A are not far from the exact value yet the values of B differ substantially. In the first case, the result equal to -1 is not even admissible based on the original assumption about the limits of B. The last value that is closest to the exact value still is 34 percent off. By rounding to any degree at the initial stage of calculation in this case provides a final result that is quite inaccurate.

Now consider the second problem where we consider the effect of the final value of A rounded to various degrees.

A	B
.2	-1
.19	.8
.194	.08
.1936	.152
.19360	.152
.193596	.15272
.1935965	.15263

In this case we can see from the table that rounding A to four decimal places (.1936) yields a value of B (.152) that is within 0.001 or 0.7 percent of the required value of the actual value rounded to the required number of decimal places (0.153). When more decimal places in A are considered the result gets closer to the actual value.

In general, the extent to which a rounded result differs from the exact result depends on the complexity of a particular problem, the magnitudes of the numbers involved, and the way in which rounding is performed at intermediate stages of computation. Whenever possible, rounding should be done at the end of the computations. A *general rule* to follow is that the final result should be rounded to the same number of nonzero digits or to one more nonzero digit that exists in the original data used to perform the calculations. If rounding is to be done at intermediate stages, it should be done to the least extent possible; in some cases, knowledge of the nature of the problem can provide a guide with respect to the degree of intermediate rounding. This can be seen in many of the examples provided in the main text where rounding is used. Some of the results, therefore, are approximate but are based on rounding that provides a result that for practical purposes is close to the actual answer.

All calculating devices employ rounding at some stage of operation. Calculators are limited in terms of the size of a number that can be used as an input and in terms of the output, and many are limited in terms of their internal operations. Scientific calculators that operate with scientific notation carry more places in the operation.

INDEX